Hurricanes
OF THE
North Atlantic

Hurricanes
OF THE
North Atlantic

Climate and Society

JAMES B. ELSNER
A. BIROL KARA

New York   Oxford

Oxford University Press

1999

Oxford University Press

Oxford   New York

Athens   Auckland   Bangkok   Bogotá   Buenos Aires   Calcutta
Cape Town   Chennai   Dar es Salaam   Delhi   Florence   Hong Kong   Istanbul
Karachi   Kuala Lumpur   Madrid   Melbourne   Mexico City   Mumbai
Nairobi   Paris   São Paulo   Singapore   Taipei   Tokyo   Toronto   Warsaw

and associated companies in
Berlin   Ibadan

Library of Congress Cataloging-in-Publication Data
Elsner, James B.
Hurricanes of the North Atlantic : climate and society /
James B. Elsner and A. Birol Kara.
p.   cm.
Includes bibliographical references and index.
ISBN 978-0-19-512508-5
1. Hurricanes—North Atlantic Ocean Region.
2. Hurricanes—United States.   3. Hurricanes—Social aspects.
I. Kara, A. Birol.   II. Title.
QC945.E46   1998
551.55'2'091631—dc21   98-16652

Printed in the United States of America
on acid-free paper

J.B.E.   To Ian and Diana,
           the uncertainty of the storm is the challenge of your future.

A.B.K.   To Selma and Veysel,
           like the forces of nature, you have sustained my creative path.

Acknowledgments

We extend special gratitude to Professor Noel E. LaSeur who encouraged us to examine non-tropical-only hurricanes as an important component of North Atlantic hurricane activity. We graciously acknowledge the help of Carl Schmertmann, Xufeng Niu, Anastasios Tsonis, Jason Hess, Todd Kimberlain, and Gregor Lehmiller on portions of the background research.

Partial financial support for this book came from the Risk Prediction Initiative (RPI) of the Bermuda Biological Station for Research (BBSR). We greatly appreciate the enthusiastic support of Anthony Michaels, Ann Close, Mark Johnson, Fielding Norton III, Dave Malmquist, Rick Murnane, Susan Howard, and Anthony Knap. Additional support came from the U.S. National Science Foundation; with particular thanks to Sankara-Rao Mopidevi. The U.S. National Oceanic and Atmospheric Administration also provided some financial assistance on this project with special thanks to Dan Smith. Some of the material for this book was compiled by the U.S. National Weather Service (NWS) at the National Hurricane Center (NHC). The project began with a sabbatical to the lead author granted by The Florida State University.

The authors prepared the entire manuscript including text, figures, and the book format. We graciously acknowledge the help of Svetoslava Elsner with proofreading and editing. Additional thanks go to Matthew Carter, Harley Hurlburt, Bethany Kocher, Mark Laufersweiler, Melody Owens, Alan Wallcraft, David Whitehead, Eric Williford, and Xungang Yin. Gratitude is extended to Joyce Berry and Lisa Stallings for their editorial assistance.

Figures 2.4, 2.5, and 2.6 are reprinted by permission of Louisiana State University Press from *The Hurricane and Its Impact*, by Robert H. Simpson and Herbert Riehl. Copyright ©1981 by Louisiana State University Press. Figure 2.1 is reprinted by permission of Rowman & Allanheld from *The Climate of the Earth*, by Paul E. Lydolph. Figure 10.12 is reprinted by permission of American Institute of Physics from *Physics of Climate*, by José P. Peixoto and Abraham H. Oort. The maps displayed in Figures 9.11, and 9.16 are reprinted by permission from *State Maps on File* by Facts On File, Inc. Copyright ©1984 by Facts On File, Inc.

The opinions, findings, and conclusions presented in this book are those of the authors and do not necessarily reflect the views of the funding sources. Any mistakes or omissions are the fault of the authors.

Preface

Often called the greatest storm on Earth, the hurricane is an awe-inspiring feature of tropical weather. Accounting for a relatively small percentage of global tropical cyclone activity, hurricanes of the North Atlantic have a tremendous impact on the people and economies of nations in and around the Caribbean Sea. When measured in terms of past loss of life and property damage, hurricanes rank near the top of all natural hazards, rivaling major earthquakes. Hurricane *Mitch* is a grim reminder of their deadly potential. Despite significant reductions in the number of deaths from hurricanes, economic costs of hurricanes affecting the United States have increased. Hurricane *Andrew* caused approximately $30 billion dollars in damage to Florida and Louisiana, making it the costliest hurricane on record. As economic development continues on islands and shorelines, our vulnerability to hurricanes will rise at an increasing rate, regardless of changes to the climate.

This book examines North Atlantic hurricanes with respect to both "climate" and "society." Our purpose is a comprehensive reference for users of hurricane information. Users include geographers, meteorologists, climate scientists, economists, and decision makers in government and industry, particularly those involved in urban planning, disaster relief, and insurance. The emphasis is on physical models to explain statistical relationships of hurricane activity with respect to weather and climate events. The better people are informed, the better they can prepare. The book is suitable for use as a reference textbook for graduate and undergraduate courses in applied climate science, physical geography, economics, risk management, urban planning, and so on. It is our intent for the book to be used as an information source for those interested in additional research into North Atlantic hurricane activity.

The difference between climate modeling and climate studies based on data and analysis is a modern theme on the ancient dichotomy between Plato and Aristotle. Plato thought of the world in terms of ideals with no fuzziness from contingencies. Aristotle, on the other hand, considered the world in terms of empirical classifications and inductive generalizations. For Aristotle, reality is derived from making sense of the observed world. This book is Aristotelian. The nature of hurricanes is understood from a careful examination of the data. Indeed, in much of climate science an Aristotelian rather than a Platonic approach to understanding is necessary. Yet we are quick to point out that the Aristotelian emphasis on causality is limiting when attempting to understand complex climate and social systems. In this regard, the term "significant" is used throughout the book with reference to the probability of occurrence. It is not meant to imply causality with respect to a physical influence.

Modern science must address problems of social relevance. Economic and societal costs of hurricanes are increasing yearly as societies become ever more intricate. Specifically, applied hurricane climate analysis involves a communication between science, industry, and government. Science provides answers to questions about hurricanes to industry while government and industry ask for specific interpretations of information useful to decision-making needs (e.g., the likelihood of a major hurricane striking New York City during the next decade). In reality a large gap exists; scientists are unaware of user needs while users are unsure of inherent limitations of climate information. The solution involves linking understanding of climate with human economic and political systems.

In the end we claim that simulation techniques designed to apply knowledge of historical events to hypothetical industry situations will help accurately evaluate risk. This integrated assessment represents an exciting research area on the interface between the natural and social sciences. We hope this book will enhance integrated assessment while providing a fresh perspective to students looking to make career choices. The level of presentation is aimed at the educated general public with an emphasis toward professionals like design engineers and actuaries of the insurance industry. Improvements in our ability to understand and predict natural fluctuations in hurricane activity will benefit both society and business.

The book has four main themes. It begins with a general description of hurricanes, including an examination of historical data sets and a presentation of various hurricane statistics. Details on the origin and tracks of hurricanes are provided. The second theme is the North Atlantic hurricane record most closely linked to people and society. Special focus is given to major hurricanes, landfalling hurricanes and the analysis of cycles, trends, and return periods. Prediction models for forecasting North Atlantic hurricane climate is the third principal theme. Included is a history of modeling efforts as well as a look at the potential to predict hurricane climate several years to a decade in advance. The fourth theme is societal vulnerability to hurricanes. Chapters include facts on changes in population and property. Ideas on risk management and catastrophe insurance are presented. The book is our attempt to present the historical hurricane record of the North Atlantic in a clear, comprehensive, and original manner. The user will decide if we have succeeded.

James B. Elsner
A. Birol Kara

Florida State University
January 1999

Contents

Hurricanes
OF THE
North Atlantic

1

Hurricane Characteristics

Owing to substantial size and intensity, hurricanes are the single costliest and most destructive of all atmospheric storms. Of the ten costliest weather disasters in the history of the United States, six were the result of hurricanes. Powerful winds of the hurricane produce tremendous damage to natural and man-made structures. Trees are blown down and properties are destroyed. Huge waves raise the potential of total destruction to coastal communities. The associated torrential rains create catastrophic flooding to both coastal and inland areas. The combination of strong winds, flooding rains, storm surge, and tornadoes raises the specter of mass casualties. Destruction from hurricanes rivals the destruction from major earthquakes. Though typically formed and nurtured deep in the tropics, hurricanes can travel great distances to devastate cities and towns that have a temperate climate.

Each year about four score tropical cyclones form over the warm oceans of our planet. Eighty-five to ninety percent of them originate between 20°N and 20°S latitude. The main breeding grounds are the tropical waters of the Pacific, Indian, and North Atlantic oceans. Hurricanes are most likely to develop when and where the oceans are their warmest. Tropical-cyclone activity over the North Atlantic basin accounts for 11% of world-wide activity. More than 50% of the North Atlantic tropical cyclones reach hurricane intensity. The strongest tropical cyclones develop over the western North Pacific. A significant percentage of all hurricane deaths and destruction occur with the intense storms.

This book focuses on hurricanes of the North Atlantic basin. The North Atlantic basin encompasses the waters between North America and the continents of Europe and Africa, and includes the Caribbean Sea and Gulf of Mexico. The book describes climatological features of North Atlantic hurricane activity from data archived over the past century and more. Specifically, the focus is hurricane activity relevant to planning and mitigation strategies in the United States and Caribbean. This first chapter serves as an introduction to hurricanes. It begins with definitions and descriptions of the salient features of these powerful tropical cyclones. The environmental conditions conducive to their growth and development are also examined.

1.1 Descriptions

A closed, generally circular, rotation of air spinning in a counterclockwise direction in the Northern Hemisphere is called a cyclone.[1] The term *tropical cyclone* refers to a circulation of air that develops over the warm waters of the tropical latitudes.[2] A tropical cyclone is a vortex of air circulating around a center of low pressure such that the force caused by the pressure gradient pushing the air toward the low pressure is balanced by the sum of the deflective force due to the earth's rotation and to the centrifugal force. The centrifugal force acts outward from the center of circulation. The deflecting force acts to the right looking in the direction of motion in the Northern Hemisphere. The rising motion associated with the cyclonic vortex is due to the frictional slowing of the winds near the surface creating a spiraling of air inward and upward toward low pressure. Rising air cools by expansion leading to condensation of moisture and the production of clouds and rain.

The National Hurricane Center (NHC), which is the forecast branch of the Tropical Prediction Center, considers three categories of non-frontal cyclones of tropical origin. These include tropical depressions, tropical storms, and hurricanes. All three tropical cyclones have surface winds that blow in a complete (closed) circulation. A hurricane is a tropical cyclone at maturity.[3] Table 1.1 lists the general features of mature hurricanes. In the North Atlantic, a tropical cyclone with 1-minute maximum sustained near-surface (10 meter) winds in excess of 33 m s^{-1} ($\geq$ 64 kt) is called a *hurricane*. Shorter period wind gusts are likely to be considerably stronger than the 1-minute maximum sustained speed. A one-second wind gust may be 1.25 times higher than a 1-minute average wind speed (Neumann 1987). Likewise, short duration wind lulls are substantially weaker. Hurricanes are characterized not only by strong surface winds but also by relatively symmetric and self-contained cyclonic[4] inflow of air near the ground and anticyclonic (opposite of cyclonic) outflow of air aloft. Hurricanes have diameters ranging between 200 and 1300 km and lifespans lasting between one and 30 days. The average lifespan of a North Atlantic tropical cyclone from its birth as a tropical depression to its demise is around ten days, but varies widely from one storm to another. Globally, the busiest month of the year for tropical cyclones is September, with May being the least active month. Hurricane activity in the North Atlantic peaks in September.

From the vantage point of space, a hurricane appears as a nearly circular mass of deep clouds called the *central dense overcast* (CDO) with a hole near the center called

---

[1]The word *cyclone* (meaning "coil of snake") was first used by Henry Piddington, president of the "Marine Courts" at Calcutta (Tannehill 1950).

[2]William Redfield in 1831 was the first to argue that winds around the tropical cyclone blow in a circular pattern.

[3]The term *hurricane* (from a Native American word pronounced in Spanish as huracán) is used for a mature tropical cyclone in the North Atlantic and eastern North Pacific east of the international dateline. In the western North Pacific the term is "typhoon" and near India, Australia, and Madagascar the term is "cyclone."

[4]*Cyclonic* refers to winds blowing in a counterclockwise direction as viewed from above in the Northern Hemisphere (clockwise in the Southern Hemisphere).

Table 1.1: General characteristics of North Atlantic hurricanes.

Characteristic	Range
Storm Diameter	200–1300 km
Surface Winds	≥ 33 m s^{-1} (one-min average)
Lifespan	1–30 days
Eye Diameter	16–70 km
Direction of Motion	Westward then northward
Energy Source	Latent heat release

the *eye* (Figure 1.1). The CDO is a shield of high-level (and cold) cloud tops resulting from the underlying showers and thunderstorms in the *eye wall*. The eye wall is a ring of cumulonimbus clouds[5] immediately surrounding the eye. The strongest winds and heaviest rain occur with the clouds inside the eye wall. Only a small percentage of the total hurricane contains intense rising motion; most of which occurs in the eye wall. The eye, whose diameter ranges between 16 and 70 km across depending on the storm, is general free of tall clouds, winds are light, and the air aloft is warm and dry. Surface air pressures are lowest in the eye of the hurricane due to sinking and warming of the air. The eye provides a convenient fix for tracking the movement of hurricanes from satellites, radar, and aircraft. Clouds outside the CDO are arranged in a banded pattern resembling a spiral. The spiral feeder bands create an appearance from space that looks similar to that of a galactic star system. Heavy rain and *squalls*[6] accompany the spiral (feeder) bands. The outermost bands can be 1000 km from the hurricane center. Hurricane weather at a particular location begins with a hard, brief rain and sudden gusts of wind, followed by a period of partial clearing until the next band arrives and squally weather returns. As the center of the hurricane approaches the frequency and intensity of the squalls increase. As the eye passes winds become light before switching directions and gaining strength as the storm pulls away.

1.2 Definitions

Changes to the form and structure of a hurricane can arise in various ways, leading to some confusion in terminology. The definitions used by Merrill (1984) are the most commonly accepted. The *intensity* of a hurricane is measured by the extreme value of either the minimum sea-level pressure (MSLP) or the maximum sustained near-surface winds. For distinguishing hurricanes from tropical storms and for classifying hurricanes as major, we follow the convention and use the maximum sustained winds as the defining standard. Hurricane intensification refers to a decrease in MSLP or

---

[5]Cumulonimbus clouds (or thunderstorm clouds) are deep, dense clouds that often have an anvil-shaped top and a towering appearance.

[6]A squall is defined as a strong wind characterized by a sudden onset and termination, and lasting generally less than five minutes.

Fig. 1.1: Hurricane *Andrew* located near the coast of Louisiana before its second landfall in August of 1992. Note the circular central dense overcast (CDO) surrounding the eye, indicative of a favorable upper-level environment for intensification. Hurricane *Andrew* is the costliest U.S. hurricane on record. The satellite picture is courtesy of the National Oceanic and Atmospheric Administration (NOAA).

an increase in maximum sustained winds. Winds in the hurricane increase from their lowest speeds within the eye to their highest velocities immediately beyond the edge of the eye, in the eye wall. Winds decrease more gradually outward from the eye wall. Figure 1.2 shows various features of an idealized hurricane. The wind speed profile (panel b) indicates a maximum speed (V_{max}) at the eye wall with a rapid drop in the speeds within the eye itself. Wind speeds in gusts are higher with the ratio of gust speed to sustained speed decreasing with height but increasing with roughness of the underlying surface.

The *frequency* (or abundance) of hurricanes refers to the number of storms over a given time interval. Typically, the frequency is expressed as the number of hurricanes in a season. Changes in the abundance of hurricanes over a year or two are linked to regional climate fluctuations. The frequency over decades to centuries is related to global climate factors. The conditions that govern the intensity of hurricanes are apparently

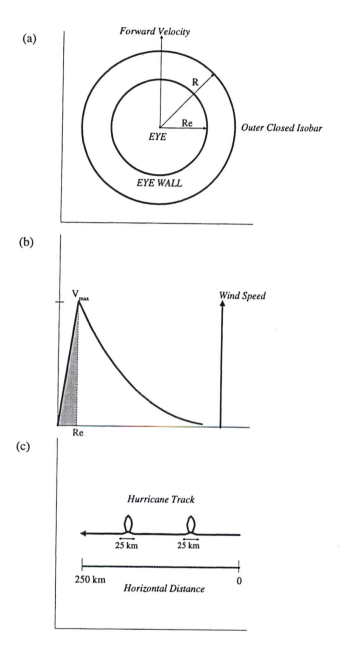

Fig. 1.2: Idealized hurricane features showing (a) the relationship of the eye to the eye wall and to the outer closed isobar (OCI), (b) the wind speed profile, and (c) the size of the cyclodial oscillations (wobbles) to the overall track distance.

quite different from those factors that influence frequency (Emanuel 1997). Indeed, the 1992 North Atlantic hurricane season saw only four hurricanes but produced the very intense hurricane *Andrew* that devastated parts of southeastern Florida.

The geometric *size* (or breadth) of a hurricane is an indication of the storm's horizontal dimension (or lateral extent) and is often measured by the average distance from the hurricane center to the region of maximum wind speed, termed the *radius to maximum winds*. Hurricane size is of interest as it indicates the spatial coverage of damaging winds and the duration that a specified location can expect to be under the influence of the destructive winds. Another measure of hurricane size is the average distance from the hurricane center to the outermost closed *isobar*.[7] Since this distance will depend to some extent on the pressures outside the hurricane, it might be more convenient to measure the distance from the center to the neutral point of the wind field. The neutral point represents the position on a streamline analysis immediately outside the hurricane where the winds are calm.[8] The various measures of a hurricane's breadth are not the same. As with intensity, the size of hurricanes varies dramatically among storms. The largest storms can be ten times the size of the smallest. Size can also vary substantially for a single hurricane over its lifespan. Growth of a hurricane refers to an increase in size. According to Merrill (1984), the largest hurricanes in the North Atlantic occur during October over subtropical latitudes. Hurricane size is only weakly correlated with hurricane intensity. This is reflected in the fact that although both hurricanes had similar intensities, hurricane *Hugo* in 1989 was a very large hurricane while hurricane *Andrew* in 1992 was a relatively compact hurricane. The distribution of the wind field around the storm is often asymmetric featuring the strongest winds in the right rear quadrant. Unfortunately, very little is known about factors and conditions that govern the geometric size of individual hurricanes.

The depth of a hurricane is constrained by the extent to which the atmosphere exhibits a significant fall of temperature with height (*lapse rate*). Lapse rates are sufficiently large throughout the troposphere, but become constant above, in what is called the *tropopause*.[9] In tropical latitudes the tropopause extends to about 16 km above the ground; consequently hurricanes do not extend much above this level. The tropopause slopes toward the surface away from the tropics reducing the depth of storms that move or originate at higher latitudes. As a first approximation, a cap on the intensity of a hurricane is set by the limit of pressure difference between the eye and the outermost closed isobar. This difference is determined by the differences in weights of a column of air in the center and a column outside the circulation. Column weight depends on the temperature. Warmth in the eye column is determined by the temperature near the tropopause and by the depth over which the subsiding air is heated through compression. A mature tropical cyclone is characterized by warm temperatures throughout its core column. Hurricanes are appropriately described as warm-core cyclones. A tropi-

---

[7] An isobar is a line connecting areas of equal air pressure.

[8] This idea was suggested by Noel E. LaSeur (personal communication 1997).

[9] The tropopause is a layer of air on top of the troposphere where temperatures are nearly uniform with height.

cal cyclone over the tropics where the tropopause is coldest and highest will generally have a maximum potential intensity exceeding that of a storm in the subtropics where the tropopause is lower. The *strength* of a hurricane is defined as the wind speed averaged around the cyclonic circulation. Strength can change even if size or intensity remain constant. Consideration of hurricane strength is important as it can have different controls and influences on storm development than either intensification or growth. Because data on hurricane size is rather limited, breadth and strength are not considered in much detail in this book. Indeed, studies on factors that control changes in hurricane size and strength are important areas for future hurricane research.

Besides hurricane morphology, it is instructive to understand the energetics of a hurricane. In short, the energy necessary to sustain a hurricane is provided by the evaporation of warm water from the ocean surface over which hurricanes are originate and develop. The heat energy of evaporation[10] is stored in the form of water vapor. The water vapor is carried upward inside the towering cumulonimbus clouds surrounding the eye of the hurricane. The rising air cools by expansion leading to condensation of the vapor to maintain the clouds and precipitation. Condensation releases a significant portion of the stored heat energy, which in turn helps to maintain the storm's circulation. This is accomplished by the latent heat reducing the rate of adiabatic cooling below that of strictly dry air and thus providing a constant upward force on the air parcels. Significant condensation also occurs in the hurricane's feeder bands.

Hurricanes move with the air flow in which they are imbedded. Weaker storms are steered by lower-level winds, while the strongest storms move with winds higher up. Hurricanes of low latitudes will track to the west, pushed along by the northeasterly *trade winds*.[11] Hurricanes of higher latitudes track more to the northwest and north steered by the anticyclonic flow around the subtropical high-pressure system. The overall track is described as a parabolic sweep. Yet the parabolic path is a simplified description of hurricane motion. For instance, late and early season hurricanes often move directly north before acquiring an eastward component of motion (Simpson and Riehl 1980). Changes in hurricane motion are often linked to extra-tropical circulation. Strong upper tropospheric troughs extending to low latitudes can steer the hurricane northward out of the tropics. The absence of deep middle-latitude troughs allow low-latitude hurricanes to maintain a more westerly motion. A hurricane imbedded in a weak upper-level flow may move very slowly or become stationary for several days. Since the movement of an extra-tropical circulation is sometimes hard to predict in itself, the secondary effect on hurricane steering is even more difficult to forecast. A change to a more northerly direction for tropical cyclones is known as *recurvature*. The most uncertain period in the movement of the hurricane is at the time of recurvature when the storm is in a region of light steering flow. Indeed, the critical component of a hurricane-track forecast is the correct anticipation of the point of recurvature.

In addition to a variety of possible large-scale motions, hurricanes may exhibit

---

[10]This energy was first formulated by Joseph Black in 1760 and is technically called *latent heat*.

[11]Trade winds (or trades) are low-level winds that blow across the tropics flowing out from the subtropical high pressures toward the lower latitude equatorial troughs (low pressures).

small-scale oscillations (wobbles) superimposed on an otherwise steady direction of motion. These oscillations, which likely arise due to asymmetries in thunderstorm activity around the center of circulation, may be as large as 8 to 32 km. The wobbles, described as cycloidal oscillations, do not represent the steady large-scale motion of the hurricane (see Figure 1.2), and real-time tracking requires careful observations so as not to confuse a wobble with an important change in direction. When the hurricane eye becomes partially obscured by clouds, a hurricane may appear to wobble (or jog) as viewed from a loop of satellite images even as it maintains a steady direction. The tracks presented in this book are based on the "best-tracks" (see Chapter 3), which technically refers to the best estimate of the smoothed path of the hurricane's eye as it moves across the earth's surface. Determination of the best-track positions are made by the folks at NHC based on a careful reanalysis of the cyclone months after it has occurred using all the available data.

As a tropical cyclone approaches the United States, the NHC issues *watches* and *warnings*.[12] A hurricane watch means that hurricane conditions are *possible* in the specified area, usually within 36 to 48 hours. This advice is to alert residents and other interests to the potential need for evacuation or other emergency preparations and to allow time for individual planning should such action be needed. A hurricane warning means hurricane conditions are *expected*, usually within 24 hours. It is meant as a warning of imminent danger for a specified segment of coastline and the urgent need for prompt action to protect life and property. Forecasts of future hurricane positions are made using a suite of dynamical and statistical computer models, with each model indicating a somewhat different projected path as the forecast lead time increases. The ensemble of forecast paths is translated into a "cone of uncertainty." Where the cone of uncertainty intersects the coastline dictates the areas subjected to hurricane watches and warnings.

Technically, hurricane landfall occurs when all or part of the hurricane eye wall crosses the coastline. Other factors being equal, the intensity of a hurricane, as measured by the central pressure, is not significantly reduced until the eye wall makes landfall, even though much of the storm may be over land. This is true despite the reduction in spiral-band convection at landfall. However, shower and thunderstorm intensity on the west side of a hurricane in the Gulf of Mexico may be weaker as the inflow of low-level air on this side of the storm originates over land. Winds along the coastline are strongest in the right front quadrant of the storm for a person standing at sea looking inland. This is because winds on the right side of the hurricane have an additional forward motion added to the speed of circulation around the storm (see Figure 1.3).

As the hurricane moves inland, the friction of land acts to increase the flow of air toward the hurricane center. This creates a greater inflow of air toward lowest

---

[12]The first hurricane warning for a North Atlantic hurricane was made by Father Benito Viñes, director of the Belen College Observatory in Havana, Cuba. His forecast warned of the Cuban hurricane of September 11, 1875. According to Fernández-Partágas and Diaz (1995a, 1995b), Father Viñes was a significant contributor to our early knowledge on many aspects of North Atlantic hurricanes.

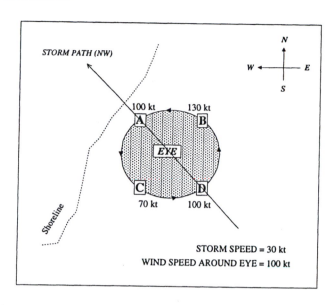

Fig. 1.3: Sketch of winds in the eye wall of a mature hurricane as it approaches the coastline. A hurricane moving toward the coastline will have the strongest winds on the right side (B) of the eye (looking toward the direction of motion) and weaker winds on the left side (C). In this case the hurricane's maximum-sustained winds are recorded as 130 kt.

pressure.[13] This in turn leads to rising surface pressures (filling) and a weakening of circulation. Surface friction may alter the movement of hurricanes at landfall. Initially there will be a difference in frictional effects between the portion of the storm over land and the portion over the water. The greater friction over land will lead to greater inflow toward the center and a corresponding increase in mass and pressure in the right-front quadrant which will cause the storm to jog to the left. More important for dissipation is the fact that over land the hurricane is removed from its source of heat and moisture (the ocean) that fuels the convection.[14] Without the rapidly rising air near the storm center, the hurricane weakens. In general, the decrease of maximum wind speed after landfall is proportional to the storm's intensity. The more intense the circulation around the eye as the storm approaches the coast, the quicker it will weaken over land. The speed of approach determines the inland penetration of hurricane force winds (Kaplan and DeMaria 1995). Tropical cyclones typically loose hurricane intensity within 12 hours of landfall, but powerful storms can remain at hurricane intensity for greater than 24 hr after landfall. Observations and model simulations indicate that tropical cyclones can maintain strength or even intensify when moving over swampy ground. Excessively heavy rainfall is often the last punch of a decaying hurricane. The associated heavy rainfall can be widespread despite winds diminishing to below tropical storm intensity.

---

[13]This is because slower rotation weakens the centrifugal force below the force of pressure gradient.

[14]Convection refers to atmospheric motions that transport heat and motion in the vertical direction. Cumulus clouds and thunderstorms are examples of phenomena associated with convection.

Since the principal focus of this book is the statistics of North Atlantic hurricanes, it is necessary to define *hurricane climate*. As used here, hurricane climate refers to the set of normal and extreme hurricane statistics as gathered from storms over the North Atlantic during the past 100 years or more. The statistics include average frequencies, mean intensities, seasonal cycles, geographic distributions of origin, and so on. The set of hurricane statistics is placed against the backdrop of climate conditions over the globe. Global climate anomalies are offered as explanations to variations in hurricane activity. Classical statistics provide a useful way to summarize the historical hurricane record, but as will be argued later, their utility may be limited when asking questions about future hurricane activity.

1.3   Environment

The average environment over the North Atlantic Ocean is not particularly favorable for tropical cyclone development (Gray et al. 1993). Of the many tropical waves (or other pre-hurricane disturbances) each season only a relatively few develop into hurricanes. Despite the lack of consensus regarding a general theory of tropical cyclone formation, partly due to insufficient upper-air observations over the oceans, there is agreement on the requirements necessary for hurricane development. These requirements include a large ocean area with sea-surface temperatures (SSTs) exceeding 26.5°C, an atmosphere that is moist (laden with water vapor), and an atmosphere for which the temperatures cool sufficiently with height. A minimum latitude of approximately 8° away from the equator, weak vertical shear of the horizontal winds, and a preexisting low-level and/or upper-level disturbance (Table 1.2) are also important conditions for hurricane development. The problem for forecasting is that, taken separately, each of these conditions is quite common in the tropics and in particular during the hurricane season, but taken together, one or more ingredients is often missing. Moreover, although they broadly define the necessary preconditions for hurricane origin and development, they are not sufficient to guarantee one will form. Let us examine each of these conditions separately in more detail.

A tropical cyclone develops and spends most of its life in the relatively homogeneous tropical air. Tropical air has only minor temperature contrasts from place to place. Warm SSTs are necessary for hurricane initiation and development. The depth to which the warm waters extend is also important. A deep warm pool of water will not easily mix with cooler water below the *thermocline*[15] as winds increase in a fledgling storm. Water evaporates from the warm ocean surface where it rises and condenses inside towering cumulonimbus clouds. The latent heat released from the condensation, together with the sensible heat from the air in contact with the warm ocean, fuels the tropical cyclone. The supply of warm, moist air must remain uninterrupted over a period of at least a day or more if the tropical cyclone is to develop. The air flow above the

---

[15]Thermocline refers to a boundary in the ocean surface layer between the relatively warm surface waters and the relatively cooler waters below.

Table 1.2: Environmental conditions favorable for hurricane development. The conditions are not all physically distinct nor statistically independent.

Condition	Criteria
Warm ocean surface	$> 26.5°C$
Unstable atmosphere	$> 7°C \ km^{-1}$
Minimum latitude	$> 8°N$
Weak wind shear	$< 10 \ m \ s^{-1}$

storm at upper levels must provide enough outflow (divergence) to allow the pressures near the surface to fall. The outflow carries the extra heat generated by the central condensation far from the storm center so that their is no warming of the air except in the core. An environment featuring marked divergent flow at high levels superimposed over a low-level cyclonic disturbance over the warm ocean is most favorable for hurricane development.

The latent-heat content of the atmosphere at a particular relative humidity increases exponentially with air temperature.[16] Observations and theory (Carlson 1971, Emanuel 1991, Evans 1993, DeMaria and Kaplan 1994a, Holland 1997) suggest that hurricane formation is sensitive to only minor increases in SSTs between 26 and 29°C because of the trade-wind *inversion of temperature*[17] located one to two kilometers above the tropical Atlantic Ocean. Penetration of the developing cumulonimbus clouds above the inversion arises suddenly when the air below (called the *boundary layer air*) becomes sufficiently warm and moist (Saunders and Harris 1997). An atmosphere that is adequately moist will produce rising parcels of air, which cool slower than their immediate environments. As such, the air parcels will find themselves considerably warmer than the undisturbed surroundings all the way from the surface to 10 or 12 km aloft. This so-called conditional instability maintains the vertical circulation inside the developing tropical cyclone. The temperature of the ocean is only half the story. Not only is it essential that the waters be warmer than 26°C to a depth, the expanse of warm ocean also needs to be substantial. Wendland (1977) found that the area of the North Atlantic Ocean with SSTs exceeding 26.8°C is directly (and nearly exponential) related to the frequency of tropical cyclones. When the mean monthly area of warm SSTs is less than $8.5 \times 10^6 \ km^2$ (roughly the area of the United States), tropical cyclones do not readily occur. Figure 1.4 shows paths of tropical cyclones across the Northern Hemisphere superimposed with average July SSTs. The ocean temperatures over the western North Atlantic are warm enough to support the development of tropical cyclones as far north as 35°N latitude. In the eastern North Pacific, SSTs are generally too cold north of 25°N latitude.

Implicit in the requirement of a warm, unstable atmosphere for hurricane formation and development is the requirement of ample moisture. Indeed, tropical cyclones

---

[16]The exact relationship is called the *Clausius-Clapeyron equation*.
[17]An inversion occurs when temperatures increase with altitude.

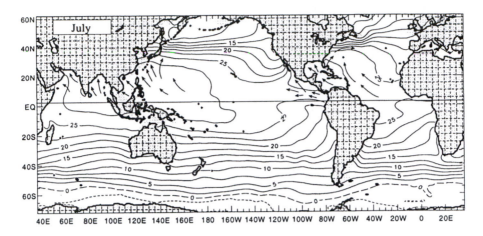

Fig. 1.4: Average sea-surface temperatures (°C) during July in intervals of 2.5°C, along with the typical paths of tropical cyclones during the tropical cyclone seasons of the Northern Hemisphere. The data are from the Comprehensive Oceanographic and Atmospheric Data Set (COADS; daSilva et al. 1994). The dotted line indicates average SSTs below 0°C.

require plenty of moisture evaporated from the ocean surface through a deep atmospheric layer. Under the influence of above-normal surface pressures over the development areas of the Atlantic basin, cooler and drier conditions typically prevail in the atmospheric boundary layer making the environment hostile for hurricane initiation and growth (Knaff 1997). Above-average sea-level pressures (SLPs) are also linked to greater vertical wind shear, so the exact physical mechanisms by which sea-level pressures influence hurricane activity remain elusive. A necessary component of organization is the effective force caused by rotation of the earth. A disturbed area containing numerous showers and thunderstorms near the equator will remain disorganized and fail to form a tropical cyclone. The so-called Coriolis force[18] keeps the air from filling the weak low-pressure area. Instead, the air is forced into a circular, cyclonic flow around the low pressure. The horizontal component of the Coriolis force is too small to be effective in organizing tropical cyclones south of about 8°N latitude.

Vertical wind shear is a term used to describe the changes in horizontal winds between two levels in the atmosphere. Wind changes can arise from changes in speed and changes in direction. For example, if the easterly wind has a speed of 5 m s^{-1} at the surface and 15 m s^{-1} at 2 km above, then there is an easterly shear of 10 m s^{-1} between the two levels. During the Northern Hemisphere winter and spring, winds several kilometers above the surface over the tropical North Atlantic are strong and have a westerly component (wind blowing from west to east). The westerlies override the low-level easterlies resulting in vertical shear. The consequence of wind shear on the development of a hurricane are complex and not fully understood. It is known

---

[18]Named after its discoverer Gaspard Gustove de Coriolis, a 19th-century French mathematician.

that a circulation consisting of low-level inflow of air coupled with upper-level out-flow through strong rising currents can be disrupted by vertical shear of the horizontal winds. Another influence of shear is to increase the area under the influence of latent heating, thereby diminishing the effect of upper-level heating on reducing the surface pressures.[19] The effect of shear on a tropical wave or depression is complex, but in general, the larger the shear the lower the probability of hurricane development. As a rule of thumb, vertical wind shears in the layer between 850 and 200 mb over an area of 350,000 km^2 that exceed 8.5 m s^{-1} will generally act to retard hurricane in-tensification and shears in excess of 10 m s^{-1} will typically inhibit tropical cyclone development (Zehr 1992).

Upper-level troughs of low pressure are responsible for fast moving air and strong vertical shear of the horizontal winds. In particular, the middle Atlantic trough sand-wiched between a ridge near Bermuda (Bermuda high) and another near the Azores (Azores high) is a semi-permanent feature of the subtropical North Atlantic during the hurricane season. This high-level feature is referred to as the *tropical upper-tropospheric trough* (or TUTT for short). TUTTs are found in both the North Atlantic and North Pacific basins. They are likely maintained by subsidence warming at high altitudes (12–16 km) which counteracts the radiation cooling associated with the sub-tropical ridges (highs). A portion of the trough might become completely isolated from the main trough creating a TUTT low. A TUTT or TUTT low will aid the development of a tropical cyclone if positioned in a way that enhances upper-level outflow above the cyclone (see Chapter 2). The motion of a tropical cyclone or hurricane can, in some cases, be influenced by the proximity of the TUTT. On average, the trough axis extends from the central subtropical Atlantic southwestward to the eastern Caribbean Sea. TUTT-related tropical cyclone genesis is more common over the western North Pacific than over the North Atlantic.

Despite the general hostility of upper-level dynamics to hurricane formation and development, there are circumstances under which such dynamics are beneficial to hurricane intensification. For instance, it has been known since the 1940s that under the right conditions, the interaction of an upper-level cyclonic circulation with its atten-dant vertical wind shear can cause a nearby tropical cyclone to develop. More recent work by Bosart and Bartlo (1991) and Molinari et al. (1995) have tried to put such interactions into a consistent theory. One of the cornerstones of this book is the dis-tinction between the "classic" deep-tropical hurricane and hurricanes that form with the aid of middle-latitude upper-level dynamics. Indeed, the North Atlantic basin is unique in that tropical cyclone formations can occur 30° or more away from the equa-tor (Figure 1.5). The percentage of high-latitude hurricane formations (north of 20°N) to low-latitude formations is greatest in the North Atlantic basin.

An important ingredient for the formation of a hurricane is a pre-existing low-level atmospheric disturbance. The majority of North Atlantic hurricanes develop from

---

[19]The relationship between upper-level heating and surface pressure changes is called the hydrostatic balance.

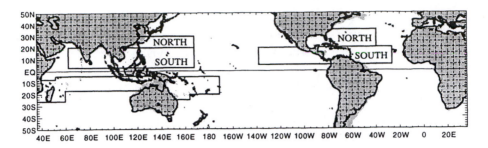

Fig. 1.5: Global tropical cyclone source regions as indicated by rectangular areas. Note that the North Atlantic region is unique in having a relatively large area north of 20°N latitude (NORTH) supporting tropical cyclone development. The ratio of the areas of formation north and south of 20°N is largest over the North Atlantic basin.

convectively active tropical waves emerging from western Africa. The waves move westward at fairly regular intervals between 10 and 20°N latitude, primarily during August and September. Approximately 40 to 70 such waves occur in a season with the vast majority never reaching hurricane strength. The waves are relatively cool in the lower atmosphere but warm aloft. The associated shower and thunderstorm activity make these atmospheric waves potential precursors to tropical storms and hurricanes. The development of a tropical wave is indicated by the occurrence of low-level westerly winds against the prevailing easterlies. The conditions necessary for hurricane development beyond the initial disturbance are detailed next.

2

Hurricane Categories and Impacts

Hurricanes are classified in a number of different ways. The most intense tropical cyclones are called major hurricanes, while the less intense ones are called regular hurricanes. Other classifications are based on developmental factors. Hybrid storms, subtropical cyclones, tropical-only hurricanes, and baroclinically-enhanced hurricanes represent categories of tropical cyclones that are distinguished by different developmental or structural characteristics. Hurricanes differ in fundamental ways from extra-tropical cyclones common in the middle latitudes. This chapter examines the development and growth of hurricanes. Consideration is given to the various categories of North Atlantic hurricanes. The impacts of hurricanes from wind, rain, and storm surge are considered.

2.1 Life Cycle

Tropical cyclones, like other atmospheric storms, progress through regular stages of development from a disturbance to a mature hurricane. The life cycle of a hurricane is divided into four stages. The formative stage begins with a closed circulation. Minimum central pressures drop to 1000 mb or lower. Technically the term *development* refers to an increase to tropical storm intensity. The term *intensification* is reserved for development beyond minimal tropical storm intensity. The early mature stage begins at hurricane intensity and continues until winds reach maximum intensity and pressures drop to their lowest values. Pressures are less than 1000 mb and the organized wind field is usually symmetric. The appearance of narrow feeder bands of showers is a feature of the early stage of hurricane intensification. The mature stage lasts from the time of maximum intensity until the storm weakens below hurricane strength or transforms into an extra-tropical cyclone. At this stage the hurricane grows in size and the strongest winds extend farther from the center of circulation. A distinct asymmetry featuring the fastest winds on the right side of the storm may develop. The fully mature phase usually begins at the time of recurvature or at a latitude considerably farther to the north than its origin. The decaying stage is characterized by a rapid decrease in winds after landfall, or by the transformation of the system to an extra-tropical storm.

Table 2.1: North Atlantic tropical cyclones. The tropical wave is typically a precursor to a tropical cyclone. Major hurricanes are called intense hurricanes.

Category	Development and Intensification Criteria
Tropical Wave	Small pressure drop (typically less than 3 mb) along a latitude. To the west of the trough surface winds are divergent (spreading apart) and air is mostly cloud free, while to the east of the trough there is enhanced cloudiness. Sometimes called easterly waves or African waves. This stage is absent in some developments.
Tropical Depression	The early stages of a tropical cyclone in which the maximum sustained (1-minute average) surface wind speed is below 18 m s^{-1}. Also, the decaying stages of a tropical cyclone in which the maximum sustained surface wind has dropped below 18 m s^{-1}.
Tropical Storm	A warm-core tropical cyclone in which maximum sustained surface winds range between 18 and 32 m s^{-1}.
Hurricane	A warm-core tropical cyclone in which maximum sustained surface winds are at least 33 m s^{-1}.
Major Hurricane	A hurricane in which maximum sustained winds are at least 50 m s^{-1}. Also called an *intense hurricane*.

The length of time a storm spends in each phase varies widely and a particular stage may be skipped entirely.

The various stages of hurricane development lead to a classification of tropical cyclones as shown in Table 2.1. Gordon E. Dunn[1] was the first to suggest the importance of easterly (or tropical) waves as a precursor to hurricane formation. He observed a regular sequence of falling and rising pressures associated with waves moving from east to west across the Caribbean Islands. The tropical wave is essentially a trough of low pressure near the ground embedded in the deep easterly flow on the equator-ward side of the subtropical high-pressure area. The classic scenario for tropical cyclone formation in the North Atlantic basin is development from a tropical easterly wave. Figure 2.1 shows a schematic of an easterly wave over the Caribbean Sea and western North Atlantic. Most of the shower and thunderstorm activity forms to the east of the wave axis in the area of convergence of low-level easterly winds. Easterly waves are a common feature of the tropical North Atlantic region between May and November, with maximum frequency occurring in August and September. Originating over Africa generally east of 20°E latitude, most waves move westward across the North Atlantic. Above the surface, the tropical wave is characterized by a wind shift (northeast

---

[1]Gordon E. Dunn was a preeminent forecaster of tropical cyclones and was the first official director of the National Hurricane Center (NHC), though it is generally recognized that his predecessor, Grady Norton, a pioneer in tropical weather prediction, was the first unofficial director, being the meteorologist-in-charge at the Miami forecast office before it was named the NHC (Sheets 1990). Former directors of NHC include Robert H. Simpson, Neil L. Frank, Robert C. Sheets, and Robert W. Burpee. Jerry D. Jarrell is current director of NHC.

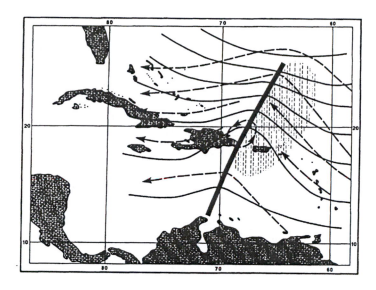

Fig. 2.1: Easterly (or tropical) wave over the Caribbean Sea. Solid lines show the isobars (lines of constant pressure) at the surface. Dashed lines are the flow of air (streamlines) at 500 mb above the surface. The heavy line shows the trough axis. Stippling indicates areas of increased convection (showers and thunderstorms). Reproduced with permission from Lydolph (1985).

to southeast) that is most pronounced near the 700 mb level (Avila and Pasch 1995). Though only relatively few easterly waves intensify into hurricanes, nearly half of all North Atlantic tropical cyclones originate from such waves (Frank 1970a). The forward speed of a tropical easterly wave is on the order of 5° of longitude per day (20 to 25 km hr^{-1}). Usually an easterly wave that is slowing down indicates the potential for development. In contrast, an accelerating wave suggests a weakening of the wave. Satellite pictures provide an effective means of tracking tropical waves as they cross the open waters of the North Atlantic.

Specific intensification criteria for tropical waves were first identified in an objective way by Robert H. Simpson in 1971. His work produced a decision tree to aid forecasters in determining the potential for intensification (or decay) of disturbances in the tropics as well as in the subtropics (Simpson 1971). The criteria used were largely adapted from factors that Gray (1968) showed to be climatologically related to tropical cyclone development. These factors were qualitatively identified in Riehl (1954). Hebert (1977) introduced a slight modification to the decision tree of Simpson (1971). In general, criteria favorable for development include a cloud mass north and east of the tropical wave axis, above normal temperature aloft, feeder bands extending south of the main cloud mass, abundant moisture at low levels, warm SSTs, weak shear of horizontal winds aloft, surface cyclonic vorticity[2], and advection of vorticity at upper levels (200 mb). These conditions are conducive to the organization of convection into

---

[2]Vorticity is a vector measure of the local rotation in moving air.

distinct mesoscale convective elements. Hebert (1977) also remarked about the favorable environment for wave development produced by the proximity of a baroclinic trough. If a trough is approaching a wave, but remains at least 10° latitude away, then upper-level outflow over the incipient wave is enhanced increasing the chance for development. This criterion is most important for disturbances moving to higher latitudes where westerly winds dominate. A climatology of tropical cyclones that develop into hurricanes with the aid of middle-latitude baroclinic factors is presented in Chapter 6.

As a tropical wave deepens (surface pressures fall), and a closed circulation is detected, the wave is classified as a tropical depression. The intensity of the depression can be estimated by the strength of the west winds to the south of the center of circulation. Tropical depressions in the North Atlantic are numbered consecutively beginning with the first depression of the year. On average, the Atlantic basin experiences about 12 tropical depressions a season. A tropical depression becomes a tropical storm when surface winds reach gale force (maximum sustained speed of 18 m s^{-1}). It takes a day or two, sometimes longer, for a depression to strengthen to a tropical storm. The tropical storm stage is marked by the appearance of a core region of deep convection. At this stage the tropical cyclone is given a name. The naming convention was introduced to avoid confusion when more than one storm was being tracked at the same time. Separate sets of tropical storm names are used for the central and eastern North Pacific as well as for the North Atlantic.[3] A separate list of names in alphabetical order is used for each new hurricane season and the same list is repeated every six years with the exception of names that are retired. Names of hurricanes causing significant damage or loss of life like *Camille* and *Gilbert* are deleted from the lists (see Chapter 18).

Figure 2.2 shows the track and the stages of development for hurricane *Andrew* in 1992. The tropical cyclone originated as a tropical depression near 10°N latitude, became a tropical storm east of 45°W longitude, and reached hurricane intensity near 25°N latitude and 68°W longitude. *Andrew* remained a hurricane as it crossed south Florida and the central Gulf of Mexico, and did not weaken to a tropical storm until after landfall in Louisiana. Further development of a tropical cyclone beyond the minimum tropical storm intensity is referred to as intensification. Understandably, hurricane research has focused largely on the intensification stage.

It is important to use *sustained winds* in characterizing tropical cyclone intensification because the actual winds are gusty and can be considerably greater or less at a given instant. Sustained winds are defined as the wind averaged over a one-minute interval. When a tropical storm intensifies so that maximum sustained winds reach 33 m s^{-1}, it is classified as a hurricane.[4] At this intensity, or slightly stronger, a trop-

---

[3]The official naming of North Atlantic tropical storms by the U.S. government began in 1950 with the phonetic alphabet (*Able, Baker*, etc.). The practice of using women's names for western Pacific storms during World War II was introduced for North Atlantic tropical storms in 1953. Alternating men's and women's names began in 1979. No names begin with the letters Q, U, X, Y, or Z. Historically, hurricanes in the Spanish islands of the Caribbean were named after the Saint's day on which the hurricane occurred.

[4]In 1806, Admiral Sir Francis Beaufort devised the Beaufort wind scale to help sailors determine how much sail a ship should carry. Based on watching the influence of the wind on the waters, his designation of winds of 33 m s^{-1} or more as hurricane force is the origin of the current definition of a hurricane. Beaufort's scale was later adapted to winds observations on land.

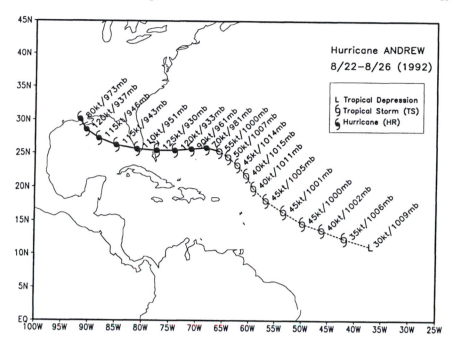

Fig. 2.2: Stages of development and intensification for hurricane *Andrew* in 1992. The tropical storm did not reach hurricane intensity until after a turn toward the west. *Andrew* devastated parts of southern Florida and Louisiana, becoming the costliest hurricane on record for the United States.

ical cyclone typically develops a relatively calm central region known as the hurricane eye, though sometimes the eye is ill-defined until intensities reach 40 m s^{-1} or more. The eye, as a warm column of air extending from the surface to 15 km or more above, is one of the most distinguishing features of the mature hurricane. Tropical cyclone intensification is more likely for slower moving storms and storms that track to the north rather than to the south. The passage of a migratory surface high pressure area to the north of the cyclone usually signals intensification. Stronger hurricanes are known to develop suction vortices, which are cores of faster winds, usually less than a few kilometers wide, within the eye wall.

Hurricanes that intensify to maximum sustained surface winds in excess of 50 m s^{-1} are considered major (or intense) hurricanes. Intense hurricanes may have a very small eye and frequently develop a concentric double eye-wall structure. As intensification peaks, the inner eye wall begins to break apart and is replaced by a larger outer eye wall. The outer eye wall begins to tighten as the hurricane goes through another period of intensification. The hurricane's *maximum potential intensity* is limited by the temperature of the ocean surface. A warmer ocean provides a greater potential for hurricane intensification. However, the actual intensity of a hurricane is limited by the presence of wind shear (see Chapter 1). Moreover, the strong winds generated by the hurricane over the ocean surface can create local ocean changes (including added sea spray) that feed back to inhibit or enhance further intensification. Rapid intensification often occurs in conjunction with an eye wall replacement cycle. Hurricanes that intensify rapidly are of particular concern to forecast and warning operations. Rapid intensification is defined as a drop in minimum central pressure of approximately 1 mb hr^{-1} sustained over a day. Necessary, but not sufficient, conditions for significant deepening of hurricanes include weak vertical shear of horizontal winds, warm SSTs (>28°C), and weak fluxes of local angular momentum at upper levels (200 mb) above the hurricane. These conditions are common over the deep tropics during the heart of the hurricane season. All other things being equal, storm decay does not begin until the eye has passed completely over land. As a first approximation, the over-land wind speed decay rate is proportional to wind speed at landfall. Faster winds diminish more quickly. The rougher the terrain over which the hurricane passes, the more rapidly the storm will dissipated. Though comparatively rare, major landfalling hurricanes are responsible for the largest percentage of all hurricane damage in the United States.

The life cycle of tropical cyclones are also marked by distinct changes in surface pressure patterns. Tropical waves are characterized by a trough of low pressure oriented north to south and located on the equator-ward side of a subtropical high-pressure region (called the Bermuda high when situated over the western subtropical North Atlantic). Pressures are at most only a few millibars lower in the middle of the trough than to the east or west of the trough axis. The tropical depression stage features the formation of a closed area of low pressure with minimum values of pressure between 1007 and 1016 mb. Central pressures inside a tropical storm are considerably lower, typically in the range of 990 to 1005 mb. Hurricanes have central pressures that are on average as low as 950 mb, but the values can go considerably lower. The lowest surface air pressure ever recorded in the Atlantic basin was 888 mb (26.17 in of mercury) during hurricane *Gilbert* during September of 1988[5] (Chapter 9 mentions the impact of hurricane *Gilbert* on the Caribbean).

Figure 2.3 shows the distribution of central pressures for all hurricane observations over the North Atlantic basin for the period 1886–1996. The distribution is asymmetric as a result of the extremely low pressures in rare hurricane cases. Half of all hurricane observations have central pressures 980 mb or lower. The central pressures at the surface inside a hurricane provide a more reliable indication of intensity than do the wind speeds. An approximate relationship between the maximum sustained wind speeds and minimum surface pressures for Atlantic basin hurricanes is empirically estimated by Kraft (1961). The formula, which is based on the cyclostrophic wind equation,[6] is given by

$$V_{max} = 7\sqrt{(1013 - P_c)}, \tag{2.1}$$

where V_{max} is the maximum sustained wind speed in m s^{-1} and P_c is the minimum

---

[5]Typhoon *Tip* in the western North Pacific had an observed central pressure of 870 mb on October 12, 1979.

[6]The cyclostrophic wind equation results from a two-way balance between centrifugal and pressure gradient forces.

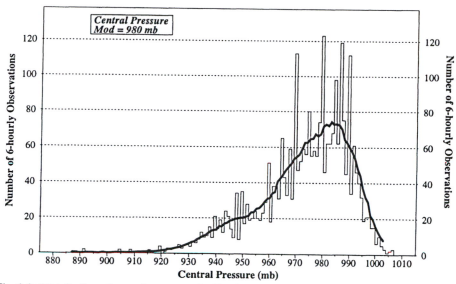

Fig. 2.3: Distribution of central pressures for North Atlantic hurricanes based on data from 1886 through 1996. Lower central pressures indicate a more intense hurricane. The median pressure is 976 mb. Data sources are discussed in Chapter 3.

central sea-level pressure in millibars. The relationship is only approximate because, in part, the wind flow around a hurricane is strictly related to horizontal gradients of pressure rather than a single, central value. However, the above formula is useful in that historical hurricane records do not always contain simultaneous pressure estimates from elsewhere around the storm.

As is true in much of the natural sciences, the strict criteria for classifying tropical cyclones hides the fact that the observations necessary to make the delineation are often subjective. For example, wind speeds for cyclones over the eastern North Atlantic are typically inferred indirectly from satellite pictures. Although users of information in this book should be aware of limitations of this sort, the large number of cases available make the results presented here quite robust. The limitations of the available data sets are discussed in Chapter 3.

2.2   The Saffir/Simpson Scale

As a tropical cyclone intensifies beyond a minimum threshold intensity, it is further classified using a disaster-potential scale. The hurricane disaster-potential scale is descriptive and was developed in the 1970s by the U.S. National Weather Service (NWS) as an experiment to give public safety officials more information on the potential for wind and *storm-surge*[7] damage from a hurricane threatening a coastal community. The

---

[7] Storm surge is the wind- and pressure-caused high tide that accompanies the landfall of hurricanes. This "wall of water" can be as high as 5 m or more above the normal tide levels in a major hurricane.

Table 2.2: The Saffir/Simpson hurricane disaster-potential scale. The numbers indicate the category, P is central barometric pressures. V is maximum sustained 1-minute averaged wind speeds and G is peak wind gusts. Surge heights are given in meters (m). An example for each category is also given. Note that normal barometric pressure at sea level is 1013.25 mb, which is 29.92 in of mercury. $1 \text{ m s}^{-1} = 1.94 \text{ knot} = 2.24 \text{ mph}$. Examples include hurricanes *Agnes* in 1972, *Cleo* in 1964, *Betsy* in 1965, *Andrew* in 1992, and *Camille* in 1969.

Cat.	Damage	P (mb)	V (m s^{-1})	G (m s^{-1})	Surge (m)	Example
1	Minimal	>980	33–42	41–53	1	*Agnes*
2	Moderate	980–965	42–50	53–62	2	*Cleo*
3	Extensive	964–945	50–58	62–72	3	*Betsy*
4	Extreme	944–920	58–69	72–86	4–5	*Andrew*
5	Catastrophic	<920	>69	>86	>5	*Camille*

scale is now referred to as the *Saffir/Simpson scale* after its authors Herbert Saffir and Robert H. Simpson. The Saffir/Simpson scale is used by the NHC and the NWS to give public officials information on the destructive potential of a hurricane threatening landfall. It represents an estimate of what a hurricane could possibly do to a coastal area if it were to make a direct hit. The scale is based on maximum sustained one-minute averaged near-surface winds and can be considered as an extension of the classification of the various stages of development up to hurricane strength (Table 2.2). Peak wind gusts are based on 1-second averages.

Much of the hurricane damage to coastal communities results from the storm surge. The magnitude of the surge depends on the wind speed as well as on the forward speed and direction of the hurricane with respect to the coastline as the eye wall moves on shore. Other factors being equal, the greater the wind velocity, the higher the storm surge. Yet the speed and direction of the approaching hurricane can also affect surge height. The faster and the more direct (more perpendicular to the coast) the approach, the greater the storm surge height (see next section). The following information is taken from an article appearing in *Weatherwise* of August, 1974 describing the effects of hurricanes. A category one (or minimal) hurricane on the Saffir/Simpson scale has winds between 33 and 42 m s^{-1}. Damage from these hurricanes is primarily to small trees, foliage, and unanchored mobile homes. Storm surge can be expected to be one meter above normal height with flooding to low-lying coastal roads and minor pier damage. Small water vessels can be ripped from their moorings. A category two hurricane has winds of 42 to 50 m s^{-1}. Damage to trees and foliage can be considerable, with some trees completely blown over. Mobile homes are vulnerable to significant damage, particularly ones in open areas unprotected from the wind. Minor damage to roofs and windows of anchored buildings and homes is common. Damage patterns may be localized depending on the strength of the thunderstorms embedded in the core region of the eye wall. The stronger the convection, the more widespread the damage. Storm surge can reach 2 meters above normal, causing flooding to coastal roads and low-lying evacuation routes 2 to 4 hours before landfall. With category two hurricanes, there is the potential for considerable damage to piers and marina flooding. Evacuation of some shoreline residences and low-lying island areas may be necessary.[8]

More than three-quarters of all hurricane damage results from hurricanes in categories three, four, or five on the Saffir/Simpson scale. These are the major (or intense hurricanes). In a category three hurricane winds are between 50 and 58 m s^{-1}. These wind speeds are capable of uprooting large trees and tearing foliage entirely. Buildings typically experience roof, window, and door damage. Structural damage is likely in smaller buildings. Mobile homes are often completely destroyed. Storm surge can reach 3 meters above average. Widespread flooding occurs at the coast and most smaller structures on the coast are destroyed. The larger structures near the coast are damaged by waves and floating debris. The low-lying evacuation routes are cut by the rising waters 3 to 5 hours before hurricane landfall. The evacuation of low-lying coastal residence within several blocks of the shore is recommended.

A category four hurricane has winds between 58 and 69 m s^{-1}. Winds of this magnitude can blow down large shrubs and trees. Extensive damage occurs to roofs, windows, and doors. The complete failure of roofs is possible on many small residences, and the total destruction of mobile homes is likely. The storm surge can reach 4 to 5 meters above normal, causing widespread flooding as far as 10 km inland. Major damage results from wave action on structures near the shore and due to floating debris. Major shifts in the sand along the beaches occur. Massive evacuation of all coastal residences within 200 m of shore is advised. A category five hurricane has winds in excess of 69 m s^{-1}. Very severe and extensive damage is likely with these winds. Most roofs and windows on residential and commercial buildings fail. Small buildings can be completely overturned or blown away. The storm surge exceeds 5 meters above normal height, causing catastrophic damage to all structures within 150 m of the shore. There have only been two category five hurricanes to strike the United States this century, the most recent being hurricane *Camille* in 1969.

Intense precipitation rates and copious rainfall amounts accompany most hurricanes. Since the hurricane resides primarily over the open ocean, this rainfall is often of little societal concern. Over the ocean the distribution of rainfall relative to the hurricane varies markedly from one storm to another. As the hurricane approaches land, heavy rain rates lead to major flooding problems particularly along the immediate coast. The intensity of rainfall generally diminishes rapidly with distance from the shoreline. However, if the circulation of winds around the hurricane stays intact as it approaches hills or mountains additional heavy precipitation falls along the windward slopes (Schoner 1968). Some of the worst floods on record in the eastern United States were caused by hurricanes. Indeed, hurricane *Agnes* in 1972 was responsible for only minor damage when it made landfall along the Gulf coast, but as the circulation moved north into the mountains of West Virginia and Pennsylvania it caused catastrophic floods. On the other hand, heavy rainfall accompanying hurricanes can be

---

[8]*Bob* in 1991 was a category two hurricane at landfall over New England.

beneficial to agriculture when the rains break a prolonged episode of drought conditions. Frequently the heaviest rainfall amounts accumulate when a hurricane stalls its forward motion.

2.3 Storm Surge

The most destructive component of the hurricane is the storm surge. The storm surge is defined as the difference between the storm tide and the normal tide. The normal tide is quasi-periodic representing a combination of the astronomical tide and the geophysical tide. The normal tide has periods of approximately 12.5 and 25 hr. Astronomical tides are caused by the gravitational pull of the sun and moon. Geophysical tides are the deviation from the mean sea level for the tidal period due to large-scale meteorological and oceanographic effects such as prevailing winds, pressure gradients, sea and air temperatures as well as variations in the salinity of the water. Storm tides are caused by hurricane-generated winds.

The astronomical tide is regular and predictable since it depends on the known motions of the earth and moon. Astronomical tides arise from the interplay of several forces. The moon exerts a pull on the oceans that is stronger on the waters facing it. A similar, but weaker, pull is exerted on the oceans by the sun. Finally, these two differential forces act on a rotating planet. The result is a pair of tidal bulges, one from the moon and a smaller one from the sun, that sweep across the earth (see Zebrowski 1997). The storm tide is more difficult to forecast; it depends not only on the size and intensity of the hurricane but also on the shape of the coastline, the angle between the direction of approach and the coastline, as well as other factors, such as depth of continental shelf (bathymetry), and the duration of strongest winds. The significance of a storm surge depends on the stage of the normal tide and the timing of the storm tide.

As a general guide, peak surge heights from a landfalling hurricane increase with a decrease in central pressure, an increase in hurricane size, and an increase in speed of approach (Simpson and Riehl 1980). Greater surge heights are found along coastlines where the bottom topography slopes more gently (continental shelves). Inundation from storm surge is most acute along the right-forward quadrant of the hurricane at time of high astronomical tides. Deviations from this pattern are produced by local land topography. Peak surges are often found away from the coast near the headwaters of an open bay or inlet. The period of high water associated with the surge is likely to last from 6 hours to several days. Figure 2.4 is a schematic of a storm surge generated from a hurricane. Heights may exceed 4 to 5 m and are greatest on the right side of the hurricane eye. Over the open ocean, sea swells propagate outward from the center of the storm with a tendency for heavier swells in front (ahead of the direction of motion) of the hurricane. In general, the wave frequency (the inverse of the time between successive swells) is constant but the amplitude diminishes with distance from the storm. The swells transport no mass horizontally as long as there is sufficient depth

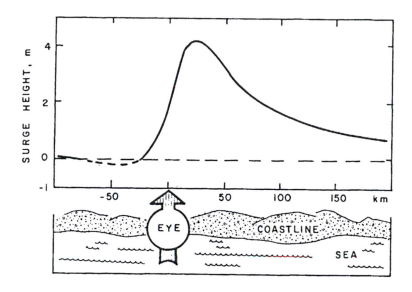

Fig. 2.4: Schematic vertical cross-section of surge heights along the coastline as a hurricane makes landfall. Reproduced with permission from Simpson and Riehl (1980).

to the ocean. However, in shallow water along the coastline, the swells crest and break, carrying huge amounts of water toward the coast and eventually onto the beach and inland.

The forward velocity in the wave crest exceeds the backward velocity in the trough transporting water forward toward the coast. Water movement by waves is variable along the coast and can produce extreme levels of water height above the surge. Sustained winds blowing over the ocean above the continental shelf produces a net transport of water to the right of the wind direction due to the Ekman spiral.[9] The shore-parallel wind component has a significant influence on water level rises at the coast. For instance, southerly winds over the eastern Gulf of Mexico piles the water up along the west-facing beaches of peninsular Florida. The suctioning of water due to the low atmospheric pressures within the hurricane eye contributes to the surge height. The ocean level rises by a third of a meter for a 30 mb drop in central air pressure. For a category five hurricane this amounts to an additional one meter above the 5 m of wind-driven surge.

Flooding by the storm surge is locally variable. It depends on the extent of shallow water near the coast, the shape of the coastline, the height of the shoreline, and the orientation of the winds with respect to the coastline. Flooding is more likely in coastal locations where the coastline is concave forming a bay or inlet. In bays and inlets the situation is complicated and often extreme. Tides are highly intricate in estuaries,

---

[9]The Ekman spiral is a change of wind-driven ocean currents with depth. The change is a clockwise turning and slowing of the currents with increasing depth from the surface.

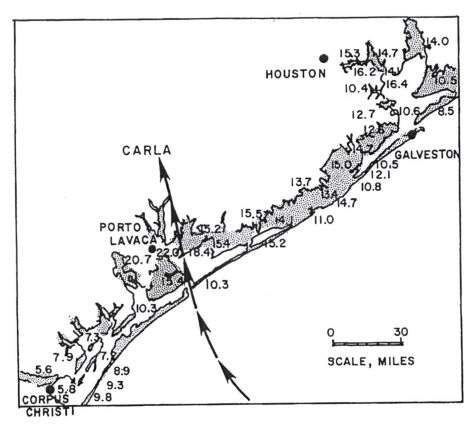

Fig. 2.5: Storm-tide levels in feet for locations along the central Texas coast during hurricane *Carla* in 1961. The arrows indicate the track of the center of the hurricane. Reproduced with permission from Simpson and Riehl (1980).

lagoons, and bays because inlets and outlets to these basins are not arranged in a regular way. Simpson and Riehl (1980) note that water-level rise in bays and estuaries may exceed those at an open coast by as much as 50% or more for slow moving hurricanes. In fact, hurricane-driven tides in bays may reach 6 to 8 meters with the storm waves on top of that. Tidal effects in an estuary are further complicated by the rate at which the river water is flowing into the estuary (river discharge rate). During rainy periods discharge rates are high, exacerbating storm-surge levels. Figure 2.5 gives the storm-tide levels in feet during hurricane *Carla* in 1961. Inland tide levels were generally higher than tide levels along the open coastline. Storm tides are recorded as historical events and projected into the future as return periods (see Chapter 11). For instance, a 100-year storm-tide map yields the area where storm-surge flooding can be expected under conditions which conspire on average once a century (an annual probability of 1%).

Table 2.3: Fujita scale for tornado intensities. Wind speeds are given in m s^{-1}. Most tornadoes are of magnitude (Mag.) F-1 or less.

Mag.	Damage	Wind Speed (m s^{-1})	Characteristic Effects
F-0	Light	18–33	Damage to small trees, signs, and chimneys.
F-1	Moderate	33–50	Roofs damaged, mobile homes pushed from foundations, moving cars swept off roads.
F-2	Considerable	50–70	Large trees uprooted, roofs torn from houses, mobile homes destroyed. Damage from flying debris.
F-3	Severe	70–92	Trees leveled, cars overturned, walls removed from buildings.
F-4	Devastating	92–116	Well-constructed homes are demolished, cars are lifted from the ground.
F-5	Incredible	116–142	Buildings lifted from foundations and obliterated in the air, cars are carried great distances. Total devastation.

2.4 Hurricane Tornadoes

The potential destruction of a hurricane is partly linked to the possibility of hurricane-spawned tornadoes. A tornado[10] is a relatively small (100 m or so), but powerful vortex of winds in the shape of a funnel extending from the base of a cloud to the ground and rotating at tremendous speeds. In brief, the deceleration of the hurricane-force winds due to friction over land, coupled with rapidly rising air inside thunderstorms, can generate tornadoes as the hurricane makes landfall. Though typically much weaker, narrower, and shorter-lived than the classic springtime tornadoes of the central United States, their destructive and life-threatening potential is an important component of the overall destructive threat from hurricanes. Hurricane-spawned tornadoes are particular threat to the Gulf coast states. Tornado intensities are rated on the Fujita scale (Table 2.3). The Fujita scale[11] estimates wind speeds based on reported damage.[12] The vast majority of tornadoes and nearly all hurricane-spawned tornadoes are of category F-0 or F-1. Damage is described as light to moderate with damage to trees, signs and roof tops. Hurricane *Isbell* in October 1964 produced four confirmed F-2 tornadoes over southwestern Florida. Only rarely is a substantially more powerful F-3 tornado observed with a hurricane landfall.

Estimates indicate most hurricanes that make landfall in the United States south

---

[10]Tornado is Spanish for "tornada," which means thunderstorm. The term comes from Latin "tornare," meaning "make round by turning" (Petak and Atkisson 1982).

[11]After Theodore Fujita of the University of Chicago.

[12]This is fundamentally different than the Saffir/Simpson scale for hurricanes, which estimates the damage potential based on observed (or estimated) wind speeds (M. Green, personal communication).

of New England generate tornadoes (Gentry 1983). Hurricane-spawned tornadoes are
responsible for up to 10% of the overall hurricane fatalities and up to 50% of the overall damage (Novlan and Gray 1974). Most of the tornadoes develop within 150 km of
the hurricane center with a strong preference for development ahead and to the right
of the direction of motion (see Figure 2.6). The preferred location results from coincident areas of strong vertical shear of the horizontal winds, rapidly rising and sinking
air, and strong local convergence of the winds (Gentry 1983). Tornado formation is
most likely when the winds above the surface ($\approx$ 850 mb) remain strong but weaken
near the ground. The majority of hurricane-spawned tornadoes occur in the outer spiral bands of convection with about a fifth occurring in the thunderstorms on the outer
edge of the eye wall (Gentry 1983). Rapidly moving hurricanes and those with more
westerly paths have fewer tornadoes. Significant tornado outbreaks over the Florida
peninsula are more likely associated with hybrid tropical cyclones (Hagemeyer 1997).
These hybrid (or non-tropical-only) systems have both tropical and extra-tropical characteristics and tend to occur early or late in the hurricane season. One of the earliest
reports of hurricane-generated tornadoes occurred over southeastern Florida during the
late September hurricane of 1929. The eye of the storm passed directly over Long Key.

2.5   Subtropical Cyclones

Development and origin mechanisms provide another way to classify hurricanes. Tropical cyclones sometimes develop in the absence of tropical waves. In fact, it is known
that tropical cyclones can originate from a polar trough or an upper-level cold-core circulation. A cold-core circulation which becomes displaced south of the high-latitude
westerly current is called a *cut-off low*. The transformation of a cold-core cut-off low
into a warm-core circulation can be quite sudden but typically takes several days to a
week or more. These cold-core systems are sometimes labeled *subtropical cyclones*.
The process of latent heat release by condensation will gradually warm the central
regions of the subtropical storm in the absence of cold-air advection.[13]

A subtropical cyclone is a low-level manifestation of a cut-off low and is formally
defined as a low-pressure system that develops over tropical waters with a non-tropical
upper-level circulation but with significant convective (cumulonimbus) clouds characteristic of tropical cyclones. Subtropical cyclones with maximum sustained winds less
than 18 m s^{-1} are called *subtropical depressions*. Subtropical cyclones with winds exceeding this threshold are called *subtropical storms*. They differ from tropical cyclones
in that the region of maximum sustained winds is usually farther from the low-pressure
center, perhaps as far away as 160 km or more. Moreover, with subtropical cyclones,
the minimum central pressures can be lower than the corresponding winds would suggest (Hebert et al. 1993). Though there is no wind-speed maximum for subtropical
storms, subtropical storms that reach or exceed 33 m s^{-1} have undergone sufficient

---

[13]Advection is the transfer of atmospheric properties, such as temperature, by the winds.

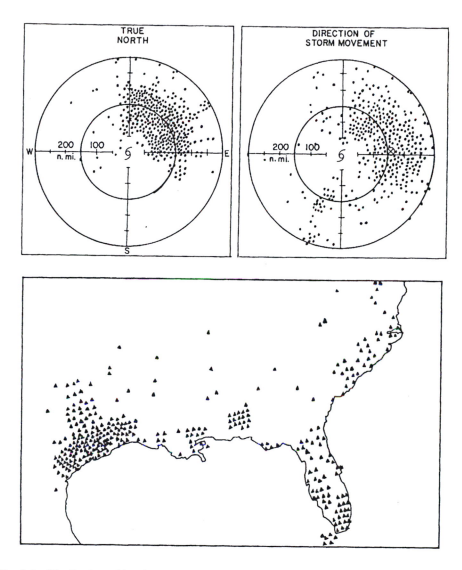

Fig. 2.6: Distribution of hurricane tornadoes with respect to the center of the hurricane and with respect to true north and the direction of movement (top). Also shown is the distribution of hurricane tornadoes along the U.S. coastline (bottom). Reproduced with permission from Simpson and Riehl (1980).

transition to a warm-core circulation to be designated as hurricanes. Hurricane *Florence* in November 1994 developed as a subtropical cyclone. Subtropical cyclones are more common during the early and later parts of the hurricane season.

2.6 Tropical Only Hurricanes

Differences in North Atlantic tropical cyclone origin and development mechanisms allow a classification that identifies hurricanes by ontogeny[14] rather than morphology. According to this classification, a "tropical-only hurricane" is defined as a hurricane that originates from a tropical wave (or disturbance) and develops to hurricane intensity devoid of enhancing middle-latitude baroclinic[15] influences. This definition does not imply that tropical-only hurricanes are necessarily exempt from interactions with baroclinic weather systems. Often adverse environmental conditions produced by middle-latitude upper-air currents hinder the development of a tropical cyclone (see Chapter 1). A developing tropical cyclone that begins as a tropical wave and reaches hurricane strength despite encounters with hostile effects of baroclinic systems is considered a tropical-only hurricane. Only if baroclinic influences are such that they *aid* formation and growth in a situation where otherwise development was unlikely, is the hurricane not considered a tropical-only hurricane. It is important to keep in mind that the definition of tropical-only hurricanes is a description based on the process of tropical-cyclone initiation and growth and is distinct from a definition based on form or structure as is the definition of a subtropical storm. The classification of tropical-only hurricanes is motivated by a desire to understand the climatology of North Atlantic hurricanes, and may not be important for forecasting individual hurricane tracks and intensity changes. Moreover, it may not be directly relevant to other tropical-cyclone basins, including the eastern North Pacific.

Since tropical-only hurricanes are defined with respect to their origin and evolution, it serves us well to note that biologists still argue about how to define "species" 135 years after Darwin (Dennett 1995). One must be careful, of course, but it may be best to consider hurricanes as having various origin and development mechanisms. Research has shown (Hess and Elsner 1994a, Hess et al. 1995, Elsner et al. 1996b) that it is useful to consider North Atlantic hurricane activity as the sum of different hurricane types. This distinction allows one to see otherwise hidden structures in the record. Moreover, as noted in Chapter 1, there are studies identifying underlying physical connections that cause some hurricanes to develop under the direct influence of middle-latitude baroclinic dynamics.

In the North Atlantic there are basically two distinct mechanisms for the development and formation of tropical-only hurricanes. In the eastern and central Atlantic, tropical-only hurricanes originate from tropical waves moving off the west coast of Africa near the islands of Cape Verde (see Figure 2.7). These waves are a product

---

[14]Ontogeny refers to the process of growth.

[15]Baroclinic refers to a state of the atmosphere where air density varies from one place to another at the same atmospheric pressure.

of instabilities in the low-level African easterly jet stream and the resultant hurricanes are traditionally referred to as *Cape Verde hurricanes*. In the western Caribbean Sea and Gulf of Mexico, hurricanes sometimes form from disturbances originating in the confluence of the northeast trade winds with air flowing northward across the equator as an extension of the southeast trades of the Southern Hemisphere. This region of confluence is known as the *inter-tropical convergence zone* (ITCZ) and is associated with widespread shower and thunderstorm activity. Areas of lower pressure within the confluence, or trough, axis (monsoon trough) can lead to a *monsoon depression*, characterized by a significant expanse of scattered deep convection and a large core (no definitive center) of light winds surrounded by bands of much stronger winds. Though more common over the western North Pacific, monsoon depressions, with their quasi-stable configuration of winds and convection, occasionally develop into North Atlantic hurricanes. The ITCZ reaches its most northerly latitude across the eastern North Pacific extending into the Caribbean in late summer and early fall. Frequently, when disturbances from the ITCZ reach hurricane strength, they are classified as tropical-only hurricanes. However, it is not uncommon to observe an enhancement of the cross-equatorial flow resulting from a cold front that has plunged southward (called a "norther") through the Gulf of Mexico. In such cases the disturbance may intensify to hurricane force due to a favorable baroclinic environment. These are non-tropical-only hurricanes.

2.7 Baroclinically Enhanced Hurricanes

Subtropical cyclones that transform into hurricanes are part of the larger category of baroclinically-enhanced hurricanes. Strictly speaking, baroclinically-enhanced hurricanes are non-tropical-only hurricanes. Actually, they come in two flavors. A hurricane born of a non-tropical wave or disturbance is considered a baroclinically-enhanced hurricane. Also, a hurricane that originated from a tropical wave or disturbance, but benefited favorably from mid-latitude baroclinic influences as it intensified to hurricane strength is considered a baroclinically-enhanced hurricane. To make a distinction between these two types of baroclinically-enhanced hurricanes and to recognize that climatologically they are likely distinct, the term baroclinically-initiated (BI) hurricane is applied to the former type. This classification also helps to emphasize that not all subtropical cyclones of hurricane intensity are necessarily hurricanes, only those that attain warm-core thermodynamic characterisitics.

As a tropical cyclone interacts with an upper-level trough (or TUTT; see Chapter 1), it usually encounters large vertical wind shear resulting in the destruction of the incipient storm. In general, the westerly shearing winds are strongest to the south of the TUTT axis (Fitzpatrick et al. 1995). In some instances, when the vertical shear is weak, *vorticity advection*[16] (Montgomery and Farrell 1993), or the transfer of angular momentum (Pfeffer and Challa 1992) associated with the upper-level trough, can lead

---

[16]Vorticity advection is the transport of vorticity by the winds.

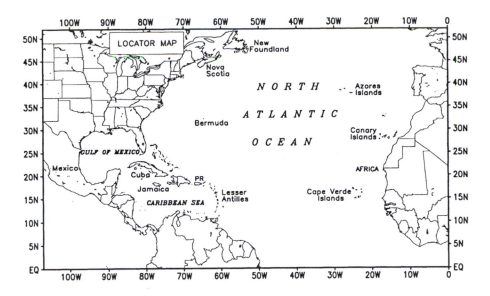

Fig. 2.7: Locator map of the North Atlantic basin. The basin consists of the North Atlantic Ocean, the Caribbean Sea, and the Gulf of Mexico. The islands of the Greater Antilles separate the Gulf of Mexico and the Atlantic to the north from the Caribbean Sea to the south. The islands of the Lesser Antilles separate the tropical Atlantic to the east from the Caribbean Sea to the west.

to an intensification of an otherwise benign tropical system. Whether or not trough interaction enhances development of a tropical cyclone depends on the breadth of the trough and its distance from the storm. Narrow troughs that remain at an upstream distance from the tropical cyclone can enhance development by providing a more favorable upper-level outflow pattern. The difference between tropical-only and some baroclinically-enhanced hurricanes can be understood as a difference between the level at which initial development takes place. Classic tropical-only hurricanes are initially most organized near the surface and develop from the bottom of the atmosphere upward (bottom-up development). In contrast, baroclinically-enhanced hurricanes (particularly the baroclinically-initiated storms) begin most energetically aloft and become hurricanes as they develop from the top downward (top-down development).

Baroclinically-initiated hurricanes originate in several different ways. The most common development occurs in conjunction with a middle latitude frontal trough, or baroclinic trough, that extends into the subtropics. If the trough is associated with divergent air flow at upper levels, the development of a non-tropical surface, low pressure system is likely. Strengthening and a transformation into a warm-core tropical depression is possible if the frontal low is situated over warm waters. Baroclinically-initiated hurricanes can also form from the transformation of an upper troposphere cold low that becomes "cut off" from the prevailing westerly wind flow. Transformation occurs slowly as shower and thunderstorm activity persists in association with the upper-level disturbance. Other, less common mechanisms include the transformation of a decaying (occluded) extra-tropical cyclone and the development of a mesoscale convective complex (MCC; Maddox 1980), which initially forms over the continental United States and drifts south into the Gulf of Mexico or the western North Atlantic.[17] In comparison with tropical-only hurricanes, baroclinically-initiated hurricanes form in closer proximity to the middle-latitude upper-air jet stream, develop over cooler waters, and strengthen in an environment of substantial thermal contrasts. Moreover, baroclinically-initiated hurricanes typically originate at higher latitudes with the result that they generally do not become as strong as tropical-only hurricanes. The idea of dividing North Atlantic hurricane activity into separate subsets, each characterized by unique origins and developmental processes, is an assumption that changes the view of North Atlantic tropical cyclone climate. As is illustrated throughout this book, our understanding of hurricane climate will likely improve by adopting this classification scheme.

2.8   On Frogs and Hurricanes

The distinction between tropical-only and baroclinically-enhanced hurricanes can be illustrated with an example. Frogs are born as tadpoles from eggs, usually laid in water. They grow and develop into adult frogs. This growth and development is called ontogeny.[18] Ontogeny refers to the "becoming" of a frog. Yet frogs belong to the class *Amphibia* which are backboned animals having moist, glandular skins and toes without claws. Amphibians include salamanders and newts, as well as frogs and toads. Along with toads, frogs belong to the order *Anura* of which there are nearly 3700 species known worldwide. The species *Rana* represents the typical frog species known commonly as "true" frogs. The class, order, and species is derived largely from morphology (though often phyletic information is extracted from ontogeny). Morphology or form refers to the structure (being) of the adult frog.

A cursory examination of adult frogs might lead to a classification based on size— small like the north Florida bog frog (*Rana okaloosae*) or large like the bullfrog (*Rana catesbeiana*). This is a categorization based on morphology (recall the size difference between hurricanes *Andrew* and *Hugo*). Another classification can be made based on length of time from egg to adult. Temperatures and other environmental factors control, to some extent, the growth rate of tadpoles so that *Rana okaloosae* might be classified with the more southern subspecies of *Rana catesbeiana*, at least with respect to this particular aspect of ontogeny. Extending this idea a little, we could classify frogs according to whether the tadpole environment is free of sub-20 °C nights. If so, we could define them as tropical-only frogs. Note this could be a very important classification when considering the possible influence of global warming on frog mortality.

---

[17]A MCC is a group of thunderstorms in close-enough proximity that mutually favorable interaction occurs between the storms enabling the cluster to persist well beyond the typical lifetime of a single storm.
[18]The term *ontogeny* is in fact borrowed from biology.

It is in this sense that we have classified North Atlantic hurricanes. As one must have a record of temperatures throughout the lifecycle of a frog to classify it as tropical only, one must examine the growth and development of a hurricane from its inception to hurricane intensity to classify it as tropical only or baroclinically enhanced. The new classification is based on ontogeny rather than on morphology. It is necessary to understand this distinction to appreciate the meaning of tropical-only hurricanes. Concerning the baroclinically-initiated classification, this is a stratification based on morphology. In other words, what is the form or structure of the original disturbance? If it is cold core, then it is baroclinically initiated. There is no requirement to examine any other aspect of development to make this classification. Note that a baroclinically-initiated hurricane is necessarily baroclinically enhanced, since the initiation stage is certainly part of the lifecycle of a tropical cyclone, just as the egg is the first stage in the lifecycle of a frog. That is, morphology unfolds chronologically as a sequence of ontogenies. However, "trajectories" made from the sequence of ontogenies may "converge" so that a tropical-only hurricane is indistinguishable in form from a baroclinically-enhanced hurricane. Examining structure and form of a fully-developed hurricane will tell us little about whether or not it is a tropical-only hurricane.

2.9   Extratropical Cyclones

An extra-tropical (or middle-latitude) cyclone is a storm that develops in association with the polar front. The polar front is a boundary between two air masses[19] differing in temperature and density. The polar front separates the cool, dry air originating at high latitudes from the warm, humid air of the tropics. The front is weaker and closer to the pole during the summer months. The source region of the cool, dry air is the arctic and alpine tundra regions of North America. The polar front takes the form of a warm front if the warm air is moving northward to displace the cool air, and a cold front if the cool air is moving southward to remove the warm air. Extratropical cyclones are the primary weather producers in the middle and high latitudes. Middle-latitude cyclones, with their large area of low surface pressures, generally travel from west to east across the United States and the North Atlantic. Mature extra-tropical cyclones will have a cold front extending to the southwest of the center of low pressure and a warm front extending to the east. The *warm sector* is the area to the south of the warm front that lies to the east of the cold front. Warm, humid tropical weather conditions prevail in the warm sector.

Extra-tropical cyclones are more numerous than tropical cyclones and may originate over the continents as well as over the oceans. They are a daily occurrence across the United States, though more frequent in winter than in summer. Extra-tropical cyclones are typically larger in diameter than tropical cyclones, but generally less intense. The energy necessary to maintain a higher latitude extra-tropical cyclone is consider-

---

[19] An air mass is a large volume of air, usually 1600 km or more across, that is characterized by similar atmospheric properties at a given height.

ably different than that of a lower-latitude hurricane. Extra-tropical cyclones are driven by extracting available potential energy stored in areas where there is strong temperature and moisture contrasts (baroclinity). In comparison, as temperature contrasts are rather small in the tropics, hurricanes derive substantial energy from latent heat release within the cumulonimbus clouds.

On satellite pictures, extra-tropical cyclones are distinguished by their "comma shape" appearance. The comma shape is a consequence of the accompanying cold front and the associated enhanced cloudiness trailing southwest from the center of the storm system. Convergence of the air toward low pressure and toward the frontal boundaries produces the associated clouds and precipitation of the extra-tropical cyclone. Mature hurricanes have a pinwheel shape as seen from space, although hurricanes within an environment of strong wind shear may appear as comma shaped.

Middle-latitude cyclones can forestall the demise of a hurricane. Hurricanes typically remain vigorous as they push northward out of the tropics, and in some situations their intensity increases. This occurrence is favored when their arrival into the middle latitudes coincides with the development and movement of an extra-tropical cyclone. If the hurricane motion is fast enough to stay east of the developing extra-tropical storm in the warm sector, the two storms will remain distinct. The air in the warm sector provides a tropical environment for the hurricane, and the hurricane provides an envelope of moisture for the extra-tropical storm. Upper-level winds to the east of the associated middle-latitude trough lead to favorable outflow for the hurricane.[20] If the hurricane catches up with the middle-latitude storm, the circulation of the hurricane becomes absorbed into the circulation of the extra-tropical cyclone. This is typical at latitudes north of the United States-Canadian border.

A hurricane can also transform into an extra-tropical cyclone. The non-tropical environment (characterized by sharp horizontal temperature contrasts and the advection of cyclonic vorticity) encountered at higher latitudes causes the winds, rainfall, and temperatures around the center of the hurricane to become asymmetric as the center moves toward the region of strongest thermal contrasts. As a result, it is common for the winds to subside a bit and the circulation to expand. Expansion of the circulation occurs in conjunction with a strengthening of winds farther from the storm's center. Winds may increase on the north and northeast side in response to a greater pressure gradient as the hurricane approaches higher surface pressures associated with a middle-latitude anticyclone. The anticyclone is maintained by a region of upper-level convergence of air located between the upstream ridge and the downstream trough in the middle-latitude westerlies. The stronger the pressure gradient, the faster the winds will blow. Even in this situation it is possible for the transforming cyclone to retain a small region of hurricane-force winds and an area of extremely heavy rainfall. This is particularly true if the convection remains wrapped completely around the center of circulation.

Transition of a hurricane is typically preceded by the entrainment of drier air into the circulation. Continental air from the middle latitudes is a source of the dry air. Dry

---

[20]The "symbiosis" may remain in place for several days.

air weakens the storm by limiting the lateral extent of moist adiabatic ascent inside the thunderstorms. The forward motion of the cyclone is often faster as transition occurs and can reach speeds of 100 km hr^{-1} or more. If the baroclinic dynamics are favorable the post-tropical storm[21] may intensify as an extra-tropical cyclone and pose a serious threat to the maritime communities of the North Atlantic, including eastern Canada and the United Kingdom. Fast-moving, vigorous extra-tropical cyclones can pound the western shores of Europe with high winds and heavy surf. On the other side of the continent, extra-tropical transition of western North Pacific tropical cyclones result in powerful middle-latitude storms that threaten Alaska and British Columbia with high winds and rough seas.

---

[21]This is the designation given to these systems by the Canadian Hurricane Centre.

3

Hurricane Climate Data

The historical record of tropical cyclones over the North Atlantic, Caribbean Sea, and Gulf of Mexico extends back more than half a millennium. Geological records provide clues to hurricane activity farther back still. This chapter identifies the sources of historical hurricane data for the North Atlantic basin. It provides a chronology of the technological advances that make it possible to accurately account for past and present tropical cyclone activity. Caveats concerning the limitation of the classifications outlined in Chapter 2 within the context of data resolution and reliability are considered. The chapter concludes with a look at the potential for improvements and extensions to the historical hurricane record through the use of proxy reconstruction methods.

3.1 Data Sources

From the perspective of European history, the first North Atlantic hurricane sighting was made by Christopher Columbus while sailing along the north coast of Cuba in June of 1494 (Millás 1968). The hurricane of July 1500 is the first known hurricane to affect the present day United States (Florida). In these earlier times, accounts were kept only of hurricanes that caused loss of life or property damage either to ships and crew or to coastal communities. The "triangular trade route" between Europe, Africa, and the Caribbean that existed from the late 16th through the middle 19th century provided an early source of information about Atlantic basin hurricanes. The southern branch of the trade route, where ships carried slaves from Africa to the new world, cut through the main breeding grounds of North Atlantic hurricanes. The apparent increase in occurrence of tropical cyclones over the last 500 years (Figure 3.1) is largely due to substantial increases in the ability to observe, report, and catalogue storms over the years. For example, not until the 1898 Treaty of Paris between Spain and the United States were regular weather stations maintained in the West Indies.

Though accounts of individual tempests date to the time of Columbus, detailed information on North Atlantic hurricanes extends back only about 120 years or so. Much of this information is maintained by the National Oceanic and Atmospheric Administration (NOAA) of the U.S. Department of Commerce through the National Hurricane

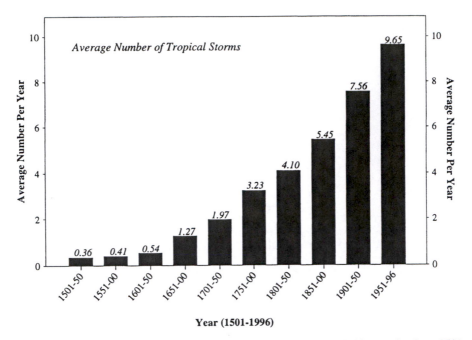

Fig. 3.1: Average number of North Atlantic tropical storms per year in half-centuries from 1501 to 1996. The year is defined as June through November. The data prior to 1901 are interpolated from Figure 61 of Tannehill (1950). Data after 1901 are from the best-track data set.

Center (NHC). The NHC, which is the forecast branch of the Tropical Prediction Center, is located on the campus of Florida International University.[1] The organization and dissemination of tropical cyclone data from around the world is done by the National Climatic Data Center (NCDC) located in Asheville, North Carolina (*NC*). Tropical cyclone data for the North Atlantic and eastern North Pacific basins are collected by the NHC. Annual summaries of hurricane activity over these two basins are routinely published in the *Monthly Weather Review*.[2]

Details on individual tropical cyclone tracks for the North Atlantic basin, including statistics for the period 1886–1958 were first provided by the U.S. Weather Bureau (now called the U.S. National Weather Service) in *Technical Paper Number 36*. The tracks and statistical analyses have been updated at various times since then, first by G. W. Cry and subsequently by the NCDC in cooperation with the NHC through the *Historical Climatological Series*. The latest version available was issued under the subtitle *Tropical Cyclones of the North Atlantic Ocean, 1871–1992* and authored by C. J. Neumann, B. R. Jarvinen, C. J. McAdie, and J. D. Elms (Neumann et al. 1993). This so-called best-track data set represents the most complete and reliable source of all

---

[1]Until 1996 the NHC was located at the Miami NWS Forecast Office in Coral Gables. Before 1943 hurricane forecasting was done from the Jacksonville Weather Bureau Office.

[2]*Monthly Weather Review* is one of several technical journals published by the American Meteorological Society (AMS).

North Atlantic hurricane information back to 1886, and is considered the official U.S. National Weather Service record of North Atlantic hurricanes.[3] Practically all research involving historical Atlantic hurricanes is based on this data set as is the majority of tables and figures presented in this book. Updates to the North Atlantic hurricane best-track data set can be obtained from the NHC.

The best-track data records are compiled from various publications and represent a rigorous, post-season analysis of all tropical cyclone intensities and tracks every 6 hours. Data reliability is not uniform throughout the period of record. Uncertainties in wind and pressure estimates are largest before aircraft reconnaissance. Before this time, only occasional ship and land reports provided information on the location and intensity of individual storms. In the period 1931–1957, 12-hourly position and intensity estimates were made from these scattered observations. Before 1931, estimates were made once per day. The positions and intensities given in the best-track data for these earlier times are interpolations based upon available estimates.

Historically, the most consistent source of track, intensity, and other hurricane information is the *Monthly Weather Review*. Though this monthly technical journal evolved in format, emphasis, and content from its inception in June 1872, regular reports of seasonal tropical cyclone activity appear in the journal during most years since 1922. Descriptions of specific storms by various authors continues to be an occasional feature of the journal. Summaries of North Atlantic tropical cyclone seasons for the period 1950–1980 are found in the U.S. Weather Bureau's publication *Climatological Data National Summary*, volumes 1–31. For the period 1951–1992 data are found in NCDC's *Storm Data*, volumes 1–34. The details of early hurricanes affecting the United States for specific states can also be found in *Climatological Data* published by the U.S. Weather Bureau.

For the time period prior to the *Monthly Weather Review* summaries and articles, North Atlantic tropical cyclone information is available from various sources. Tropical cyclone activity over the period 1871–1880 was chronicled by Elias Loomis in his papers entitled "Contributions to Meteorology" published in the *American Journal of Science*, volumes 14–29. Various authors are responsible for hurricane track records since the early work of Elias Loomis including I. M. Cline, O. L. Fassig, E. B. Garriott, C. L. Mitchell, and I. R. Tannehill. Using the above-cited primary sources, as well as numerous other secondary sources, G. W. Cry published "Tropical Cyclones of the North Atlantic Ocean" as *Technical Paper No. 55* of the U.S. Weather Bureau in 1965. This 148-page paper is the primary reference for hurricane tracks and intensities over the period 1871–1963.

According to Neumann and McAdie (1997), there are both systematic and random errors in the best-track data set mostly confined to the years before 1968. The random errors were introduced largely through mistakes made as the data were entered on punch cards. The systematic errors are a result of the interpolations necessary to provide 6-hourly reports from 12- and 24-hour observations. These errors include winds

---

[3]It is also known by the acronym HURDAT for HURricane DATa.

that are not always consistent with pressures; some pressures being peripheral rather than central; and forward motion estimated to be too fast at the beginning and end of the hurricane track, among others. For instance, before 1900 when there were few direct measurements, many hurricanes were assigned an 85 kt wind. Storm intensity climatologies that include these data will have a bias toward category two hurricanes. NOAA has recently begun a project to improve the quality of the best-track data. The reanalysis project may, for instance, determine that hurricane *Andrew* in 1992 was a category five storm at landfall over southeastern Florida. Neither the errors nor the future alterations in the best-track data will significantly impact many of the hurricane climate statistics presented in this book.

Fernández-Partágas and Diaz (1995a, 1995b, 1996a, 1996b) examined numerous ship logs and newspaper reports to enhance the historical hurricane record back to 1851. They found additional information on known hurricanes and discovered previously undocumented storms. The additional information was used to slightly revise a few hurricane tracks reported in the best-track data set. Since the Fernández-Partágas and Diaz data are not in easily accessible form, they are not used in our general climatology of North Atlantic hurricanes. However their updates are incorporated in the statistics on U.S. hurricanes starting with Chapter 8.

3.2 Technological Advances

Improvements in the ability to log hurricane occurrences and detail individual tracks are dependent on technological progress as well as on a modern economy. This is because hurricanes spend most of the time over the open ocean waters where regular and routine observations are difficult to obtain. Before the era of aircraft reconnaissance and weather satellites, the detection of tropical cyclones was largely a chance event. A hurricane that failed to affect a population or was not detected by a merchant ship never made it into the record logs. A famous example of a chance encounter with an early American hurricane occurred in 1743 when Benjamin Franklin was unable to observe a much-anticipated lunar eclipse due to cloud cover from the approaching storm. Yet over the North Atlantic basin, coincidence of typical tropical cyclone paths with shipping lanes and populated islands made it unlikely that a significant hurricane would have escaped detection, even as far back as the middle 19th century (Neumann et al. 1993). The weaker, shorter-duration tropical cyclones are more likely to have gone unnoticed.

The invention of the telegraph in 1835 ushered in the electric age of meteorological observations in the United States. In 1844 the construction of the first telegraph line (from Washington to Baltimore) was completed. By 1846 Joseph Henry, as the first secretary of the newly formed Smithsonian Institution in Washington, initiated a weather telegraphy program to study U.S. storms (Hughes 1994). In the same year, hurricane pioneer William Redfield suggested the telegraph might be used to warn Atlantic seaports of approaching storms. The wind-speed anemometer was invented in the middle 1800s. By the late 19th century anemometers were used at weather observation sites throughout the United States and elsewhere.

A proclamation by President Ulysses S. Grant gave weather warning and forecasting responsibility to the U.S. Army Signal Corps in 1870. One of its first responsibilities was the issuance of tropical storm warnings. The collection and collation of data from tropical cyclones received by land stations via telegraph and by ships through marine logs were first routinely carried out under the direction of the Chief Signal Officer of the Army. The earliest reports came in from Havana, Santiago de Cuba, and Kingston in August of 1873. The first storm forecast was issued for an extra-tropical cyclone near the Great Lakes on November 8, 1870 by Increase A. Lapham, assistant to the Chief Signal Officer in Chicago. The first Signal Corps weather map of a hurricane was drawn on September 28, 1874.[4] A Hurricane Warning Service was organized in 1898 with the establishment of a forecast center in Kingston, Jamaica. The center was transferred to Havana in 1899 and to Washington in 1902. The first radio weather observation received by the U.S. Weather Bureau from a ship at sea occurred on December 3, 1905 from the *S.S. New York* located near 40°N latitude and 60°W longitude.

The era of radio communication to report tropical cyclones began on August 26, 1909 when the *S.S. Cartago* in the southern Gulf of Mexico reported a hurricane near the coast of the Yucatan Peninsula (Tannehill 1950). The amount and quality of ship reports increased over the following decades, so that by 1959, the number of tropical cyclone observations over the North Atlantic Ocean exceeded 64,000 (Neumann et al. 1993). The warning service was decentralized in 1935 with centers in Jacksonville, New Orleans, and San Juan, but Washington continued with its warning duties. A continuous 24 hr monitoring of the hurricane season was instigated at this time. By 1937 a fully-established network of radiosondes were in place for tracking upper-air changes related to changes in movement and intensity of hurricanes. The Jacksonville center was moved to Miami in 1943. The U.S. Navy began the use of moored oceanic buoys in the late 1960s. The first buoy system to support hurricane analysis and forecasting occurred in 1972. It expanded over the years to include important installations over the central and northern Gulf of Mexico as well as along the eastern U.S. seaboard (Sheets 1990). The buoys provide a dependable source of data for daily analysis and are a vital part of the current hurricane warning system.

The era of major technological advances in tropical cyclone detection began after World War II. Colonel Joseph P. Duckworth and Lieutenant Ralph O'Hair made the first intentional aircraft flight inside a hurricane on July 27, 1943 using a single-engine plane (Hughes 1987). Ever since, it has been routine to fly reconnaissance aircraft missions at frequent intervals into tropical cyclones to update storm location and intensity information. The position of the hurricane eye and details of the wind and temperature fields surrounding the storm are determined with fine accuracy using aircraft surveillance. Dropsondes are released throughout the storm by the plane in

---

[4]The storm was located off the coast between Savannah and Jacksonville.

order to measure the storm structure. Hurricane reconnaissance by aircraft is one of the most important tools the modern hurricane forecaster uses. Both U.S. Air Force and Navy aircraft were initially used, but the Navy discontinued operations after the 1974 hurricane season. In September of 1944 Major Harry Wexler participated in the first research reconnaissance flight into a hurricane. NOAA currently operates several aircraft loaded with instruments for the collection of research and operational hurricane information. NOAA's *Gulfstream G-IV jet* is the newest of the hurricane-research aircraft, capable of flying faster and higher than the previous generation of hurricane hunters. Modern aircraft are equipped with Global Positioning System (GPS) dropsondes. The dropsonde system is capable of providing data throughout the hurricane at high frequency sampling rates.[5]

The post–World War II use of radar to track tropical cyclones near the coast further improved observational capabilities. By the middle 1950s, radar systems were available at some U.S. Weather Bureau sites along the Gulf and Atlantic coasts, and by the early 1960s the Weather Surveillance Radar-1957 (WSR-57) system was fully established providing coverage from Brownsville, Texas to Eastport, Maine (Sheets 1990). Radar is particularly useful as the hurricane approaches to within 400 km of the coast and can be used to detect sudden and often subtle shifts in the track of an approaching storm. New, better resolution, Weather Surveillance Radar-1988 with Doppler (WSR-88D or NEXRAD) were installed throughout the United States during the late 1980s and early 1990s. A network of these radars currently covers the entire United States including the coastlines vulnerable to hurricane landfalls.

Perhaps the single most important technological advance for hurricane detection and monitoring is the weather satellite. Satellites provide a global view of the weather, which is particularly essential in areas lacking in ground-based weather stations such as over the open waters of the North Atlantic Ocean and in the tropics in general. The original U.S. weather satellites were an extension of experimentation with rocket photography in the 1940s and 1950s by the U.S. National Aeronautics and Space Administration (NASA). The first pictures of a tropical cyclone (located to the east of Australia) were taken in 1960 from the TIROS-I, launched on April 1st. The first completely operational satellite (ESSA-I) was launched in 1966. Both the prototype TIROS-I and the ESSA series were polar-orbiting, low-altitude ($\approx$ 850 km) satellites that took pictures of the tropical Atlantic waters once per day.

The prototype geostationary orbiting satellite was the Applications Technology Satellite (ATS) launched on December 6, 1966. It contained a "spin-scan" camera invented by Verner E. Suomi at the University of Wisconsin-Madison. Geostationary satellites orbit at an altitude of approximately 36000 km over the equator and travel at the same speed as the earth rotates. This keeps them fixed over the same location at all times. With the development and deployment of geostationary satellites, the ability for continuous daytime monitoring became available by the late 1960s. The Geosta-

---

[5]The first GPS dropsonde was deployed in the eye of hurricane *Guillermo* over the eastern North Pacific during September 1997.

tionary Operational Environmental Satellite (GOES) series, initially launched in 1975 completed the surveillance capability of the satellite network by providing continuous nighttime pictures and today constitutes the primary support for tropical meteorological observations. The original spin-scan camera has been replaced by the three-axis stabilized imager on the GOES-8 launched in April of 1994. The new camera provides higher resolution pictures and a lens that continuously points at the earth.

Pictures from geostationary and polar-orbiting satellites are developed from cameras (radiometers) that measure the amount of electromagnetic radiation emitted and scattered by the earth and its overlying atmosphere. Radiation is grouped into various ranges depending on wavelengths. Radiation having wavelengths in the range of 0.4–0.7 micrometers[6]corresponds to visible light; wavelengths of 10–12 micrometers correspond to the infrared radiation (IR) range; and wavelengths of 6–7 micrometers to the water vapor range. A visible image measures the amount of reflected sunlight from the ground and cloud tops and is only available during the daylight hours. Infrared radiation images provide information on the temperature of the clouds and ground, while water vapor images show regions of high and low atmospheric moisture content.

Because of the conspicuous shape of the attendant cloudiness, visible and IR satellite imagery allow for unmistakable detection of a hurricane. Cirrus clouds extend a great distance from the storm center and are readily observed from satellites. A careful examination of a satellite picture makes it possible to specify with accuracy the center of the storm and make a good estimate of its intensity. Techniques for intensity estimates of tropical cyclones from satellite pictures are outlined in Dvorak (1973, 1975), and for subtropical cyclones in Hebert and Poteat (1975). This represent an important advance for hurricane analysis and forecasting. Cloud patterns as observed from satellites are routinely consulted for estimating tropical cyclone intensities for tropical cyclones over the far eastern North Atlantic where the logistics and economy of aircraft reconnaissance is prohibitive. Objective measures to estimate hurricane intensities from cloud images are presently being investigated. Water vapor pictures provide a valuable source of information concerning the movement of air at middle levels in the atmosphere. The movement of fronts and low-pressure troughs in advance of the hurricane often provide clues as to the hurricane's future track.

In addition to photographs, satellites provide measurements of wind, temperature, and rainfall associated within tropical cyclones. The movement of high cirrus clouds above the hurricane give a good indication of the upper-level steering currents. Wind estimates are made directly by tracking individual cloud elements on successive images. In fact, the ability to animate or "put the pictures in motion" is the product of major technological advances in computer hardware and software systems. The first satellite viewing, processing, and analyzing software system was the Man-computer Interactive Data Access System (McIDAS) developed at the Space Science and Engineering Center (SSEC) at the University of Wisconsin-Madison in the 1970s (Smith 1975). The McIDAS is used operationally at the NHC (Sheets 1990). Temperature and rain-

---

[6]1 meter = 10^6 micrometers.

fall fields associated with the hurricane are estimated using algorithms that convert meteorological values from radiation profiles.

3.3 Classifications

3.3.1 Major Hurricanes

A distinction is made between regular hurricanes and major hurricanes. A regular hurricane is a tropical cyclone with maximum sustained winds between 33 and 50 m s^{-1} (65–100 kt). Major hurricanes have maximum sustained winds of at least 50 m s^{-1}. The best-track data set gives the maximum sustained wind speeds in 2.5 m s^{-1} (5 kt) intervals. A finer resolution of hurricane strength is not practical with the uncertainties of the various estimation and observation techniques (Landsea 1993). In science and engineering the units of meters per second are used for wind speeds, however knots are used throughout this book when describing statistics based on the best-track data. According to the Saffir/Simpson scale (Table 2.2, Chapter 2), which is recognized as the best overall measure of hurricane intensity, categories three through five are major hurricanes. Approximately 97% of all hurricane destruction is from major hurricanes (Gray and Landsea 1992), although a significant proportion of hurricane-related deaths occur from flooding events generated from weaker tropical cyclones. Statistics on major hurricanes of the North Atlantic are provided in Chapter 7.

The simplicity of defining major hurricanes hides some difficulties in distinguishing them, particularly from incomplete historical records. There is little doubt that subjective judgment was needed in many cases for prescribing the tropical cyclone intensity of the earliest hurricanes in the data record (Jordan and Ho 1962). Moreover, there are likely biases in the hurricane data record. For instance, before the era of aircraft reconnaissance which began in the middle 1940s, some tropical cyclones may have been incorrectly classified with respect to their precise intensity. Furthermore, Landsea (1993) notes that data from a few hurricanes over the period 1944–1969 are inconsistent with respect to wind and central pressure estimates, suggesting a slight exaggeration of major hurricane activity during this period (Table 3.1). These errors and biases influence the statistics to some extent.

Figure 3.2 shows the relationship of recorded maximum sustained winds and central pressures for all North Atlantic hurricane observations over the period 1886 through 1996. The 95% confidence lines indicate the points farthest from a linear fit. Winds are generally reported in 5 kt intervals. Despite the large variability, a strong relationship exists between maximum sustained winds and minimum central pressures. For a few weaker hurricanes, the wind speeds appear to be somewhat underestimated for the measured central pressure. For more intense hurricanes (sustained winds between 50 and 70 m s^{-1}) an overestimate of the wind speeds is noted for several observations. For the most intense hurricanes an underestimation of wind speed occurs for extremely low pressures. There is not a large systematic bias in the best-track data set, at least for reports where both winds and pressures are provided.

Table 3.1: Major hurricane frequencies. Values are from the best-track data and from Landsea (1993) for years in which there is a difference. No differences are noted after 1969.

Year	Best-track	Landsea (1993)	Difference
1945	3	2	1
1950	8	7	1
1951	5	2	3
1953	4	3	1
1955	6	5	1
1958	5	4	1
1961	7	6	1
1962	1	0	1
1964	6	5	1
1969	5	3	2
Total	*50*	*37*	*13*

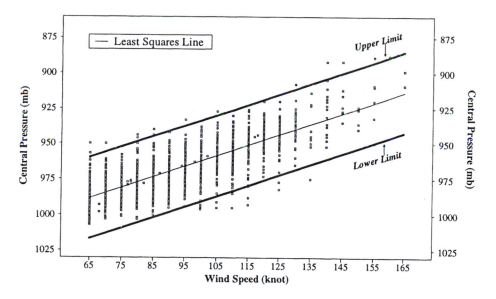

Fig. 3.2: Maximum sustained wind speeds versus central pressures. Values are for all North Atlantic hurricane observations over the period 1886–1996 from the best-track data set. The 95% confidence lines (upper and lower limits) are drawn using the student t distribution.

As wisely noted in Henry (1929), no two people will agree on the allocation of a large number of hurricanes to *any* common scale of intensity. Even today, with the abundance of hurricane surveillance, the delineation of hurricanes is sometimes difficult to make. Considering the substantial effort and great care that went into the development of the best-track data over the years, there is a good deal of confidence on the completeness, consistency, and accuracy of the information. These data are used throughout the book in compiling a comprehensive statistical survey of North Atlantic hurricane activity. It is likely that some changes to the data will be forthcoming with NOAA's reanalysis project.

3.3.2 Tropical Only Hurricanes

The distinction between tropical-only and baroclinically-enhanced hurricanes provides an interesting way to understand the vagaries of North Atlantic hurricane statistics. For instance, Hess et al. (1995) showed that an improved empirical prediction algorithm of season Atlantic basin hurricane activity is obtained by treating tropical-only and baroclinically-enhanced hurricanes separately. To develop a forecast model they first subjectively classify hurricanes over the period 1950–1993. Stratification is done by examining the summaries of past tropical-cyclone seasons published regularly in *Monthly Weather Review*. These reviews describe the lifecycle of each tropical storm by season. Typically, the descriptions are adequate to decide whether baroclinic influences were an enhancing factor for development to hurricane strength or not. Other sources include daily synoptic weather maps and consultation with Professor Noel E. LaSeur who flew on many of the pre-satellite-era reconnaissance missions and who was director of the National Hurricane Research Laboratory (NHRL)[7] from 1975–1977. The weather maps are part of the *Daily Series: Synoptic Weather Maps* prepared by the U.S. Weather Bureau in cooperation with the U.S. Army, Navy, and Air Force. The charts include both surface and 500 mb analyses for the Northern Hemisphere. The once-per-day maps provide detail on the positions of tropical waves, upper-air troughs, and shear lines, although analyses in the tropics are necessarily incomplete. Figure 3.3 shows a reproduction of the surface map from this series for September 15, 1953. Numerous surface reports from land stations and ships allow an analysis of weather systems in the tropics.

The method of Hess et al. (1995) is subjective in that it was done by an individual (in consultation with others) looking at each storm separately. This should not be interpreted to mean that there are no underlying physical mechanisms supporting the classification. On the other hand, some other individual looking at the same resources might produce a somewhat different selection of tropical-only hurricanes. As with the distinction between regular and major hurricanes, there will always be debate about the classification of some hurricanes. In order to extend the subjective classification to include hurricanes prior to the time of detailed descriptions, Elsner et al. (1996a)

---

[7]NHRL is now called the Hurricane Research Division (HRD).

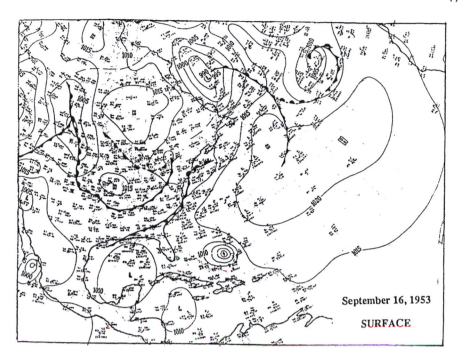

Fig. 3.3: Surface analysis for September 16, 1953 reproduced from the *Daily Series: Synoptic Weather Maps* prepared by the U.S. Weather Bureau in cooperation with the Army, Navy, and Air Force. Originals of these maps are used in Hess et al. (1995) as an aid in categorizing North Atlantic storms into tropical-only and baroclinically-enhanced hurricanes.

develop a set of rules that identify tropical-only hurricanes in an objective manner based on initial tropical storm and hurricane locations and on the time of year. The classification is done using a partially adaptive classification tree (PACT) developed at the University of Wisconsin-Madison. The PACT algorithm provides a series of decision rules through a combination of statistical methods. The procedure begins with an initial decision node, and adds further nodes as constrained by the tree-growth parameters. Since it is always possible to create a 100% classification accuracy by complete partitioning of the variable space (time and location of the storm), a criterion is used to determine the optimal tree size. This is achieved by using a direct-stopping rule and maximizing the cross-validated classification (see Chapter 14) accuracy as a function of the stopping rule. A direct-stopping rule halts the tree growth once the number of observations remaining within a terminal node falls below a percentage threshold of the total number of observations. In other words, suppose a direct-stopping rule of 6% is chosen. Then if the number of observations in a particular node is less than this percentage of the total number of observations, the growth is stopped for that node and the node is designated a terminal. The direct-stopping rule is applied to all terminal nodes of the decision tree.

Overall the algorithm functions as follows: First, if the initial or a subsequent node has sufficient number of observations, the algorithm performs an analysis of variance (ANOVA) on each potential variable and selects the variable that has the most significant F statistic. To avoid ignoring variables that have a large degree of nonfunctional group separation, a Levine's test is conducted to identify which variable has the largest inequality of variances caused by group classification. The F statistics for this test are also obtained. PACT selects the splitting variable for the decision node based on the variable having the largest F value over both test procedures. Next, the algorithm performs a one-dimensional linear discriminant analysis (LDA), using the variable selected above. The decision rule for the decision node in question is created from the linear discriminant function (LDF), which partitions this node into two additional nodes. Finally, each of these nodes is checked to see if it has a sufficient number of observations, and the process repeats until all of the remaining nodes become terminal nodes, thus completing the tree. The classification tree, once completed, allows for rather straightforward group classifications. Since LDA is technically a Bayesian classifier (Mardia et al. 1979), prior probabilities (see Chapter 14) can be utilized in the LDFs for each decision node.

In applying the algorithm to hurricane records after 1950, the rules correctly classify at better than 90% accuracy providing guidance for classifying all North Atlantic hurricanes (see Chapter 5). Figure 3.4 shows the pre-hurricane tracks of the tropical-only hurricane *Hugo* in 1989 and the baroclinically-enhanced hurricane *Arlene* in 1967. Large differences are noted. In general, tropical-only hurricanes originate from tropical easterly waves that track near the Islands of Cape Verde and maintain a westward motion, like *Hugo*. Baroclinically-enhanced hurricanes, on the other hand, may come from waves that reach hurricane strength only after a swing to the north (recurvature), like hurricane *Arlene*.

3.3.3 Landfalling Hurricanes

Hurricanes that remain at sea are of concern only to shipping interests. Landfalling hurricanes directly impact people and property. Information on landfalling storms is more accessible. Hieroglyphics from the Mayan civilization of ancient Mexico are possibly the earliest human record of Atlantic hurricanes (Konrad 1985). Using historical accounts it is possible to extend the record of hurricane landfalls with confidence. The most intense hurricanes have left indelible marks on the local history making it possible to use historical storm accounts from local newspapers and other sources as additional information on these early hurricanes. As a noteworthy example (Rappaport and Fernández-Partágas 1997), a fleet of ships carrying passengers from England to Virginia was struck by a hurricane in 1609. Some of the ship's fleet landed in Bermuda and become the island's first permanent residents.[8]

The hurricane warning service was established by President William McKinley

---

[8]Bermuda got its name from a Spaniard named Juan Bermudes who was another hurricane shipwreck victim fortunate to land on the island (Hughes 1987).

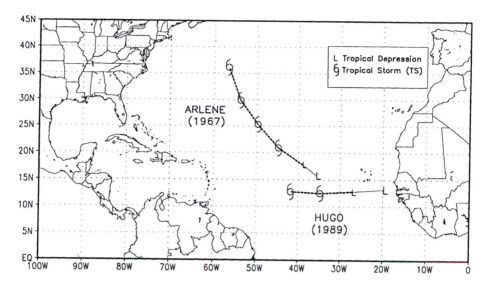

Fig. 3.4: Pre-hurricane tracks of *Hugo* (1989) and *Arlene* (1967). The tracks are shown from their initial tropical depression stage through their tropical storm stage. *Hugo* is classified as a tropical-only hurricane and *Arlene* as a baroclinically-enhanced hurricane according to the classification of Hess et al. (1995).

at the turn of the 20th century after Willis Moore of the U.S. Weather Bureau had shown that hurricanes and other storms had sunk more ships than all the wars in history. McKinley reportedly said, "I am more afraid of the West Indian Hurricane than I am of the entire Spanish Navy," (Gedzelman 1980). The definitive source on hurricanes affecting the U.S. during earlier times is provided by David M. Ludlum. His treatise is entitled "Early American Hurricanes, 1492–1870" published by the AMS in its series called "The History of American Weather" (Ludlum 1963). More details on landfalling hurricane data are given in Chapter 8.

3.4   Proxy Data

Besides written historical records, there are other sources of data on past hurricanes. Proxy data are from indirect sources that are used where actual observations are lacking. Indirect evidence of North Atlantic hurricanes predates historical accounts. Evidence of past hurricane landfalls and their dramatic effects are etched in the local biota and geology. Recent developments in paleotempestology permit the reconstruction of past histories of catastrophic hurricane landfalls. Proxy data are categorized by effects from wind, storm surge, and rainfall (Table 3.2).

Concerning effects from wind, intense hurricanes that rip through a forest will kill or drastically reduce an entire plant species. Large trees are particularly vulnerable to extremely high winds. The New England hurricane of 1938 produced significant

Table 3.2: Proxy reconstruction methods. Proxy approaches are divided into methods based on the effects of wind, storm surge, and rainfall.

Effect	Proxy Approach	Limitations
Wind	Changes in species composition of forest alters pollen types in wetland sediments.	Other factors alter pollen records; expensive.
Storm surge	Storm-generated lake/ocean sediments contain sand layers.	Representation of a single core.
Rain	Hurricane rain is isotopically lighter.	Isotopic fractionation within ground water.

damage to the forest canopy of the region. The widespread destruction of tall trees can abruptly change the species composition of a forest resulting in alterations in the pollen types deposited in layers of ground sedimentation. Thus soil cores taken from the forest can be used to reconstruct the history of catastrophic wind damage. The problem with this method is that other conditions, such as widespread diseases, also alter the pollen record. Moreover, detecting subtle changes in ground sediments can be expensive.

The effects of the storm surge are used for extracting information about hurricane frequencies. The most useful proxy for intense hurricane landfall activity have come from sand layers in coastal lakes and marshes. Hurricane generated sediment deposits on the continental shelf, beaches, and lakes are generally sandier, less organic, and coarser grained than deposits under less stormy conditions. Moreover, the thickness of the inorganic layers of sediment are proportional to storm intensity. Liu and Fearn (1993) demonstrate that a chronologically and stratigraphically accurate record of prehistoric hurricane strikes can be obtained from the sediments of coastal lakes. They find sediment cores removed from the center of Lake Shelby in Alabama contain multiple layers of sand suggesting that major hurricanes of category four or five have an average return interval (see Chapter 11) of approximately 600 yr. These data also indicate no major hurricanes before about 3400 years ago suggesting the possibility of a substantial climate shift over the region. Similarly, sediment cores from the west coast of Florida indicate hurricane-induced extreme rainfall events more than a thousand years ago (Davis et al. 1989). The main problem with inferring hurricane information from sediment cores is the problem of point measurements. For example, a major hurricane that skirts the shoreline will cause the same sedimentation signature as a minor hurricane that makes a direct hit on the beach. The problem is overcome by making multiple measurements around a single site to construct a more comprehensive climatology.

Another interesting proxy method is based on the fact that rain from tropical cyclones is isotopically lighter than rain produced by other types of summertime thunderstorms (Lawrence and Gedzelman 1996). The reason remains speculative but is likely related to the fact that the abundance of atmospheric moisture surrounding a hurricane limits the possibility for widespread evaporation of cloud water. Punctuation of isotopically light groundwater runoff are recorded within the calcium carbonate of cave deposits and karst formations. A problem with extracting hurricane data from these records is that isotopic breakup can occur by the ground-water runoff process itself, thus contaminating the signal recorded in the ground. These proxy reconstruction methods are valuable for improving our understanding of long-term hurricane variations as they provide independent estimates of hurricane probabilities at a given location.

4

North Atlantic Hurricanes

Hurricanes are a regular feature of the North Atlantic basin during late summer and early autumn. But, what is the average hurricane frequency and where are hurricanes most likely to form? These are some of the questions addressed in this chapter. Specifically, a climatology of North Atlantic hurricanes (historically referred to as West Indian hurricanes) is presented based on the best-track data set. The focus is on frequencies, duration, origins, tracks, and dissipation of the known North Atlantic hurricanes over the period 1886 through 1996 (111 years). The North Atlantic basin includes the Gulf of Mexico and the Caribbean Sea. The recent updates to the best-track data made by José Fernández-Partágas and Henry F. Diaz are not included in this chapter as they constitute rather minor adjustments to the data set.

North Atlantic hurricanes form, move, and die over a large part of the tropical and subtropical waters of the North Atlantic basin. The official hurricane season runs from June through November, but hurricane occurrences are strongly concentrated in the months of August, September, and October. By various measures the peak of the season occurs during the second week of September. Late- and early-season hurricanes originate farther west compared to the hurricanes of late August and September. Hurricane dissipation occurs at high latitudes and over land. The most reliable hurricane data extend back only to about 1944 (see Chapter 3). Statistical characteristics of major (or intense) North Atlantic hurricanes and those hurricanes striking the United States (landfalling hurricanes) are subjects for Chapters 7 and 8.

4.1 Abundance

The number of hurricanes occurring in any given season is referred to as the *seasonal abundance* or *frequency*. The term "seasonal activity" is often used interchangeably with the term "seasonal abundance" or "seasonal frequency." The abundance of hurricanes is determined by the prevalence of necessary environmental conditions for tropical cyclone development and intensification. However, since every hurricane develops from a tropical storm, which in turn comes from a depression or some other disturbance, the abundance of hurricanes is also a function of the frequency and vitality of the pre-hurricane disturbances. Although there is a strong seasonal cycle to North Atlantic hurricane activity with a pronounced peak during September, it is convenient to consider annual statistics based on the calendar year (January through December). Over the 111-year period from 1886 through 1996, a total of 554 North Atlantic hurricanes are noted in the best-track record. Hurricane activity of the North Atlantic basin represents only a small percentage of global tropical cyclone activity. The western North Pacific, for instance, generates about 20 typhoons per year. Typhoons of the North Pacific are generally stronger than hurricanes of the North Atlantic.

4.1.1 Interannual Variability

Interannual variability refers to fluctuations over a year or more. Here we examine the interannual variability of hurricane abundance by considering semidecadal, biennial, and annual frequencies over the period 1886 through 1996.

Figure 4.1 (top) is the time series of hurricane abundance in 5-year intervals beginning with the period 1886–1890. Hurricane activity in these 22 consecutive pentads ranges from a low of 14 hurricanes for the 1911–1915 period to a high of 37 in the 1951–1955 period. The maximum in *any* 5-year interval is 39 for the period 1950–1954. The average (or *mean*)[1] number of hurricanes in these 5-year periods is 24.8, with a standard deviation of 6.2 hurricanes. Except for the slight drop in activity during the early part of this century, the number of hurricanes is relatively consistent over this period. The *coefficient of variation*,[2] which is 25% for the pentad series, is 37% for the biennial series [Figure 4.1 (bottom)], and 50% for the annual series (Figure 4.2). Thus the variations increase as the length of the period decreases (see Dunn and Miller 1964). Stated another way, as the number of hurricanes sampled increases, the coefficient of variation diminishes. The biennial variation of North Atlantic hurricane activity is shown in Figure 4.1 (bottom). The 2-year hurricane numbers range from as few as three during 1913–1914 to as many as 20 during 1995–1996. The expected number of hurricanes in these 2-year periods is 10.0 with a standard deviation of 3.7 hurricanes. The very active 1995 and 1996 hurricane seasons are examined in more detail in Chapter 12.

The annual variation of hurricanes is depicted in Figure 4.2 and in Table 4.1. Annual activity ranges from no hurricanes in 1907 and 1914 to 12 hurricanes in 1969. There are eleven hurricanes in 1916, 1950, and 1995 and only one hurricane in the years 1890, 1905, 1919, and 1925. Table 4.1 indicates, in bold, the years of high hurricane activity (nine or more hurricanes) and, in italics, the years of low activity (three or

---

[1]The mean (or expected value) is a statistic that measures the central tendency of the observations and is computed as $\bar{x} = 1/n \sum_{i=1}^{n} x_i$, where in the above case the x_is are the hurricane frequencies over a specified interval. The standard deviation is a measure of the spread (or dispersion) of observations around and the average, and is computed as $s.d. = [1/(n-1) \sum_{i=1}^{n} (x_i - \bar{x})^2]^{1/2}$. Under the assumption of normality, approximately 68% of the observations fall within a range of ± 1 $s.d.$ of the average; approximately 95% fall within 2 $s.d.$ of the average.

[2]The coefficient of variation (or coefficient of dispersion) is a relative measure of spread and is computed as $V = s.d./\bar{x}$. It is typically expressed as a percentage.

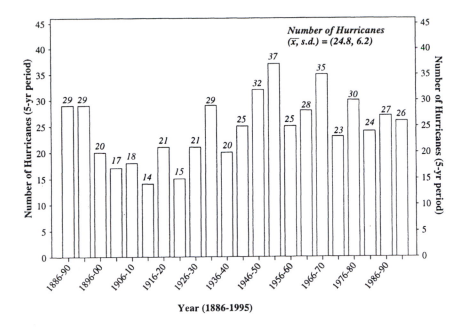

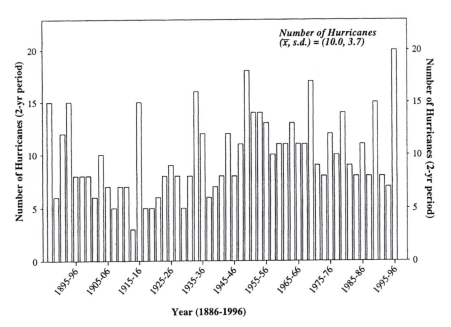

Fig. 4.1: Time series of the abundance of North Atlantic hurricanes in 5-year intervals (top) and 2-year intervals (bottom). The average (or mean) number of hurricanes ($\bar{x}$) along with the standard deviation (*s.d.*) are given in the upper-right corner of each graph.

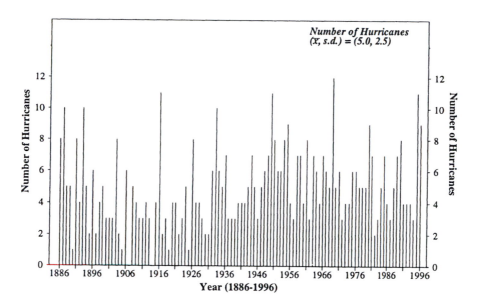

Fig. 4.2: Time series of the annual abundance of North Atlantic hurricanes over the period 1886–1996. The average number of hurricanes ($\overline{x}$) along with the standard deviation (*s.d.*) are given in the upper-right corner.

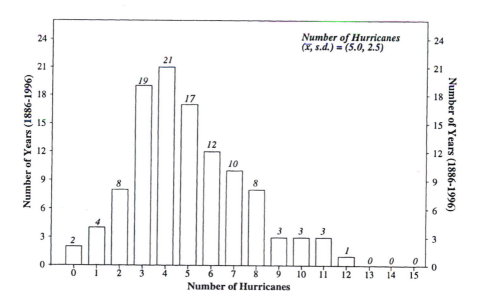

Fig. 4.3: Distribution of the annual abundance of North Atlantic hurricanes over the period 1886–1996. The most common annual occurrence is four hurricanes.

Table 4.1: North Atlantic hurricanes over the period 1886–1996. Y denotes the last digit of the year. Years of nine or more hurricanes are indicated in bold and years of three or fewer hurricanes are indicated in italics. Ten-year totals are given in the bottom row. Partial totals are indicated for the 1880s and 1990s. The 1950s and 1960s were particularly active.

Y	188	189	190	191	192	193	194	195	196	197	198	199
0	.	*1*	3	3	4	2	4	**11**	4	5	**9**	8
1	.	8	3	3	4	2	4	8	8	6	7	4
2	.	4	3	4	2	6	4	6	*3*	*3*	*2*	4
3	.	**10**	8	*3*	3	**10**	5	6	7	4	*3*	4
4	.	5	*2*	*0*	5	6	7	8	6	4	5	*3*
5	.	*2*	*1*	4	*1*	5	5	**9**	4	6	7	**11**
6	8	6	6	**11**	8	7	*3*	4	7	6	4	**9**
7	**10**	*2*	*0*	*2*	4	*3*	5	*3*	6	5	*3*	.
8	5	4	5	3	4	3	6	7	5	5	5	.
9	5	5	4	*1*	3	3	7	7	**12**	5	7	.
	28	47	35	34	38	47	50	69	62	49	52	43

fewer hurricanes). Except for 1995 and 1996, years of high activity are typically separated by years of lower activity. Extreme years of ten or more hurricanes are generally separated by 16 or more years.[3] In contrast, years of low activity are more numerous and tend to cluster in consecutive or nearly consecutive years (e.g., 1917–1919, 1929–1931). Furthermore, there does not appear to be any large trend in the annual number of North Atlantic hurricanes. This subject is considered in detail in Chapter 10. The distribution of annual hurricane counts is given in Figure 4.3. The *mode*[4] is four hurricanes with 21 years having exactly this many storms. For comparison, the average number of hurricanes in a year is five, and there are 17 years with this many storms. The distribution is asymmetric with a skewness[5] value of 0.49. The skewness is positive since the longer tail of the distribution is to the right of the central maximum (i.e., $\bar{x} >$ mode).

4.1.2 Seasonal Variability

Seasonal (or intraannual) variability refers to fluctuations in time that are within a single season (year). Here we examine the seasonal hurricane variability by considering monthly abundance. Figure 4.4 shows the monthly distribution of hurricanes over the 111-year record. A tropical cyclone reaching hurricane intensity in July, for example, is considered a July hurricane regardless of whether it remained a hurricane into August. Moreover, a tropical cyclone that reaches hurricane intensity more than once in its lifes-

---

[3]The 10 hurricanes in 1998 makes the 1990s the first decade with 3 years featuring 9 or more hurricanes.

[4]The *mode* is a measure of central tendency of a group of observations, and is defined as the most common occurrence in a distribution.

[5]The skewness is a measure of the departure of the distribution from symmetry, and is computed as skewness $= \frac{\bar{x} - \text{mode}}{s.d.}$.

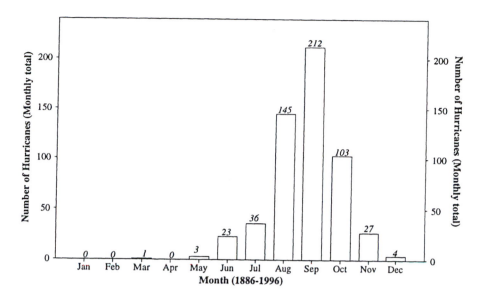

Fig. 4.4: Monthly distribution of North Atlantic hurricanes over the period 1886–1996. September is clearly the most active month followed by August and October. June and July are considered early-season months, with November considered a late-season month.

pan [there are thirty such hurricanes ($\approx$ 5%) over the period of record] is counted only once based on its first occurrence. Here we see the seasonal cycle of North Atlantic hurricanes with a pronounced peak in activity during August and September. Only 17% of the seasonal activity occurs outside the 3-month period of August through October. This late summer into early autumn maximum in hurricane frequency corresponds to the time of year when critically warm sea-surface temperatures (SSTs) cover the largest expanse of the North Atlantic Ocean. It also coincides with the time of year in which the vertical shear of horizontal winds (see Chapter 1) in the tropical Atlantic are typically at a minimum (Gray 1979). Approximately 50% more hurricanes occur in July than in June, and about the same percentage more occur in August than in October. October has four times as many hurricanes as November.

According to Hebert and Taylor (1979a), the earliest observed North Atlantic hurricane occurred on March 7, 1908, while the latest occurred on the last day of 1954 and continued into the first few days of 1955. The observed frequency with which a particular number of hurricanes occurs in different months of the season along with the probability of occurrences for various class intervals is given in Table 4.2. September is clearly the most active month with a 90% chance of observing at least one hurricane and a better than 50% chance of observing at least two. The probability of four or more hurricanes during September is 11%. In comparison, the probability of at least four hurricanes in August is only 4%. Over the 111-year period there are three years in which September was visited by exactly five hurricanes. Although there are slightly

Table 4.2: North Atlantic hurricane frequencies and probabilities over the period 1886–1996.
The whole numbers represent years with that many hurricanes. The probability of four or more
hurricanes developing in a single month peaks in September at 11%.

Number of Hurricanes	Jun	Jul	Aug	Sep	Oct	Nov
0	90	82	28	11	40	88
1	19	24	43	38	48	19
2	2	3	24	27	17	4
3	0	2	12	23	4	0
4	0	0	2	9	1	0
5	0	0	2	3	1	0
Observed Probability	Jun	Jul	Aug	Sep	Oct	Nov
At least 1 hurricane	0.19	0.26	0.75	0.90	0.64	0.21
2 or more hurricanes	0.02	0.05	0.36	0.56	0.21	0.04
3 or more hurricanes	0.00	0.02	0.14	0.32	0.05	0.00
4 or more hurricanes	0.00	0.00	0.04	0.11	0.02	0.00
5 or more hurricanes	0.00	0.00	0.02	0.03	0.01	0.00

more years with exactly one hurricane in October compared to August, the probability
of multiple hurricanes is substantially higher in August than in October. The probability of two or more July hurricanes is greater than the probability of multiple June
hurricanes. Hurricanes are more common in November than in June.

It is interesting to compare monthly distributions of hurricane activity for years
having above and below average activity (Jordan and Ho 1962). Table 4.3 shows these
distributions expressed as a percentage of total activity. High activity is defined, as
in Table 4.1, as years with nine or more hurricanes and low activity as years with
three or fewer hurricanes. The seasonal cycle of hurricane activity differs between the
two extremes. Years of high activity have more hurricanes and a greater percentage
of annual activity during August than during September. For low-activity years the
percentages of total annual activity during August and September are roughly the same
as the percentages for all years. Differences between the extreme years also extend to
the early- and late-season months. June is more active during low-activity years.

Thus an early start to the hurricane season seems to indicate fewer hurricanes overall. This is in contrast to what happens in the case of drought, where rainfall deficits at
the start of a season create conditions that lead to continued dry conditions throughout
the rest of the season (Changnon and Changnon 1990). July, October, and November
are more active during high-activity years. A characteristic of above-average hurricane
seasons is a late season surge in activity. Not surprisingly, it appears as if the "window of opportunity" for hurricane formations is more narrowly restricted during years
of low activity. This is likely due to widespread subcritical environmental conditions
during the peak months of the hurricane season. Low-activity years are characterized
by relatively few late-season storms.

Table 4.3: Monthly distribution of North Atlantic hurricanes over the period 1886–1996 for years of high and low activity (Table 4.1). Frequencies are expressed as a monthly total and as a percentage of the season total. N is the number of hurricanes in a single year while n is the monthly total. Totals by season (June through December) are also included.

Month	→	Jun	Jul	Aug	Sep	Oct	Nov	Dec	Total
High Activity	n	3	8	34	30	24	7	1	107
($N \geq 9$)	(%)	2.8	7.5	31.8	28.0	22.4	6.6	0.9	100
Low Activity	n	6	2	22	32	13	2	1	78
($N \leq 3$)	(%)	7.7	2.6	28.2	41.0	16.6	2.6	1.3	100
All Years	n	23	36	145	212	103	27	4	550
(1886–1996)	(%)	4.2	6.6	26.3	38.6	18.7	4.9	0.7	100

4.2   Season Length

North Atlantic hurricane activity occurs from June through November. These six months are designated as the "official" North Atlantic hurricane season by the U.S. National Weather Service (NWS). The warm tropical oceans supply heat and moisture in abundance during these months. This decreases the stability of the atmosphere and creates an environment favorable for organized convection as a precursor to tropical cyclones. Instability is greatest when SSTs over the North Atlantic are warmest during late summer and early autumn. The most active period for hurricane activity is August through October. Yet the season length for particular years can vary widely. For instance, in 1979 the first hurricane formed on July 10th and the last one dissipated on September 16th. Five years later, in 1984, the first hurricane formed on September 10th and the last one of the season dissipated in late December.

The begin and end dates of North Atlantic hurricanes during the official season are shown in Figure 4.5. A 9-day running average is used to smooth the large daily variability. The pronounced seasonality in hurricane occurrence is clearly evident. Most hurricanes occur between August 15th and October 20th. The likelihood of observing a hurricane form on any day from June 1st through July 31st is roughly constant and not very high. Starting in August there is a dramatic rise in the probability of hurricane formation peaking between September 1st and the 12th. The most common date for hurricane formation is September 5th. Fifteen hurricanes over the 111-year record formed on this date alone. The seasonal decline in hurricane activity is less dramatic. Two local minimums occur; one between September 16th and September 18th and the other, more pronounced, between October 4th and October 9th. Half of all North Atlantic hurricanes form on or before September 9th. The seasonal distribution of end dates mirrors the distribution of begin dates, though the frequencies are shifted later by a few days. The most common end date is September 15th and the median end date is September 14th. The secondary peak in hurricane activity during the middle of October is also evident in this figure.

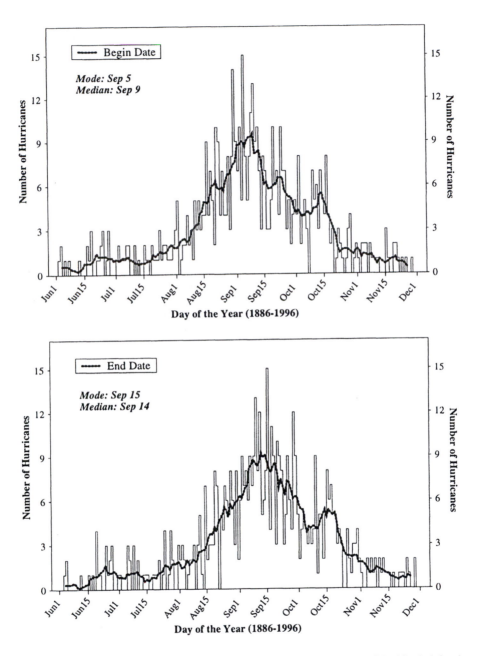

Fig. 4.5: Distributions of begin (top) and end (bottom) dates for hurricanes of the North Atlantic over the period 1886–1996 (Jun–Nov only). The dotted line is a 9-day running average. The season peaks during the second week of September.

The begin date of the season's first hurricane defines the *start of the season* for the year, while the end date of the last hurricane defines the *end of the season*. Based on this description, the interval between the start of the season and the end of the season defines the *season length*. Hurricane-season length for each year is plotted in Figure 4.6 (top). Variability in the date on which the hurricane season begins is larger than the date on which it ends. As such, the season length is dictated to a greater extent by the date on which the first hurricane forms. In general, a season which sees its first hurricane in June or July will be longer than one that does not see its first hurricane until August. Indeed, for the 6 years with the longest hurricane season, three began before June 1st and all six began before July 1st.[6]

The start and end dates of the hurricane season are examined as cumulative frequency graphs in Figure 4.6 (bottom). For the season onset, May 1st is plotted as the origin. For each date the number of years in which the first hurricane began on or before that date is summed. The same is done for the end of season.[7] The horizontal line represents the 50th percentile, which identifies the median begin (August 8th) and median end (October 16th) dates of the North Atlantic hurricane season. The two cumulative distributions show substantial differences although both distributions display regions of linear scaling signifying constant probability of onset or termination.

A scaling region is noted for the season onset curve during the months of June and July. Over these 2 months the probability of the hurricane season starting before a particular day ranges from a few percent on June 1st to 45% by the end of July. That is, by early August there is a 50% probability that the hurricane season is in progress. The linear relationship in the cumulative distribution graph is a consequence of a constant likelihood of season onset throughout June and July and suggests a consistent forcing over the North Atlantic during this time period. In other words, the conditions favorable (or unfavorable) for hurricane formation in early June are roughly the same as they are at the end of July, at least in a climatological sense. On average, conditions are not considerably more favorable for the season to begin in July as they are in June.

Climatological conditions change drastically in August as seen by a second scaling region appearing in the onset curve for this month. Again, this part of the curve can best be represented by a straight line, although the slope is approximately doubled from what it is during June and July, indicating substantially more favorable conditions for hurricane formations. The cumulative frequency for the end of season shows several scaling regions. The first extends from early September until early October. The second is a rather short, roughly 2-week period centered on the median end date of October 16th. A third scaling region extends from late October through November. It is interesting to note that this last scaling region has approximately the same slope as the initial scaling observed in the onset curve, suggesting that the dynamics operating to increasingly discourage activity at the end of the season might be similar to those acting to increasingly encourage activity at the beginning of the season. It is speculated

---

[6]The 1954 hurricane season was unusual in that the last hurricane dissipated in January of 1955.

[7]End date refers to the end of hurricane-force winds regardless of whether the storm became extra-tropical.

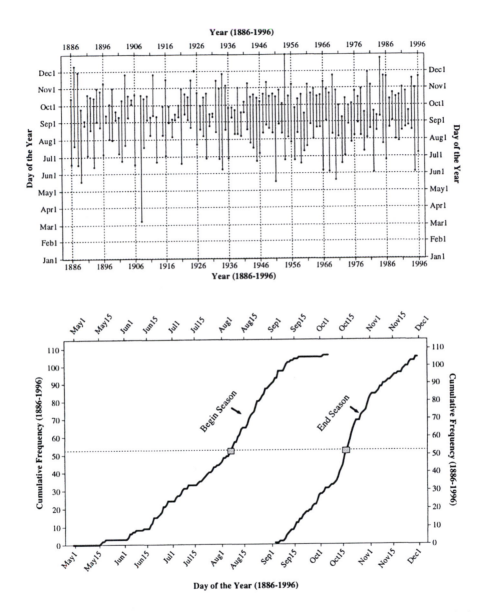

Fig. 4.6: Start and end dates of the North Atlantic hurricane season for each year in the period 1886–1996 (top) and the cumulative distribution of season onset and termination dates (bottom) based on data from 1886-1996. The horizontal line in the cumulative distribution figure identifies the median dates.

that these dynamics are tied to the seasonal warming and cooling of the ocean waters of the North Atlantic Ocean. Also of interest is the fact that the rounded average number of hurricanes occurring in seasons that begin on or after August 15th is 4, with only two late-starting seasons (1949 and 1981) having more than 6 hurricanes.[7]

4.3   Intensity and Duration

The intensity of a hurricane is measured by its 1-minute maximum sustained near-surface wind speed. For each storm there is a maximum intensity. Over an entire season the average maximum intensity is defined as the maximum intensity over all hurricanes. Figure 4.7 shows the average maximum intensity for each season over the period 1886–1996 and the annual distribution of years by maximum intensity. The overall mean average maximum intensity for all seasons is 92.4 kt. As seen, most seasons have an average maximum intensity between 85 and 90 kt. Five years have average maximum intensity greater than 115 kt, however in 3 of the 5 years the average is based on only two or three hurricanes. Average maximum intensities are generally higher during the 1940s and 1950s and clearly lower since about 1965. As noted in Chapter 3, caution must be exercised in interpreting the intensity estimates prior to about 1900, when there were few, if any, direct measurements of hurricane wind speeds.

Duration of North Atlantic hurricanes is examined by considering the average number of days of existence for all hurricanes in a season and for all years in the period (Figure 4.8). The total duration defined as the sum of all hurricane hours (or days) in a season is another component of seasonal activity. Tropical cyclones may reach hurricane intensity more than once in their lifespan. This fact is considered in these figures. Neglecting the years of no hurricanes, the average duration ranges from a low of 1 day to a maximum of 10 days with an overall average of 4.7 days and a standard deviation of 2.0 days. Since the average duration of a tropical cyclone from birth to dissipation is 10 days, the hurricane stage represents about 50% of the storm's lifespan. There is a trend toward shorter duration hurricanes with time. The question of significance of this trend is treated in Chapter 10. The distribution of hurricanes as a function of duration is given in the bottom panel of Figure 4.8. The most common duration is 1 or 2 days, but there are eight hurricanes with lifespans longer than 2 weeks. The distribution is skewed with a positive skewness value of 1.6. The skewness results because hurricanes that form in the Gulf of Mexico and the western North Atlantic travel only a short distance before dissipation over land. An apparent local minimum of hurricane duration is noted in the interval of approximately one week.

Major factors that determine hurricane duration are the place of origin and the environmental conditions at the time of occurrence. Most hurricanes persist as long as they remain over the warm waters of the tropical and subtropical ocean. The exception is when dry air, often from the north, gets entrained into the hurricane circulation. The dry air tends to enhance the associated sinking motion (down draughts), which inhibits

---

[7]The 1998 hurricane season featured 10 hurricanes, with the first occurring after the middle of August.

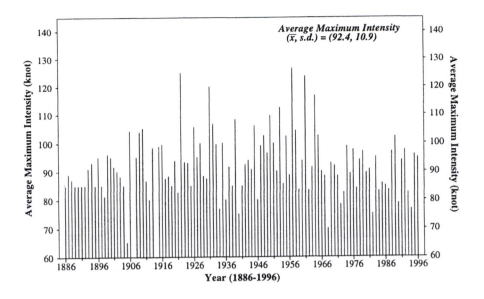

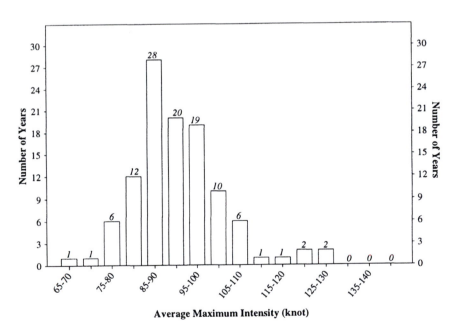

Fig. 4.7: Annual average maximum intensity of North Atlantic hurricanes for each season over the period 1886–1996 in kt (top), and the distribution of years by average maximum intensity in 5 kt intervals (bottom). During most years the average maximum intensity is in the range between 85 and 90 kt.

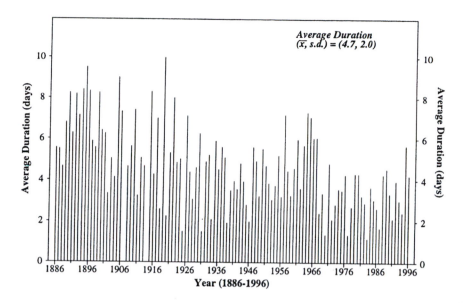

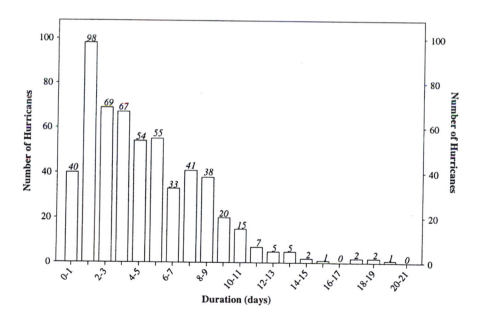

Fig. 4.8: Annual average duration of North Atlantic hurricanes for each season over the period 1886–1996 (top) and the distribution of hurricanes by duration in 1-day intervals (bottom). Note the overall trend toward shorter average duration. A majority of North Atlantic hurricanes do not last more than 4 days.

latent heat flux. Latent heat released into the atmosphere ultimate drives the circulation of a hurricane (Chapter 1). A hurricane that moves into a region of strong vertical shear of horizontal winds will tend to weaken and dissipate as the region of latent heating gets pushed away from the center of circulation. The record number of hurricanes in the North Atlantic at one time is four, which occurred on August 22, 1893.[8] September 11, 1961 saw three hurricanes at one time with a fourth tropical cyclone of near hurricane strength in the vicinity. The 1961 hurricane season was unusual in that there were no tropical cyclones during August.

Hurricanes originating in the central or eastern North Atlantic and remain at low latitudes will typically have the longest duration. Those that originate in the Gulf of Mexico or the western Caribbean Sea will have a shorter lifespan simply as a result of their proximity to land. A hurricane traveling in almost any direction out of the Gulf of Mexico will eventually encounter land and dissipate. Hurricanes that move to higher latitudes over the central and western North Atlantic rapidly loose their intensity as they make a transformation to an extra-tropical cyclone. These hurricane-initiated or hurricane-enhanced extra-tropical cyclones may last for weeks as they traverse large distances across to Europe and beyond.

It is interesting to note that the average duration during high activity years is 4.8 days (s.d. = 1.4 days), which is nearly identical to the duration during low activity years (4.5 days with a s.d. = 2.4 days). This suggests a possible difference between climatological factors responsible for tropical cyclone ontogeny and those factors responsible for maintaining the storm. The critical factors leading to hurricane origin and development may be somewhat irrelevant for hurricane sustenance. If this is the case, it is speculated that whereas growth is dominated by positive feedback mechanisms between the storm and its environment, hurricane survival may depend on an interplay of both positive and negative feedback in a situation of near equilibrium.

Intraseasonal variations are examined by tabulating average hurricane duration by month (Table 4.4). On average hurricanes during September last the longest. Late, as well as early, season hurricanes tend to be of shorter duration than storms of midseason. October and November hurricanes are longer-lived than hurricanes of July and August. December is the exception to this rule, having hurricanes of duration 25% longer than the average. However, this is based on a sample size of only four hurricanes. The monthly variability in hurricane duration can be explained in part by the different places where hurricanes originate and by the differences in tracks for hurricanes occurring at various times of the year.

4.4   Origin

A hurricane's origin is defined as the location where the tropical cyclone first reaches hurricane intensity (Dunn 1956). Though the location where the tropical cyclone is first identified as a wave or a disturbance is important, this information is less precise. We

---

[8]The 1998 season featured 4 simultaneous hurricanes on September 26.

Table 4.4: Average duration of North Atlantic hurricanes by month (June through December) over the period 1886–1996. N represents the number of hurricanes in each month and the average duration ($\bar{t}$) is given in days. Note that on average September and December support the longest-lasting hurricanes. The December average is based on only four hurricanes.

Month	→	Jun	Jul	Aug	Sep	Oct	Nov	Dec	Total
All Hurricanes	N	23	36	145	212	103	27	4	550
Avg. Duration	$\bar{t}$	4.1	3.4	3.9	5.5	4.8	4.4	5.9	32.0
Percentage	(%)	12.8	10.6	12.2	17.2	15.0	13.8	18.4	100

begin by examining the points of origin of all hurricanes throughout the North Atlantic basin over the period 1886 through 1996.

4.4.1 General Characteristics

Figure 4.9 shows the points of origin of all North Atlantic hurricanes over the period 1886 through 1996. Save for a few storms, the vast majority of Atlantic tropical cyclones reach hurricane strength between 10° and 35°N latitude. The lack of a significant Coriolis component to the winds blowing toward low pressure precludes the development of hurricanes farther south than about 8°N (see Chapter 1). In the North Pacific basin tropical cyclones sometimes form closer to the equator. The difference over the Atlantic is likely due to the northward extension of South America preventing formation south of 8° latitude. A weak Coriolis force inhibits development of tropical waves near the equator. At latitudes poleward of 30°N, colder SSTs and stronger wind shears inhibit the formation of hurricanes. Yet, at those higher latitudes otherwise benign tropical disturbances may flourish under certain baroclinic conditions. Once a hurricane develops it may move well north of tropical latitudes. The warmer waters of the western North Atlantic associated with the gulf stream allow for hurricane formation at latitudes farther north compared to formations over the eastern North Atlantic. A separate area of higher latitude origins is noted.

A distinct longitudinal variation in points of origin is apparent over the Caribbean Sea. The eastern and western Caribbean Sea appear to be source regions for hurricane formation, but there is a conspicuous lack of formations over the central Caribbean to the south of Hispaniola. This observation has been noted by others including Colón (1953) and Dunn (1940). The central Caribbean is located near the climatologically favored region of the tropical upper-tropospheric trough (TUTT). It is also near the mountains of Venezuela. It is speculated that the mountains keep the low-level air relatively dry and the TUTT acts to shear developing convection. Both factors contribute to the dearth of hurricanes in this region. Over the Gulf of Mexico hurricanes can originate anywhere, with a slight preference for formation over western regions and the Bay of Campeche. It is remarkable that some tropical cyclones become hurricanes north of 40°N latitude in the zone of prevailing westerlies.

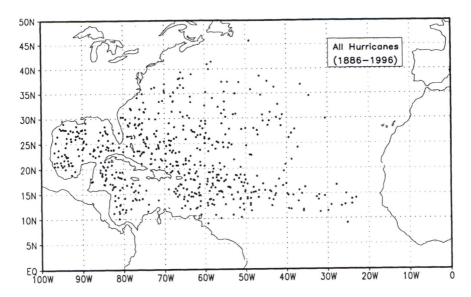

Fig. 4.9: Points of origin for the 554 hurricanes of the North Atlantic over the period 1886–1996. Preferred regions of hurricane formation include the Gulf of Mexico, the western Caribbean Sea, and near the Lesser Antilles. Note the large number of hurricane origins at high latitudes over the western North Atlantic.

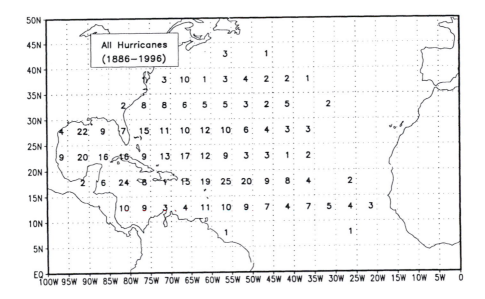

Fig. 4.10: Total number of North Atlantic hurricane origins in 5° latitude-longitude grid boxes based on the points plotted in Figure 4.9. Note the near-absence of hurricane formations south of 10°N latitude.

Table 4.5: Preferred regions of hurricane origins as defined in 5° latitude-longitude grid boxes used in Figure 4.10. Hurricane formations are most likely from the central tropical North Atlantic westward through the Caribbean Sea to the Gulf of Mexico.

Location	Total Frequency
Central Tropical North Atlantic	25
Western Caribbean Sea	24
Gulf of Mexico	22
Northwestern North Atlantic	10

A quantitative look at the spatial distribution of preferred areas of formation is given in Figure 4.10, which shows the number of hurricane formations in 5° latitude by longitude grid boxes. The area of the boxes decreases with increasing latitude. Here we can again see the preference for formations over the central and eastern Atlantic as well as over the western Caribbean Sea, Gulf of Mexico, and the higher latitudes of the western North Atlantic. Also, we can define a particular box as a region of local maximum under two conditions: The box must contain five or more points of origin and the surrounding eight grid boxes all must have a lower number of origins. In this way we identify four distinct regions of hurricane formation (Table 4.5) including the central tropical Atlantic, the western Caribbean Sea, the Gulf of Mexico, and the northwestern North Atlantic. The region of absolute maximum hurricane formations as defined in these 5° grid boxes occurs in the central tropical Atlantic, followed closely by the western Caribbean Sea and Gulf of Mexico.

A bi-modality in the distribution of hurricane origins by latitude results from the lack of formations between the active regions of the tropics and the relatively active region of the subtropics. As shown in the next chapter, these two active regions result from differences in origin and development mechanisms. The distribution of hurricane formations by longitude indicates a maximum of occurrence between 55 and 70°W longitude. This region includes the central tropical Atlantic near the Lesser Antilles. No hurricanes have formed east of 15°W longitude.

4.4.2   Monthly Activity

Substantial seasonal variations occur in the location of hurricane formation. Figure 4.11 shows the points of hurricane origin by month for June through November. In June, hurricane formation is limited to regions of the western Caribbean Sea and to the Gulf of Mexico. The early season formation is coincident, and likely linked, with the start of the eastern North Pacific hurricane season. By July, areas of hurricane formation spread eastward to include the western North Atlantic between 25–40°N and the tropical central Atlantic between 10–20°N. In August, the points of origin are considerably more numerous with many of them concentrated in the western and central North Atlantic. An increase in the area of formation to include the eastern tropical At-

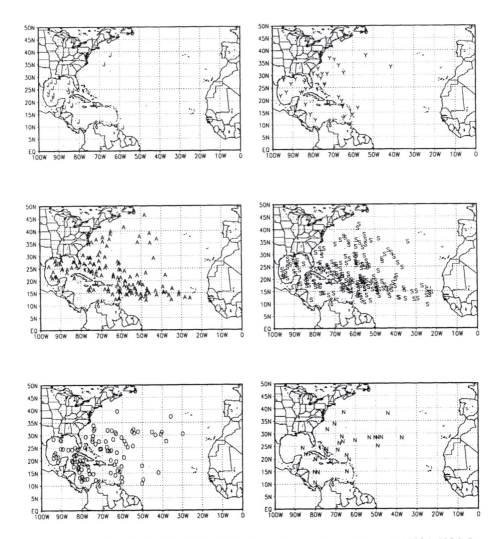

Fig. 4.11: Points of origin for North Atlantic hurricanes by month over the period 1886–1996. In the panels, J denotes June, Y July, A August, S September, O October, and N November. Early season hurricanes are concentrated over the Gulf of Mexico and western North Atlantic. Note the relative abundance of hurricane formations in the western Caribbean Sea during October, especially compared to August. By November the most likely area of hurricane formation is over the central North Atlantic between 25 and 30°N latitude.

lantic west of the islands of Cape Verde is noted at this time of year. This corresponds to the first half of the Cape Verde season (see Chapter 5). September's spatial distribution is similar to August with a preponderance of hurricane formations in the tropical central and eastern Atlantic. However, there is also an increase in hurricane origins at higher latitudes in the western North Atlantic at this time. In October, the focus of hurricane formation shifts back to the western Caribbean with substantially fewer hurricanes developing over the central and eastern tropical Atlantic. During November hurricane formations are all but absent over the tropical belt south of 20°N with the exception of the western Caribbean. Higher latitude formations are noted over the western and central Atlantic.

Points of hurricane origin are plotted by longitude and latitude in Figure 4.12 for each month of the season. The circle is the average centroid for origin points. In general the centroid does not define the most likely location of hurricane origin. The spread of points toward the east between June and August is clearly evident in this figure as is the burst of activity in August and September over much of the Atlantic. The seasonal shift in areas favorable for hurricane formation has been noted by numerous authors, including Mitchell (1924), Dunn (1958), Haggard (1958) and Cry and Haggard (1962). The reduction in overall activity and the re-emergence of formations in the western Caribbean during October is also evident.

More subtle features of seasonal variability are also indicated. For instance, as hurricane activity increases and shifts eastward with the progression of the season, it also fans northward and southward to cover a larger area of the open waters of the North Atlantic. August hurricane formations over the central North Atlantic tend to cluster into two groups; one located at low latitudes primarily south of 20°N and the other located at high latitudes generally north of 35°N. In September this clustering is less obvious. The seasonal variations in activity over the western Caribbean is quite pronounced showing a maximum in June and a minimum in August followed by another maximum in October.[9] The late season activity of November extends southwest to northeast from the western Caribbean to the central North Atlantic which is in contrast to the southeast-to-northwest orientation during the active months of August and September.

Seasonal variability of hurricane formations by latitude and longitude is depicted in Table 4.6. Only small variations in the average latitude occur throughout the season with formations during the active months of August through October located counter-intuitively farther south than during the early and late season months. Whereas the early season formations are confined to a narrow latitude belt because of the confines of the Gulf of Mexico, the mid-season formations spread both northward and southward across the North Atlantic, so that the average latitude of formation changes only slightly. In contrast, the longitudinal variation over a season is dramatic. The average longitude of hurricane formation shifts from 86°W during June to about 62°W by

---

[9]This coincides with the appearance of two distinct hurricane seasons in Jamaica and Bermuda (see Chapter 9).

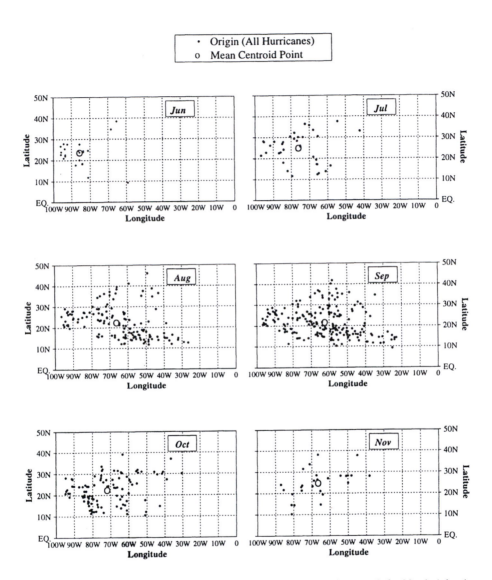

Fig. 4.12: Latitude versus longitude distribution of points of origin by month for North Atlantic hurricanes over the period 1886–1996. The circle locates the average centroid of hurricane origins. In general, the centoid does not reflect the most likely location of hurricane formation. June and July hurricanes form predominantly to the west of 60°W longitude. The latitudinal distribution of origins is broadest during September, however the centroid is located farthest to the south during this month.

Table 4.6: Average latitude and longitude of points of hurricane origin by month. Number of cases (N), average value ($\bar{x}$) and standard deviation is ($s.d.$) are also given.

Month	$\rightarrow$	Jun	Jul	Aug	Sep	Oct	Nov
	N	23	36	145	212	103	27
Latitude (°N)	$\bar{x}$	23.4	24.8	22.2	21.5	22.2	24.6
	$s.d.$	6.3	7.5	7.4	6.7	6.6	6.7
Longitude (°W)	$\bar{x}$	85.7	78.0	64.7	62.5	72.0	66.6
	$s.d.$	9.9	12.5	16.6	17.1	14.6	13.2

September. During October the activity shifts back to the west, although in November the average longitude is again farther to the east.

4.5 Tracks

There is no average or mean Atlantic hurricane track and no two hurricanes follow the exact same path. Yet many North Atlantic hurricanes move along paths having common characteristics, particularly if they form at the same time of year. For instance, hurricanes generally move around the middle atmospheric ridges of higher pressure with a tendency for a storm to move northward through a weakness in the ridge. Westward displacement of storms decreases with increasing latitude. The strongest hurricanes are steered by winds at higher altitudes. There is also a tendency for hurricanes to move toward warm, moist air in the middle levels of the atmosphere. A climatology of hurricane tracks provides a first guess for predicting paths of future hurricanes. Here general characteristics of past hurricane tracks, including their seasonal variability, are examined.

4.5.1 General Characteristics

Initially, most hurricanes move westward or northwestward pushed along by the large-scale atmospheric flow. A northward drift results from the latitudinal variation of the Coriolis force across the breadth of the storm. As they approach the western North Atlantic and the coast of the United States hurricanes are typically far enough north to be pulled more strongly in that direction. The paths of all North Atlantic hurricanes over the period 1886–1996 are shown in Figure 4.13. The general westward movement of hurricanes over the eastern and central North Atlantic with a change to a more northwesterly direction over the Gulf of Mexico and western Atlantic is clearly evident. Farther north and west the tracks indicate a northeasterly movement. The overall curvature in path (known as *recurvature*) is explained by the average position of the subtropical Bermuda high pressure with its attendant anticyclonic (clockwise) steering winds at middle and high levels in the atmosphere. The initial east-to-west movement in the tropics reflects the direction of the easterly trades to the south of this

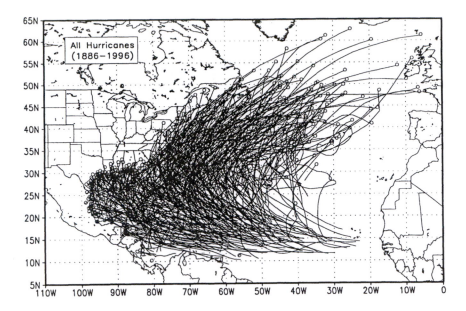

Fig. 4.13: Tracks of North Atlantic hurricanes over the period 1886–1996. Tracks are plotted from the location where the tropical cyclone first gained hurricane intensity (*point of origin*) until the location where the hurricane diminishes to tropical cyclone intensity.

semi-permanent high pressure ridge. Poleward of 30°N latitude hurricanes encounter westerly winds (westerlies), which steer them northward and eventually eastward. The westerly currents move the hurricanes faster and more directly than the easterly trade wind flow. If the subtropical high extends through a significant atmospheric depth the hurricane will take a wide turn during recurvature. If the high slopes toward the south or southeast with height recurvature will be sharp. Recurvature of individual storms depends on a number of factors. Principal among them are the position and movement of nearby upper-level troughs of low pressure. A large amplitude trough moving toward the hurricane from the west will signify recurvature. If a portion of a trough drops to a latitude south of the hurricane, then recurvature is likely. On the other hand, with a strong subtropical high to the north and a long-wave trough well west of the storm recurvature is unlikely. Recurvature often spares coastal areas of the United States a direct hurricane hit but poses a particular threat to Bermuda.

The movement of a hurricane over land or over cold waters results in its transformation. Over land the transformation is characterized by a rapid reduction in its intensity as winds subside and central pressures increase. Hurricanes may rejuvenate after crossing back over warm waters from land areas. This occurs frequently with hurricanes that cross the peninsula of Florida from the western North Atlantic. Metamorphosis to an extra-tropical cyclone may occur as the hurricane moves to higher latitudes and encounters strong temperature contrasts. The strong, concentrated center of fast winds around the hurricane decays, but the precipitation area expands. Hurricanes that become extra-tropical storms are directed by the upper-tropospheric westerlies. There is some tendency for tropical cyclones to move under cut-off upper-air lows at high latitudes. Hurricanes weaken rapidly at latitudes north of 40°N.

The recurving hurricane represents one class of North Atlantic hurricane tracks.[10] Recurving hurricanes tend to threaten the United States or Canada, whereas hurricanes that do not recurve (or do so only slightly) take a path into the Gulf of Mexico threatening the east coast of Mexico or the gulf coast of the United States. Another class of nonrecurving hurricanes are those that originate over the western North Atlantic, north of about 30°N latitude. These storms move directly toward the northeast and away from the United States. Late season hurricanes originating over the western Caribbean Sea tend to track northward and northeastward.

A picture of the average motion of hurricanes crossing the North Atlantic is obtained by averaging forward speeds and directions in 5° latitude-longitude grid boxes based on the differences in 6-hourly best-track positions. Figure 4.14 (top) gives the number of observations used to compute the averages and provides an indication of the most vulnerable areas for hurricane occurrences. Note that a single hurricane moving slowly through a grid box will generate several observations for that box at successive 6-hour periods. The average is less reliable for boxes where the direction of motion is highly variable. The region most menaced by hurricanes is over the Bahamas stretching eastward to 65°W longitude. Much of the Gulf of Mexico is also significantly threatened by hurricanes.

Figure 4.14 (bottom) gives the average forward speed and direction of hurricanes in boxes for which there are at least three observations. The figure provides a synopsis of the average steering flow across the North Atlantic during the hurricane season. The basin-wide clockwise circulation is clearly evident. Westerly-moving hurricanes at low latitudes curve to the northwest then to the north and eventually to the northeast as they reach northern latitudes. For each longitude band, average hurricane motions veer (turn clockwise) with increasing latitude. The average motion helps explain why certain coastal stretches are more likely to get hit by a hurricane compared to others (see Chapters 8 and 11). For instance, along the northern Gulf of Mexico the average hurricane motion is nearly perpendicular to the shoreline, whereas hurricane motion is nearly parallel to the northeastern coast of Florida.

Hurricanes north of 35°N generally move faster than those to the south, largely due to stronger upper-tropospheric winds at higher latitudes. The greater forward speeds for hurricanes at high latitudes adds to their destructive potential in the right-forward quadrant of the circular vortex. Despite the tendency for rotational winds to slacken as the hurricane moves over cooler waters, a somewhat weak storm making landfall along the U.S. northeast coast can still produce damage comparable to a major hurricane. Faster motion will generally reduce the likelihood of devastating floods from rainfall. We note a region of faster moving hurricanes near the Lesser Antilles extending into

---

[10] Another class of tracks is represented by hurricanes that remain at low latitudes.

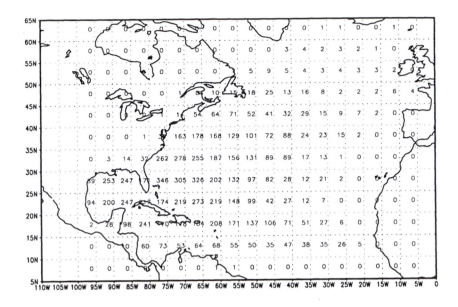

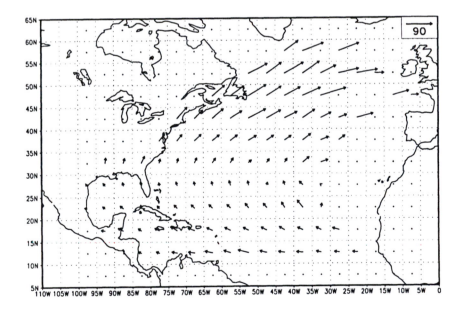

Fig. 4.14: Total number of hurricane observations made in each 5° latitude-longitude grid box based on the 6-hourly best-track data set locations (top). Average forward speed and direction of hurricanes passing through each of the 5° latitude-longitude grid boxes for boxes with three or more observations. Length of the arrow is proportional to forward speed with units in km hr^{-1}.

Table 4.7: Looping hurricanes. Values are year, begin and end dates, and direction of loop over the period 1886–1996. A track is looping if the storm retained hurricane strength through most of the loop and the loop covered an area of at least the size of a 1° × 1° latitude-longitude box. The numbers (1 through 15) refer to Figure 4.15. Dates refer to the begin and end date of hurricane strength, and the direction refers to a clockwise or counterclockwise loop.

Number	Name	Year	Begin –End Dates	Direction
1	.	1896	10/28 – 11/9	Counter
2	.	1900	9/9 – 9/19	Clockwise
3	.	1903	10/1 – 10/9	Clockwise
4	.	1908	9/30 – 10/6	Counter
5	.	1910	10/11 – 10/18	Counter
6	.	1912	10/5 – 10/9	Counter
7	.	1926	9/4 – 9/21	Clockwise
8	.	1926	9/22 – 9/29	Counter
9	.	1929	10/17 – 10/21	Counter
10	.	1943	9/15 – 9/18	Counter
11	Charlie	1950	8/28 – 9/4	Clockwise
12	Ginny	1963	10/20 – 10/29	Counter
13	Carol	1965	9/20 – 10/1	Clockwise
14	Inga	1969	9/30 – 10/10	Clockwise
15	Ginger	1971	9/11 – 9/30	Clockwise

the eastern Caribbean Sea south of 15°N. This region tends to be somewhat void of hurricane formations so perhaps factors that push hurricanes through this region rather fast are also responsible for inhibiting hurricane formations.

Though the motion of North Atlantic hurricanes is fairly smooth [after elimination of the "wobbles" (see Chapter 1)], tracks of some hurricanes are erratic, particularly in regions of light steering winds. For example, hurricanes occasionally stop and reverse directions. On rare occasions a hurricane makes a looping track defined as a path that crosses back over a particular location twice. A looping path is more likely when the middle latitude flow is more meridional with a blocking anticyclone. Figure 4.15 shows the tracks of North Atlantic hurricanes that have made a substantial loop. Looping hurricanes are most common during September followed closely by October (Table 4.7).[11] In general, counterclockwise loops occur over the western and eastern North Atlantic while clockwise loops occur over the central Atlantic. Most loops occur between 25 and 35°N latitudes in an area where average forward speeds are slowest (Figure 4.14). It is noteworthy that all clockwise loops occur in the latitude-longitude box of 25–40°N by 30–70°W, whereas the counterclockwise loops largely occur outside this region. Hurricane Ginger of 1971 was the last significant looping hurricane in the record.

Occasionally two hurricanes will interact. This occurs when a fast-moving hurricane approaches a slower moving hurricane. The interaction produces erratic tracks

---

[11] Only one looping hurricane occurred during August.

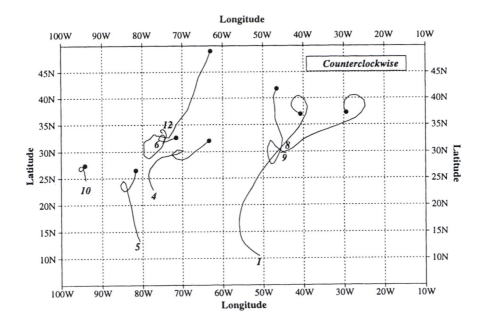

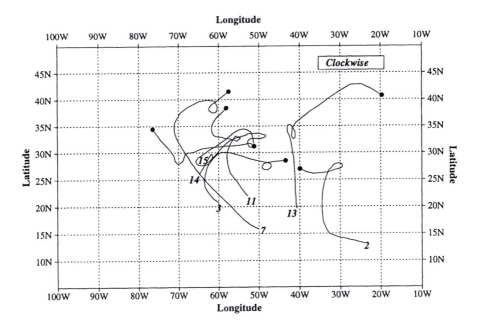

Fig. 4.15: Tracks of North Atlantic hurricanes over the period 1886–1996 showing a substantial loop. Counterclockwise loops are displayed on top. Numbers refer to the hurricanes listed in Table 4.7. The circle shows the dissipation point for each hurricane.

for both storms including equatorward deflections, rapid recurvature, and rotation. A mutual advection model (Fugiwhara 1931) is frequently used to explain the interacting process. In this model a pair of storms will rotate about a common point located on a line joining the centers of the two tempests. The point will be displaced toward a particular storm in reverse proportion to their intensities (Fugiwhara effect). More recently, Carr et al. (1997) define four alternative models of tropical cyclone motion for storms in binary interaction. They suggest the situation is more complex than the mutual advection hypothesis admits.

4.5.2   Seasonal Variability

Seasonal variability in hurricane tracks is shown in Figure 4.16. Most outstanding is the difference in hurricane paths between the middle of the season and the early and late season. In general, middle season tracks start at low latitudes over the central Atlantic and extend westward then northward and eastward over the western North Atlantic. Some August and September hurricanes track almost due westward through the Caribbean Sea into the Gulf of Mexico before striking land in Mexico or the United States. Early and late season tracks, in general, extend to the north out of the western Caribbean or western Atlantic. An important fraction of the early and late season hurricanes originate in the western North Atlantic near 30°N and track north and northeastward.

In particular, June tracks are limited to the Gulf of Mexico, the western Caribbean and the extreme western North Atlantic. There is a tendency for the early-season hurricanes to landfall along the Texas coastline (see Chapter 8). By July the tracks spread eastward at both high and low latitudes, however only a rare hurricane makes it north of 45°N. The northern Gulf and southeast coasts are likely targets for July hurricanes. In August hurricane tracks extend even farther to the east and to the north. At this time of the year it is possible for a hurricane to reach 60°N latitude. By September many hurricanes track north of 40°N particularly over the central North Atlantic between 30° and 50°W longitude.

Yet, September is also noted for low-latitude tracking storms as was the case with hurricane *Fifi* which traversed the Caribbean during the middle of September during 1974 before devastating Honduras with winds of category three intensity and flooding rains. October is conspicuous by the near absence of tracks over the tropical central Atlantic and eastern Caribbean Sea. Also at this time of year, fewer hurricanes make it into the western Gulf of Mexico although the eastern Gulf is active. Hurricanes originating over the western Caribbean during October tend to track northward toward Florida. In November hurricane tracks are largely confined to the subtropical latitudes between 20° and 45°N over the western and central North Atlantic.

The so-called hurricane alley extending from the Lesser Antilles westward to the north of the Caribbean Islands is conspicuous during July and particularly so in August and September. This region is favorable for hurricane development and intensification. Many tropical cyclones moving through this region reach the Bahamas before

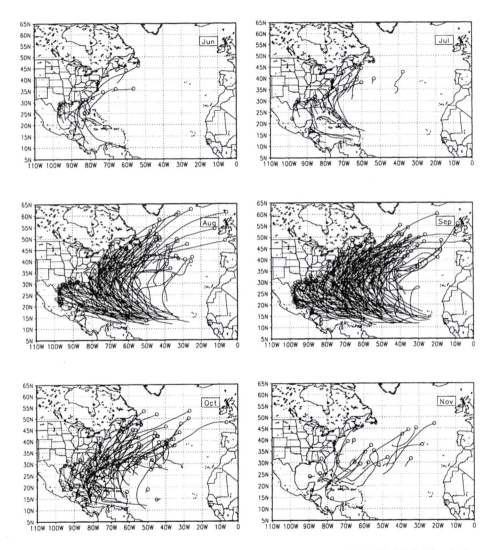

Fig. 4.16: Tracks of North Atlantic hurricanes by month over the period 1886–1996. The circle denotes the dissipation location. Many of the June and July hurricanes travel on a path that takes them near the U.S. coast. Note that by October the Texas coastline is relatively free of hurricane activity. The northeastward movement of November hurricanes usually takes them away from the U.S. coastline.

recurving to the north and northeast. Somewhat less frequently they continue west-ward, crossing southern parts of Florida before entering the Gulf of Mexico. With respect to the climatological circulation patterns, hurricane alley is located to the south of the Bermuda high on the western side of the 200 mb middle-Atlantic trough (TUTT; see Chapters 1 and 2), which is oriented northeast to southwest over the North At-lantic extending from Puerto Rico to Europe during the heart of the hurricane season (Michaels 1973).

Another preferred region for hurricane tracks is over the western Caribbean extend-ing northwestward through the Yucatan Straits and into the Gulf of Mexico. Between these two development regions, the islands of Hispaniola with its relatively high moun-tains (> 3000 m above sea level) tend to weaken tropical cyclones creating a narrow "zone of avoidance" for significant hurricane activity. For example, hurricane *David* of 1979 weakened to a tropical storm as it moved over the Dominican Republic and Haiti. *David* regained hurricane strength after moving northward out of the Windward Passage.

4.6 Dissipation

Hurricane dissipation occurs when sustained near-surface winds decrease to below 33 m s^{-1}. Typically a hurricane dissipates over land and over cold ocean waters. Dissipation over land results from increased friction and the loss of a heat and mois-ture source. In general the rate at which winds diminish is proportional to the wind speed at landfall. The surface friction associated with hurricanes over the ocean is not negligible due to rough seas, but over a warm ocean the energy source significantly compensates the frictional loss. Over cooler water the energy source is diminished and the frictional influence is proportionally more important. Dissipation is likely if the hurricane encounters strong shearing winds. Circulation changes of the hurricane's large scale environment that decrease convergence at low levels, or advect drier, cooler air into the storm will lead to dissipation. A decrease in the upper-level divergence due to vertical shear-possibly brought about by intrusion of middle-latitude westerlies into the tropics-will also lead to storm decay. Dissipation is recorded by the NHC when a sufficient transition to an extra-tropical cyclone occurs.[12]

The final dissipation point for all 554 North Atlantic hurricanes are shown in Fig-ure 4.17. Many hurricanes dissipate completely upon striking the coasts of Mexico or the United States. The exception is peninsular Florida. Hurricanes that weaken upon impact with Florida will tend to regenerate after moving back over the Gulf or Atlantic waters. Hurricanes that move northward into the southern United States tend to dissi-pate more slowly than those that move westward into Mexico. The rugged terrain over Mexico promotes rapid dissipation of landfalling hurricanes due to significant frictional effects. Hurricanes north of 35°N generally dissipate over the open waters.

Figure 4.18 shows the number of dissipation points in 5° latitude-longitude boxes.

---

[12]Dissipation occurs several times for a reintensifying hurricane.

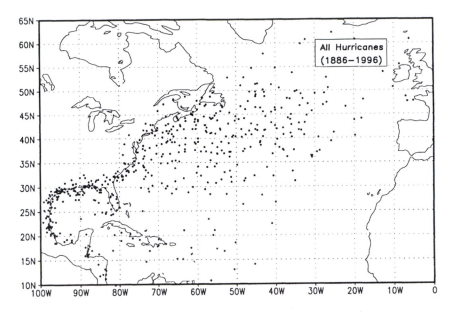

Fig. 4.17: Dissipation points for all 554 North Atlantic hurricanes over the period 1886–1996. Note that dissipation points cluster near the coastline at low latitudes and occur over open waters farther north.

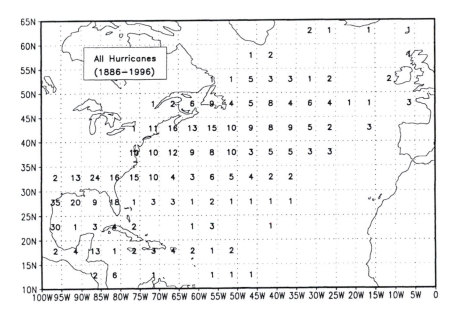

Fig. 4.18: Total number of North Atlantic hurricane dissipations in 5° latitude-longitude grid boxes based on the points plotted in Figure 4.17. Note the relative lack of hurricane dissipations over the open waters south of 30°N latitude.

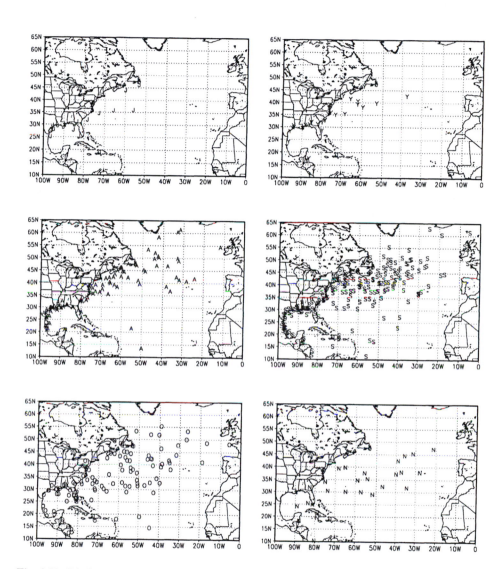

Fig. 4.19: Dissipation points for North Atlantic hurricanes by month over the period 1886–1996. In the panels J denotes June, Y July, A August, S September, O October, and N November. Early and middle season hurricanes tend to dissipate at low latitudes only when they strike the coast. Open-water dissipation is rare south of 35°N until September. Note that hurricane dissipation over Florida in October is more likely than in August.

Principal regions of dissipation include the western Caribbean, the western and northern Gulf of Mexico, Florida, and the eastern U.S. seaboard. These areas are the first significant landmasses encountered by hurricanes traveling northwestward out of the tropical Atlantic. The northern North Atlantic is also a region where hurricanes die largely as a consequence of stronger upper-level winds and cooler ocean temperatures. Storms traveling westward through the Caribbean Sea dissipate as they encounter the Yucatan Peninsula. Interestingly, hurricane dissipation is not likely in a broad area to the north of the Greater Antilles extending northward through the Bahamas. SSTs in this region are quite warm throughout the hurricane season. Hurricanes moving through the Caribbean Sea are also unlikely to dissipate. However, there is a tendency for hurricanes to dissipate if they make landfall over Hispaniola or Puerto Rico. Recall this is an area where hurricane genesis is also relatively rare due to the large landmass of the Greater Antilles. Dissipation points are uniformly distributed over the western half of the North Atlantic north of 30°N latitude.

Dissipation points by month are plotted in Figure 4.19. June and July dissipation occurs along the northern Gulf coast, along the southeast coast and over the northwestern Atlantic Ocean. By August and September, dissipation points extend over a broad expanse of the North Atlantic with some as far north as 60°N. Hurricane dissipation along northern Mexico, the Gulf coast states and along the eastern U.S. seaboard is also quite common. October continues the pattern of August and September but with fewer hurricanes. The Florida peninsula appears to be a focus for dissipation during this month. In November most hurricanes dissipate before reaching the cool waters north of 40°N latitude. Extra tropical transitions of late-season hurricanes occur farther to the south than transitions occurring during the middle of the season.

5

Tropical Only Hurricanes

To help understand the climatology of North Atlantic hurricanes it is useful to distinguish between two types of hurricanes. From the point-of-view of ontogeny, tropical-only hurricanes originate and develop free of any *enhancing middle latitude* baroclinic influences. Non-tropical-only hurricanes are baroclinically-enhanced. Motivations for considering tropical-only hurricanes as a separate category of North Atlantic basin hurricanes were presented in Chapter 3. In several ways the origin and development of a tropical-only hurricane represents the maturation of a "classic" (or textbook) tropical cyclone. Many of these hurricanes form from easterly waves traveling at low latitudes across the North Atlantic during August and September. In this chapter a fresh perspective on North Atlantic hurricanes is provided by considering various statistics of tropical-only hurricanes. Comparisons are made between tropical-only and non-tropical-only hurricane activity.

5.1   Designation

A subjective designation of tropical-only hurricanes is made by consulting reports on individual storms (see Chapter 3). An objective classification tree is applied to the subjective designation. The classification tree uses data on the latitude, longitude, and Julian date of the initial tropical depression and the initial hurricane stage of the storm. The classification produces eight rules. Elsner et al. (1996a) find an accuracy of greater than 90% when using the rules to indicate whether a hurricane is tropical-only with respect to the original subjective designation. They note that 32% of the hurricanes are classified as tropical-only by applying only rules one and two, while 36% are classified as non-tropical-only by applying rules one and three. Thus, nearly 70% of North Atlantic hurricanes are classified by invoking only three rules. Hurricane *Hugo* of 1989 is an example of a tropical-only hurricane. Hurricane *Arlene* of 1967 is an example of a baroclinically-enhanced hurricane. Figure 3.4 shows the pre-hurricane tracks of *Hugo* and *Arlene*. Tropical-only hurricanes form at low latitudes, while baroclinically-enhanced hurricanes form farther north. Rules four through eight are needed to classify the remaining 32% of the North Atlantic hurricanes.

Table 5.1: Rules for determining a tropical-only hurricane (TO) based on tropical storm and hurricane data. LATH and LONH are the latitude and longitude at which the system reaches hurricane strength, respectively, and LATS and LONS are the latitude and longitude at which the disturbance reaches tropical storm strength, respectively, and DAYS is the Julian day on which the system reaches tropical storm strength. For leap years subtract one day from DAYS. BE refers to non-tropical-only (baroclinically-enhanced) hurricanes.

Rule Number	Decision Rule	Action Taken	
		YES ?	NO ?
1	Is LATH $<$ 23.5°N ?	Rule 2	Rule 5
2	Is DAYS $<$ 303 ?	Rule 3	H=BE
3	Is LONS $<$ 79.0°W ?	H=TO	Rule 4
4	Is DAYS $<$ 166 ?	H=BE	H=TO
5	Is LATS $<$ 25.5°N ?	Rule 6	H=BE
6	Is DAYS $<$ 286 ?	Rule 7	H=BE
7	Is DAYS $<$ 203 ?	H=BE	Rule 8
8	Is LONH $<$ 73.4°W ?	Rule 9	H=TO
9	Is LATH $<$ 26.8°N ?	H=TO	H=BE

It is interesting to speculate on the physical nature of these rules. The fact that a judicious screening of storms by an individual is reduced to a relatively few objective rules, with a relatively high degree of accuracy, suggests some underlying order to the diversity of hurricane development processes. In particular, the importance of middle-latitude baroclinic disturbances in generating tropical cyclones is emphasized. Baroclinic disturbances penetrate the tropics in a variety of ways. The increase in hurricane activity during October is partly a consequence of the return of middle-latitude disturbances tracking over the still-warm waters of the North Atlantic.

Where detail information on storm processes is lacking, the classification rules are used as a guide in distinguishing hurricane types from the best-track record. However, before 1951 data on the initial depression stage of hurricanes is not available. Therefore, Elsner et al. (1996a) develop a classification tree using only data after the system reaches tropical storm strength (18 m s^{-1}). The results are similar in that the algorithm returns a classification tree with nine decision rules (Table 5.1). This second tree is accurate at better than 90%, though there is a slight bias toward tropical-only hurricanes. On the other hand, there is no time-dependent trend in the bias. The most important variables for identifying tropical-only hurricanes in order of importance are, initial hurricane day, initial hurricane latitude, and initial tropical storm latitude.

The decision rules are applied to the historical best-track data over the period 1886–1950. There is a total of 279 hurricanes over the period. Because a few of these storms were not detected until hurricane strength, the initial storm and initial hurricane data are identical (e.g., this is the case for the only hurricane of 1890). While co-linearity (correlation among the independent variables) is not a problem with classification trees as it is with other statistical models, the missing data add a bit of uncertainty to the decision in grouping these particular hurricanes. The combination of rule-based hurricane designations prior to 1950 and subjective designations after provides a complete delineation of tropical-only hurricanes in the best-track record. The separation of hurricanes into tropical-only and non-tropical-only is used to compute various statistics presented throughout the book.

5.2 Abundance

Based on the above classification scheme, a total of 358 North Atlantic tropical-only hurricanes are noted in the best-track record over the period 1886 through 1996. This represents 64.6% of all North Atlantic hurricanes. We begin a look at the climatology of tropical-only hurricanes by considering variations in their frequency of occurrence. Comparisons are made to non-tropical-only hurricanes.

5.2.1 Interannual Variability

Multi-year variation in tropical-only hurricane activity is examined by considering the ratio of the number of tropical-only hurricanes (H_T) to the total number of hurricanes (H). The ratio, H_T/H, is likely more robust than the actual number of tropical-only hurricanes against improvements in hurricane detection methods over the past century of observations. This is because some early hurricanes may have gone undetected, a situation not likely in the more recent years. It is recognized that a potential bias may exist toward one type of hurricane or the other over this time period. While the ratio may prove more robust against inconsistent biases, it is impossible to determine whether or not the pre-1951 hurricane record contains a bias toward hurricane type. Yet, no strong evidence exists to indicate that it is in fact biased. The link between low-frequency variability of tropical-only hurricanes and intense hurricane activity described later gives some support to the conjecture that the ratio is not biased over this period.

Figure 5.1 (top) shows the time series of the ratio of North Atlantic hurricanes using 5-year intervals over the period 1886 through 1996. A striking feature is the reversal in ratios during the middle 1960s. Between 1886 and 1965 tropical-only hurricanes dominate annual totals, averaging nearly three-quarters of all storms. In stark contrast, over the period 1966 through 1994 baroclinically-enhanced hurricanes account for the largest percentages in the annual totals. This reversal occurs during the reliable portion of the hurricane record, after the advent of aircraft surveillance. This finding is consistent with the observation of Hebert and Frank (1973). They note that the late 1960s into the early 1970s was a period in which hurricanes of baroclinic origin were common.

The biennial variation of the relative tropical-only hurricane abundance is shown in Figure 5.1 (bottom). Here we see more variability. In fact, despite the overall preponderance of tropical-only hurricanes early in the record, a few biennial periods before 1960 have ratios of 0.5 or less. Only one of the three hurricanes during the two-year period of 1913–1914 was tropical-only. The majority of hurricanes during the 1940s

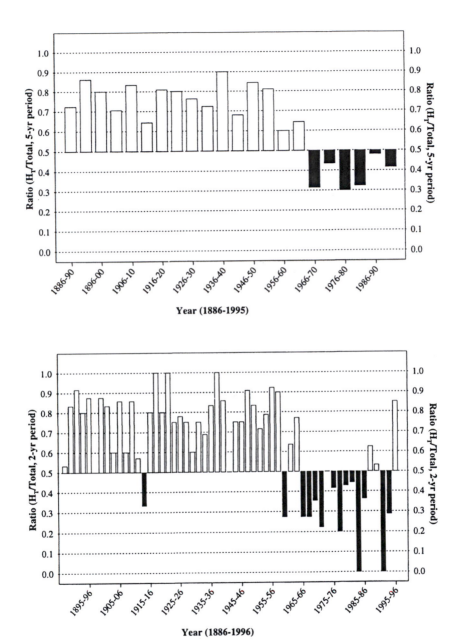

Fig. 5.1: Ratio of the number of tropical-only hurricanes to the total number of North Atlantic hurricanes in 5-year (top) and 2-year (bottom) intervals over the period 1886–1996. Tropical-only hurricanes were determined subjectively over the period 1950–1996 and objectively using rules similar to those in Table 5.1 over the period 1886–1949.

Table 5.2: Annual number of North Atlantic tropical-only hurricanes over the period 1886–1996. Y denotes the last digit of the year. Decade totals are also given. Note the relatively few tropical-only hurricanes during the 1970s and 1980s.

Y	188	189	190	191	192	193	194	195	196	197	198	199
0	.	1	3	2	3	2	3	8	2	1	4	3
1	.	7	2	3	4	2	2	4	7	2	4	0
2	.	4	3	1	2	4	2	6	0	0	0	0
3	.	8	4	1	2	10	4	6	5	1	0	1
4	.	4	2	0	4	1	5	5	5	3	0	1
5	.	2	1	4	0	4	4	9	1	4	4	9
6	8	5	5	8	7	6	2	3	2	1	0	8
7	5	2	0	2	3	3	5	2	2	0	1	.
8	3	2	3	3	3	3	5	7	1	2	4	.
9	4	4	4	1	1	3	7	1	5	2	5	.
	20	39	27	25	29	38	39	51	30	16	22	22

and 1950s were tropical-only. In contrast, the period from the middle 1960s through the middle 1980s was dominated by non-tropical-only storms. The ratios during the late 1980s and for the 1995 and 1996 seasons indicate a recent tendency for more tropical-only hurricane activity.

Annual variation in abundance of tropical-only hurricanes is shown in Table 5.2 and in Figure 5.2. Annual frequencies range from a maximum of 10 in 1933 to none in twelve different years, with 1991 and 1992 being the most recent years with no tropical-only hurricanes. The average annual number of tropical-only hurricanes is 3.2 with a standard deviation of 2.2 hurricanes. Noticeable is the relative abundance of tropical-only hurricanes during the late 1800s and during the middle 20th century. Also remarkable is the large number of tropical-only hurricanes during 1995 and 1996. The unique characteristics of the 1995 and 1996 hurricane seasons are examined in Chapter 12.

The annual distribution in the frequency of tropical-only hurricanes is shown in Figure 5.3. The distribution is multi-modal with peaks at two and four hurricanes per year. A majority of the years have five or fewer tropical-only hurricanes; 17 years have six or more tropical-only hurricanes. Only three years have exactly six tropical-only storms. The record-setting 1933 hurricane season had ten tropical-only hurricanes.

5.2.2 Seasonal Variability

The seasonal variability of tropical-only hurricanes is shown in Figure 5.4. The tropical-only season is confined to the months of August through October. The distribution is more restricted than the distribution of all hurricanes. By a significant margin the majority of tropical-only hurricanes occur during August and September. Tropical-only hurricanes during June, July, and November collectively account for less than 8%

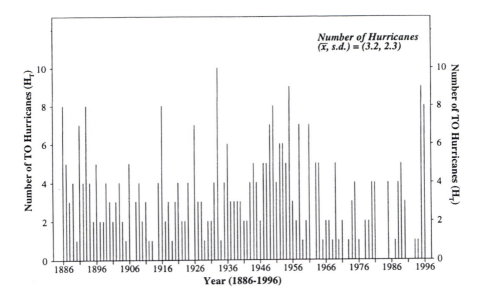

Fig. 5.2: Time series of the annual frequency of North Atlantic tropical-only hurricanes over the period 1886–1996. The average ($\bar{x}$) and standard deviation ($s.d.$) is given in the upper-right corner.

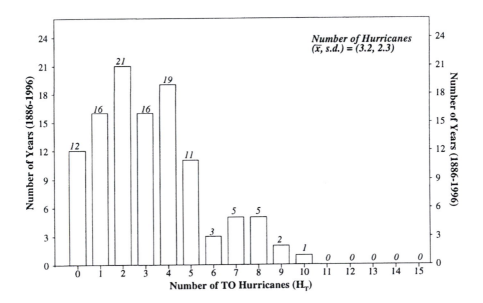

Fig. 5.3: Annual distribution of North Atlantic tropical-only hurricanes over the period 1886–1996. The distribution is multi-modal with peaks at two and four tropical-only hurricanes per year.

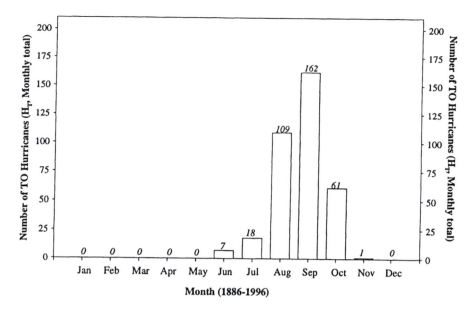

Fig. 5.4: Monthly distribution of North Atlantic tropical-only hurricanes over the period 1886–1996. The tropical-only hurricane season peaks in September.

of total tropical-only activity. Since many of the tropical-only hurricanes develop from tropical easterly waves that move west from the coast of Africa, the tropical-only season is nearly coincident with the *Cape Verde season*. The Cape Verde season runs from August through September. It is characterized by the development of tropical cyclones and hurricanes from waves passing near the islands of Cape Verde off the west coast of Africa. These storms typically remain at low latitudes and become hurricanes devoid of middle-latitude baroclinic influences. The Cape Verde season begins and ends abruptly with changes in SSTs and upper-level winds in the vicinity of the west coast of Africa. Hurricane *Marco* in 1996 is the only tropical-only hurricane to develop during November.

5.3   Season Length

As was done in Chapter 4 for all hurricanes, here the tropical-only season length is considered from various perspectives. Figure 5.5 (top) gives the distribution of begin and end dates for tropical-only hurricanes. The season begins after June 15; however, not until August 1 does the season become active. The median date of tropical-only hurricane origins is September 6th and the mode date is September 5th. Though tropical-only hurricane formations reach a peak in early September, two other surges in activity are noted; one occurs in late September and the other during the middle of October. It is unlikely that a tropical-only hurricane will form after late October.

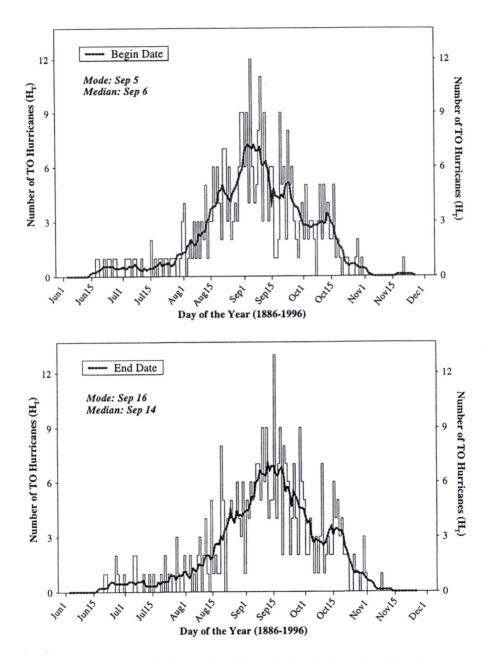

Fig. 5.5: Distributions of begin (top) and end (bottom) dates for tropical-only hurricanes over the period 1886–1996 (Jun–Nov only). The dotted line is a 9-day running average. Note the local maximum in tropical-only activity during the second week of October.

As one would expect, the distribution of hurricane end dates, shown in Figure 5.5 (bottom), is similar to the distribution of begin dates. The most common end date is September 16th and the median date is September 14th. An increase in tropical-only hurricane activity during the middle of October is clearly marked on these figures. These late-season storms sometimes originate in connection with the monsoon trough.

Season length as defined by the difference between the first origin of the first hurricane and the last dissipation of the last hurricane, is plotted for each year in Figure 5.6. Since tropical-only hurricanes form primarily during August through October, the season length is considerably more restricted compared to all hurricanes. In fact, in only 11 of the 111 years (10%) does the tropical-only season last longer than 3 months. This compares with 32% of the seasons lasting longer than 3 months when all hurricanes are considered. The March storm in 1908 was not likely tropical only.

Cumulative distributions of tropical-only hurricane season begin and end dates are shown in Figure 5.6 (bottom). The median begin date of August 16 is more than a week later than the median begin date for the season as a whole. Most interesting are the scaling regions (indicated by straight lines). The first scaling occurs from the middle of June through July, while a secondary scaling occurs from the start of August until the middle of September. The straight lines indicate that conditions favorable for the onset of the tropical-only season are relatively constant. Scaling implies that conditions do not become increasingly favorable through the time period. However, conditions are quantitatively more favorable for hurricane formation after the start of August. The discontinuity in the cumulative distribution of onset dates is remarkable. In contrast to the cumulative frequency of onset, frequency of end dates is sigmoidal indicative of a normal distribution of random events. The median end date for the tropical-only hurricane season is October 6th. This is 10 days before the median end date for the entire season.

5.4 Intensity and Duration

The average maximum intensity of tropical-only hurricanes for each season and the annual distribution of average maximum intensity are shown in Figure 5.7. Recall that a storm's highest intensity is its maximum intensity, and the average is taken over all tropical-only hurricanes in a year. In comparison to all hurricanes, maximum intensities are substantially higher for the subset of tropical-only hurricanes. The overall average maximum intensity for tropical-only hurricanes is 97.2 kt which compares with 92.4 kt for all hurricanes. Like all hurricanes, there does not appear to be a strong trend over time. Although for hurricanes prior to 1900, maximum intensities were commonly assigned a value of 85 kt. Thirteen years have average maximum intensities for tropical-only hurricanes at or exceeding 115 kt.

Duration of North Atlantic tropical-only hurricanes is examined by considering the average number of hurricane days in a season. The duration of each hurricane is averaged over all hurricanes in the season. Figure 5.8 (top) shows the average duration

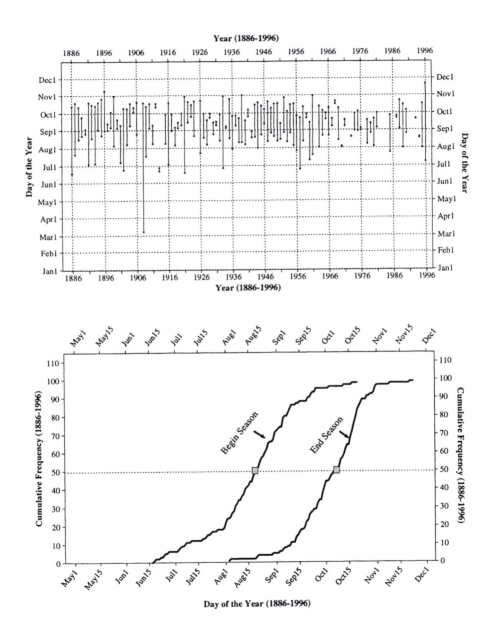

Fig. 5.6: Start and end dates of the tropical-only hurricane season for each year in the period 1886–1996 (top) and the cumulative distribution of season onset and termination dates (bottom). The horizontal line in the cumulative distribution graph identifies the median dates.

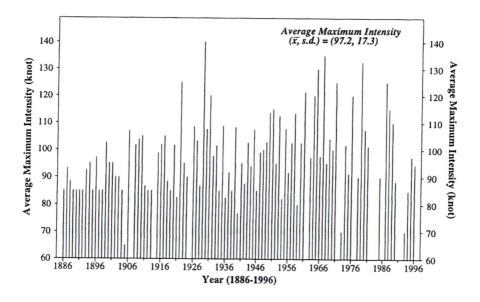

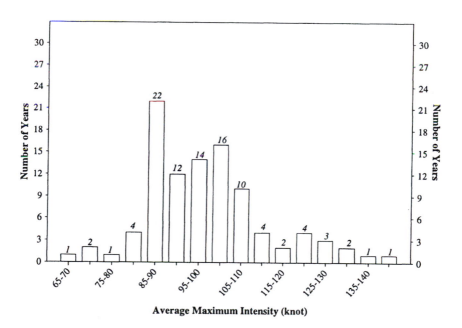

Fig. 5.7: Annual average maximum intensity (in knots) of tropical-only hurricanes for each year over the period 1886–1996 (top) and the distribution of years by average maximum intensity in 5 knot intervals (bottom).

for each year in the record. Variation in the number of tropical-only hurricane days is large. Neglecting years with no hurricanes, the average duration ranges from a low of less than one day to a maximum of 14 days with an overall average of 5.6 days and a standard deviation of 2.5 days. Because they commonly form over the central Atlantic, tropical-only hurricanes have an average lifespan that is 1 day longer than the average over all hurricanes. Similar to all hurricanes, there is a trend toward shorter duration tropical-only hurricanes over the 111-year period of record. The distribution of tropical-only hurricanes as a function of duration is given in Figure 5.8 (bottom). Here we see the much larger variation in lifespans. The most common durations include the intervals of 2 to 3 days, 3 to 4 days, and 5 to 6 days. Seven of the eight North Atlantic hurricanes with lifespans of 2 weeks or more are classified as tropical-only hurricanes; although the vast majority of tropical-only hurricanes exist for less than 11 days. The two tropical-only storms that remained at hurricane intensity for longer than 18 days were the September hurricane of 1893 and the August hurricane of 1906.

5.5   Origin

Tropical-only hurricanes form at low latitudes across the North Atlantic basin. In contrast, non-tropical-only hurricanes form at higher latitudes and farther to the west. In this section, genesis regions are examined by considering points of origin for these two types of hurricanes. The location in a tropical cyclone's track where hurricane intensity is first reached is designated its point of origin. Points of origin for tropical-only and non-tropical-only (baroclinically-enhanced) hurricanes are plotted together for comparison. General characteristics of genesis regions are considered first.

5.5.1   General Characteristics

Points of origin for tropical-only and non-tropical only hurricanes over the period 1886–1996 are plotted in Figure 5.9. Tropical-only hurricanes originate at latitudes south of 25°N. Non-tropical-only hurricanes originate north of this latitude. The division line is most pronounced over the central and eastern North Atlantic and corresponds to the latitude of recurvature for many hurricanes. In contrast, over the western Caribbean, the Gulf of Mexico, and the Bahamas both tropical-only and non-tropical-only hurricanes are likely to form. Over these regions there is no simple latitude division that separates the two classes of hurricanes.

   Whereas it is rare for a tropical cyclone to reach hurricane intensity free from enhancing baroclinic influences north of 28°N, the presence of baroclinic factors often contribute to the origin and development of tropical cyclones deep into tropical latitudes especially over the Gulf of Mexico and western Caribbean Sea. A narrow latitude zone between 20 and 23°N over the central North Atlantic, where relatively few hurricanes originate, divides the genesis regions of tropical-only and baroclinically-enhanced hurricanes. Baroclinically-enhanced hurricanes do not form south of 20°N east of 60°W.

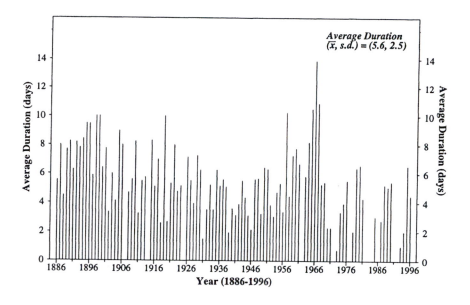

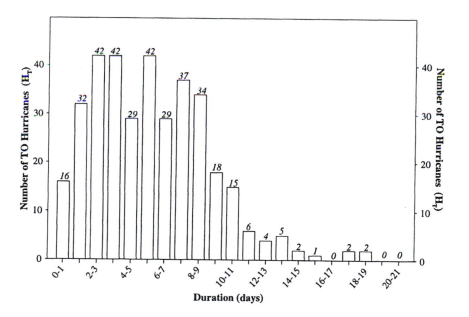

Fig. 5.8: Annual average duration of North Atlantic tropical-only hurricanes for each year over the period 1886–1996 (top) and the distribution of tropical-only hurricanes by duration in 1-day intervals (bottom).

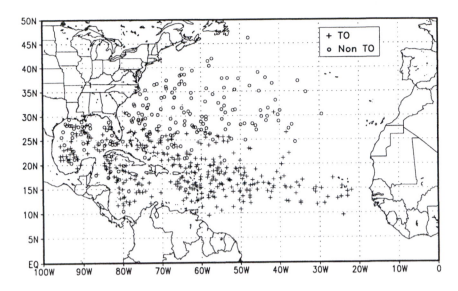

Fig. 5.9: Points of origin for tropical-only (+) and non-tropical-only (o) North Atlantic hurricanes over the period 1886–1996.

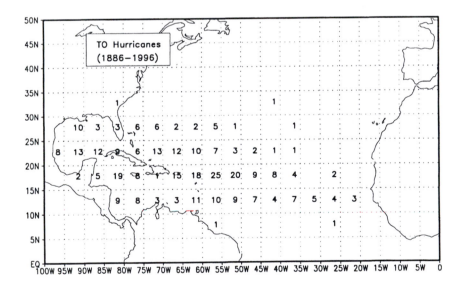

Fig. 5.10: Number of origin points for tropical-only hurricanes in 5° latitude-longitude grid boxes over the period 1886–1996. Preferred regions of tropical-only hurricane formation include the central North Atlantic, the western Caribbean Sea, and the Gulf of Mexico.

Table 5.3: Preferred regions of tropical-only North Atlantic hurricane formations as defined in 5° latitude-longitude grid boxes shown in Figure 5.10.

Region	Location	Total Occurrence
I	Central North Atlantic	25
II	Western Caribbean Sea	19
III	Gulf of Mexico	13
Total	All three regions	57

The frequency of origin points in 5° latitude-longitude boxes are plotted in Figure 5.10. Regions of local maxima in tropical-only hurricane genesis are evident. These areas include the central North Atlantic, the western Caribbean, and the Gulf of Mexico (Table 5.3). The region most favorable for tropical-only formation is the area to the east of the Lesser Antilles over the central tropical Atlantic Ocean. Only two tropical-only hurricanes have originated north of 30°N. The average latitude of tropical-only hurricane formation is 19°N. Most tropical-only hurricanes originate between 15 and 20°N. Most tropical-only hurricanes (97%) form south of 27°N. Tropical-only hurricanes are distinguished by their relatively low latitude of genesis. This contrasts with baroclinically-enhanced hurricanes which tend to form significantly farther north. Longitudinally, about half of all tropical-only hurricanes originate east of 65°W longitude.

5.5.2 Monthly Activity

Points of origin by month for both classes of hurricanes are shown in Figure 5.11. June is dominated by baroclinically-enhanced hurricanes, primarily over the Gulf of Mexico. By July, non-tropical-only hurricane formations spread northward and westward over the Bahamas and the western North Atlantic, while tropical-only hurricanes originate over the Caribbean Sea. The mid-season months of August and September continue these tendencies. September and October are characterized by non-tropical hurricane formations across the central Atlantic north of 25°N latitude. The latitude belt where relatively few hurricanes originate is conspicuous during September. By October, tropical-only genesis is largely confined to areas over the Caribbean Sea and Gulf of Mexico. In November a few non-tropical-only hurricanes originate in the western and central North Atlantic, generally north of 20°N.

Monthly tropical-only and non-tropical-only hurricane activity is summarized in Table 5.4 with respect to points of origin. In general tropical-only hurricanes originate considerably farther south (18.7°N) than baroclinically-enhanced hurricanes (29.0°N). Non-tropical-only points of origin are farther south in the early and late season compared with the middle of the season, as expected. Both hurricane types indicate a easterly shift from early to mid-season and a shift back to the west during the late season. Overall, larger variations in latitude are noted for non-tropical-only hurricanes,

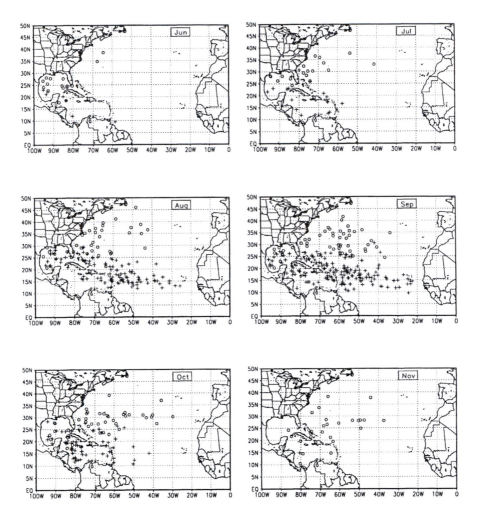

Fig. 5.11: Points of origin of tropical-only (+) and baroclinically-enhanced (o) North Atlantic hurricanes by month over the period 1886–1996. The 25°N latitude line roughly divides hurricane origin points by class, especially over the central North Atlantic. The early season is dominated by baroclinically-enhanced hurricane activity, particularly over the Gulf of Mexico and the western North Atlantic. Many of the October tropical-only hurricanes originate in the western Caribbean.

Table 5.4: Tropical-only (TO) and non-tropical-only (BE) points of origin averaged by month. The number of cases (N), the average value ($\bar{x}$), the standard deviation ($s.d.$), the linear correlation of longitude on latitude [r(lat.,lon.)], the t-statistic, and the p-values are given. "All" refers to hurricanes over the entire season. A p-value of less than 0.001 is indicated as $<$.

		Tropical-Only Hurricanes (1886–1996)					
	Month	All	Jun	Jul	Aug	Sep	Oct
	N	358	7	18	109	162	61
Lat. (°N)	$\bar{x}$	18.7	17.9	19.3	19.0	18.7	18.2
	$s.d.$	4.6	5.3	5.5	4.8	4.4	4.6
Lon. (°W)	$\bar{x}$	65.7	84.8	75.8	63.2	61.8	75.4
	$s.d.$	17.2	12.6	11.6	17.0	17.5	12.1
Correlation	r	+0.47	+0.91	+0.49	+0.70	+0.48	+0.25
	t-stat.	+10.08	+4.93	+2.26	+10.08	+6.87	+2.00
	p-value	$<$	0.002	0.036	$<$	$<$	0.050
		Baroclinically-Enhanced Hurricanes (1886–1996)					
	Month	All	Jun	Jul	Aug	Sep	Oct
	N	196	16	18	36	50	42
Lat. (°N)	$\bar{x}$	29.0	25.8	30.3	31.7	30.6	28.0
	$s.d.$	5.5	5.1	3.8	5.8	4.5	4.5
Lon. (°W)	$\bar{x}$	69.2	86.0	76.2	69.2	64.8	70.0
	$s.d.$	13.2	8.9	13.7	14.5	15.7	16.6
Correlation	r	−0.43	−0.59	−0.66	−0.54	−0.34	−0.53
	t-stat.	−6.50	−2.77	−3.54	−3.76	−2.50	−3.93
	p-value	$<$	0.014	0.002	$<$	0.016	$<$

and larger variations in longitude are noted for tropical-only hurricanes. Both hurricane types show a northward trend in areas of formation as the season advances.

Points of origin by longitude and latitude are shown again for both hurricane types in Figure 5.12. In general tropical-only hurricanes form south of 25°N latitude, while non-tropical-only occur north of this latitude. The largest exception is noted over the western part of the North Atlantic basin late in the hurricane season when non-tropical-only hurricanes form considerably farther to the south. The average point of origin by month for the two classes of hurricanes is plotted in Figure 5.13. The seasonal progression of genesis regions is clearly indicated. Regions of tropical-only hurricane formation spread to the east with the season. The spread is toward the northeast for formation regions of baroclinic hurricanes. The westward retreat of hurricane activity in October for both classes is apparent. The greatest distance between tropical-only and baroclinically-enhanced average point of genesis occurs during August. The 5° latitude belt between 20 and 25°N separates the average points of origin between tropical-only and non-tropical-only hurricanes.

The dichotomy in hurricane classes is further illustrated in Table 5.4, which gives statistics for regressions of the longitude point of origin on the latitude point of origin

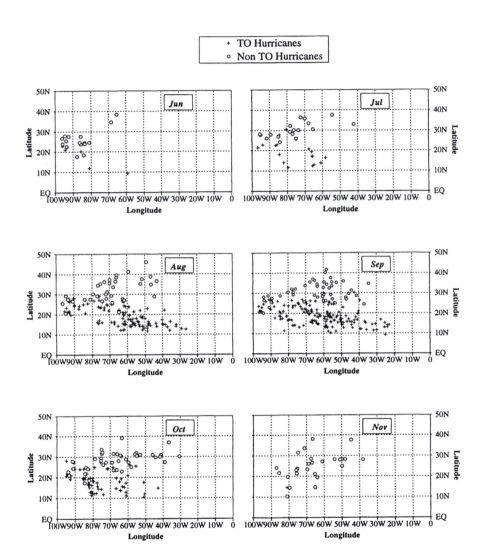

Fig. 5.12: Latitude versus longitude distribution of origin points for North Atlantic hurricanes by month over the period 1886–1996 for the categories of tropical-only (+) and non-tropical-only hurricanes (o). Note that the hurricane season begins and ends with non-tropical-only hurricanes. Midseason activity is characterized by low-latitude tropical-only formation and high-latitude baroclinically-enhanced hurricane genesis. The separation of tropical-only and baroclinically-enhanced hurricane origins is most pronounced over the central and eastern North Atlantic. The Gulf of Mexico supports the genesis of both hurricane types.

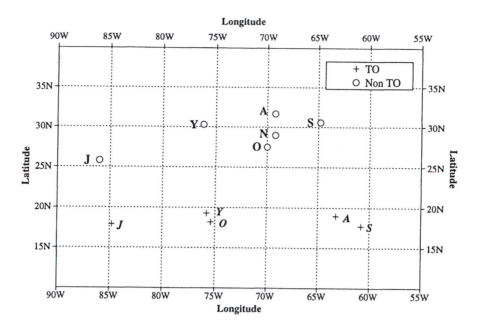

Fig. 5.13: Average centroid of the points of origin for North Atlantic tropical-only (+) and baroclinically-enhanced (o) hurricanes by month over the period 1886–1996. In the figure J denotes June, Y July, A August, S September, O October, and N November.

for each month and for both classes. As indicated by the Pearson's product-moment correlation,[1] the latitude of formation explains a portion of the longitude of formation during each month. Overall the relationships are statistically significant as indicated by p-values generally less than 0.01.[2] It is tested using the t-statistic with N degrees of freedom, where N is the number of hurricanes in each category. The squared t-statistic is equivalent to the F-statistic with 1 and N degrees of freedom. The relationship is strong during August for tropical-only hurricanes, explaining nearly 50% of the variance in longitude. The relationship is actually stronger in June, but the number of hurricanes during this month is small. The relationship between latitude and longitude of formation for non-tropical-only hurricanes is strongest in July, explaining more than 40% of the variance. The striking contrast between origin locations of tropical-only and baroclinically-enhanced hurricanes is in the sign of the latitude versus longitude correlation. For tropical-only hurricanes the correlation is positive, indicating that hurricanes originating farther to the north tend to originate farther to the west. The situation is reversed for baroclinically-enhanced hurricanes where the correlation is negative,

---

[1] Pearson's product-moment linear correlation is given by $r = \sum xy / \sqrt{(\sum x^2)(\sum y^2)}$, where x and y are differences from the mean value. The product-moment formula is used for linear correlation throughout the book except where noted.

[2] Here the null hypothesis is that latitude of formation has no linear relationship with longitude of formation.

indicating that hurricanes originating farther to the north tend to originate farther to the east. This observation is consistent with the idea that the two types of hurricanes represent contrasting origin and development mechanisms. Tropical-only hurricanes originate from low-latitude easterly waves, in contrast to baroclinically-enhanced hurricanes which develop from baroclinic disturbances along the U.S. coastline.

5.6 Tracks

Tropical-only hurricanes typically develop at low latitudes move westward before recurving to the northwest and north across the western North Atlantic basin. Tropical-only hurricanes that remain at low latitudes pose a serious threat to the United States, Mexico, and the nations of the Caribbean. These storms often intensify into major hurricanes as they remain for long periods of time over the warmest waters of the North Atlantic. Here we consider the average characteristics of tropical-only hurricane tracks, including a look at their seasonal variations.

5.6.1 General Characteristics

Figure 5.14 shows the paths of all tropical-only hurricanes over the period 1886–1996. Here we see the classic recurvature corridor as the low-latitude westerly path over the eastern and central North Atlantic becomes a northwesterly to northerly path over the western North Atlantic. Another corridor for tropical-only hurricanes is through the Caribbean Sea into the Gulf of Mexico. These storms tend to make landfall somewhere in Mexico or the United States. Tropical-only hurricanes at higher latitudes track to the north and northeast. Coastal New England is threatened by recurving tropical-only hurricanes. At latitudes north of 40°N these hurricanes obtain a significant easterly component.

5.6.2 Seasonal Variability

Variations in the tracks of tropical-only hurricanes occur within a season (Figure 5.15). June tracks are confined to the western Caribbean Sea and Gulf of Mexico while July hurricane tracks begin to extend into the western North Atlantic along the U.S. east coast. Early-season tropical-only hurricanes sometimes track through the western Gulf of Mexico and threaten Texas. By August and September—the peak of the hurricane season—much of the tropical Atlantic, Caribbean Sea, Gulf of Mexico, and the northern Atlantic are susceptible to tropical-only hurricanes. Many of the August and September tropical-only hurricanes originate from tropical waves that pass near Cape Verde. During October tropical-only hurricane paths are limited to the western half of the North Atlantic. These late-season hurricanes take a more northerly track. Florida and the Bahamas are particularly vulnerable to late-season tropical-only hurricanes. After September the western Gulf of Mexico is largely free of tropical-only hurricane activity.

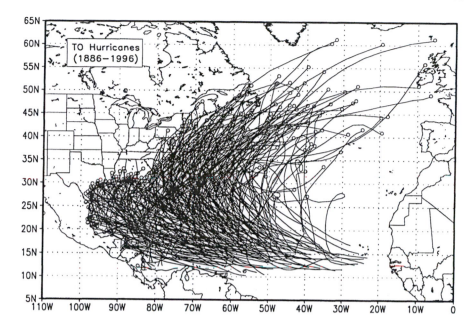

Fig. 5.14: Tracks of North Atlantic tropical-only hurricanes over the period 1886–1996. Note the general low latitude of formation and recurvature to the north and east.

5.7   Dissipation

Tropical-only and non-tropical-only points of dissipation are shown in Figure 5.16. In contrast to points of origin there are no distinct patterns that emerge which can delineate the two classes of hurricanes. There are, however, some subtle differences. For instance, near Florida tropical-only hurricane dissipation tends to occur over the peninsula or just to the east, whereas non-tropical-only dissipation occurs to the west over the eastern Gulf of Mexico. There also appears to be a split in tropical-only dissipation regions, one to the north and one to the south. This is more evident by considering tropical-only hurricane dissipation regions in 5° latitude-longitude boxes (Figure 5.17). Hurricanes that recurve over the North Atlantic tend to dissipate at high latitudes in particular over and near the northeast United States and eastern Canada (Nova Scotia, New Brunswick, and New Foundland). Warm, moist, and unstable tropical air ahead of an extra-tropical storm will allow a hurricane to maintain itself at high latitudes. However, many of these storms combine with, or transition to, extra-tropical cyclones. Tropical-only hurricanes that take a more direct path and remain at lower latitudes dissipate over the western Caribbean or over the coastal areas of the Gulf of Mexico. The average latitude at which tropical-only hurricanes dissipate is 34°N. The average longitude of dissipation is 70°W.

Seasonal variations in points of dissipation for both classes of hurricanes are shown

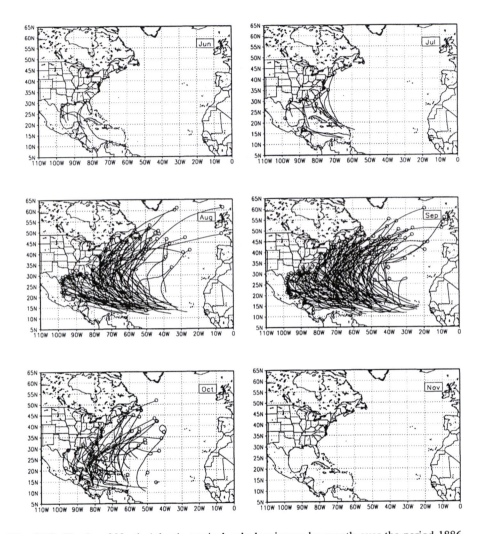

Fig. 5.15: Tracks of North Atlantic tropical-only hurricanes by month over the period 1886–1996. Early-season tropical-only hurricanes have a tendency to make landfall along the U.S. coastline, whereas a good majority of the late-season storms track away from the United States. A significant portion of the midseason tropical-only hurricanes track northward out of the tropics. Over the eastern half of the North Atlantic tropical-only hurricanes are most likely during September.

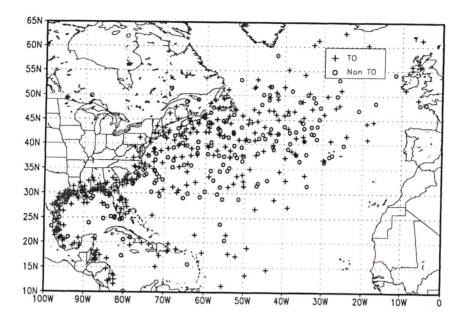

Fig. 5.16: Dissipation points for North Atlantic tropical-only (+) and non-tropical-only (o) hurricanes over the period 1886–1996.

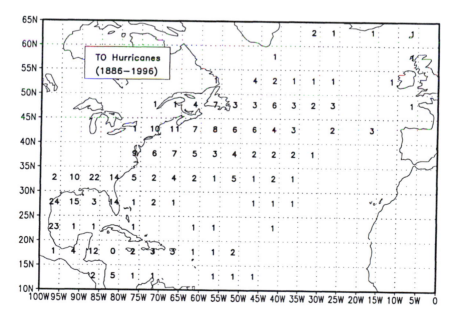

Fig. 5.17: Number of dissipation points for tropical-only hurricanes in 5° latitude-longitude grid boxes over the period 1886–1996.

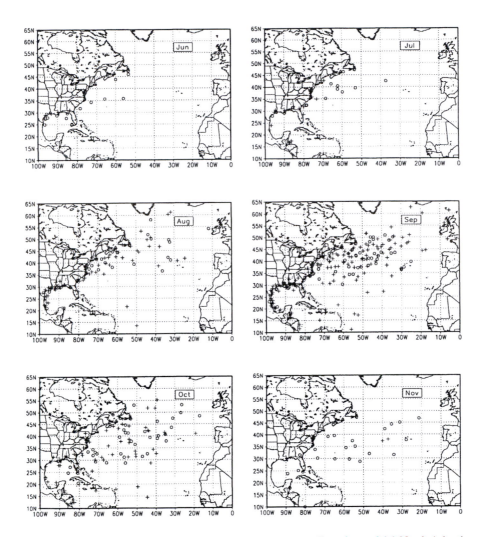

Fig. 5.18: Dissipation points of tropical-only (+) and baroclinically-enhanced (o) North Atlantic hurricanes by month over the period 1886–1996. Tropical-only hurricanes are most likely to make landfall in the United States. compared to baroclinically-enhanced hurricanes. This tendency is most pronounced during October. Note that late-season dissipation and transition tends to occur over the open waters of the North Atlantic away from land.

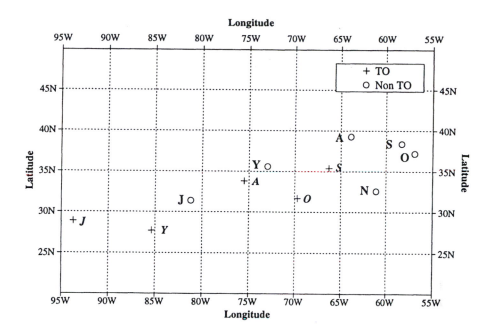

Fig. 5.19: Average centroid of the dissipation points for North Atlantic tropical-only (+) and baroclinically-enhanced (o) hurricanes by month over the period 1886–1996. In the figure J denotes June, Y July, A August, S September, O October, and N November.

in Figure 5.18. During June and July, tropical-only dissipation tends to be closer to or directly over land, whereas outside the Gulf of Mexico non-tropical-only dissipation occurs over the open waters of the North Atlantic. By August and September, both classes of hurricanes show dissipation at high latitudes in the Atlantic and along coastal regions of eastern North America. A subtle difference is that tropical-only hurricanes have a tendency to reach more extreme northerly latitudes compared to the non-tropical-only hurricanes. Because of the geography of the western North Atlantic basin, there is a noticeable dependence of dissipation longitude on dissipation latitude. That is, higher latitude dissipation occurs farther to the east. The relationship is strongest for non-tropical-only hurricanes. Dissipation at low-latitude tends to occur as a result of landfall, whereas at higher latitudes hurricane dissipation occurs over the open ocean. Preferred regions of tropical-only hurricane dissipation include the Texas and southeast U.S. coastlines.

Seasonal variations in dissipation regions are re-examined in Figure 5.19, which shows the average point of dissipation by month for tropical-only and non-tropical-only hurricanes. During June and July average dissipation points for tropical-only and non-tropical-only hurricanes are located over the northern Gulf of Mexico and off the southeast U.S. coast, respectively. By August and September hurricane dissipation shifts north and east to the western North Atlantic for both tropical-only and baroclinically-enhanced hurricanes. Later in the season, hurricanes tend to dissipate farther south over the western North Atlantic. Interestingly, tropical-only dissipation occurs to the south and west of baroclinically-enhanced dissipation for each month of the hurricane season. Regions favorable for non-tropical-only dissipation in a given month tend to be favorable regions for tropical-only dissipation in the following month.

6

Baroclinically Enhanced Hurricanes

Hurricanes that form as a consequence of favorable middle-latitude baroclinic influences are termed *baroclinically-enhanced*. There are two sources of baroclinically-enhanced hurricanes. A tropical depression or tropical storm may strengthen to hurricane intensity under favorable baroclinic dynamics at high latitudes or the initial disturbance leading to the hurricane may be baroclinic. In the later case, the hurricane is called *baroclinically-initiated*. It is important to qualify baroclinic enhancement with the term "middle latitude" because the tropical easterly waves—the seedlings of most tropical-only hurricanes—are frequently associated with north-south atmospheric temperature contrasts across western Africa.

This chapter examines the climatology of middle-latitude baroclinically-enhanced North Atlantic hurricanes. In particular, we consider the climate characteristics of both baroclinically-enhanced and baroclinically-initiated hurricanes. Baroclinically-enhanced hurricanes comprise 35% of all North Atlantic hurricanes. Baroclinically-enhanced hurricanes are relatively more frequent during the early and late parts of the hurricane season and they generally form at higher latitudes. Baroclinically-initiated hurricanes represent a subset of baroclinically-enhanced hurricanes that initiate from middle-latitude disturbances. These hurricanes tend to form over cooler waters compared to tropical-only hurricanes. The frequency of baroclinically-initiated hurricanes peaks in the month of October. Because the idea of presenting the climatology of baroclinically-enhanced hurricanes is new, we begin this chapter with a look at some early studies that investigated the influence of middle-latitude processes on tropical cyclone formation.

6.1   Middle Latitude Influence

The idea that hurricanes can develop from baroclinic disturbances[1] is not new. For example, Dunn (1940) was aware that the spread of cool, dry air into the tropics behind a cold front inhibits the development of tropical waves. On the other hand, he mentions

---

[1]Baroclinic disturbances extract energy from horizontal temperature gradients.

the importance of middle-latitude fronts penetrating the southwest North Atlantic and western Caribbean Sea to occasional hurricane development. Late summer and early autumn fronts with their surface wind shift and favorable upper-level divergent winds aloft can generate a hurricane under the right circumstances. When significant low-level cyclonic vorticity[2] attends the surface front development is more likely. Other studies suggested the importance of a constant interaction between the middle-latitude westerly winds (westerlies) and the tropical easterly trades (easterlies) for hurricane development (Riehl and Shafer 1944, Riehl and Burgner 1950). In particular, Gray (1968) noted formation mechanisms of hurricanes that begin south of about 20°N latitude in the deep tropics tend to be different than the mechanisms responsible for hurricane formation northward of this latitude where interactions with middle-latitude disturbances are common.

The role of subtropical cyclones in the initiation of hurricanes has also been previously investigated. For instance, Palmén (1949) devoted some effort to understanding the formation of upper-tropospheric cold lows in the subtropics of the North Pacific Ocean near the Marshall and Hawaiian Islands. He found that these upper-level lows are sometimes associated with a subtropical cyclone at the surface. Since the cyclones are detached from the main stream (jet stream) of middle-latitude westerlies, they tend to move slowly over the warm ocean water and can gradually acquire tropical characteristics from the latent heat released inside the surrounding showers and thundershowers. Simpson (1952) recognized the importance of these cyclones (*kona* storms) in producing showers over Hawaii particularly during the springtime. In particular, he noted the similarity of kona storms in the Pacific to subtropical cyclones in the North Atlantic and gave evidence that the region between 15–30°N latitude and 30–60°W longitude is the principal breeding ground for North Atlantic subtropical cyclones. Moreover, he acknowledged that subtropical cyclones can become hurricanes. He suggested the violent *Yankee Storm* of November 1935 that affected Florida and the Bahamas was a subtropical cyclone turned hurricane. Other early investigations including Moore and Davis (1951) and Colón (1956) examined hurricanes originating from non-tropical disturbances.

Dunn (1964) commented that the character of some hurricanes in a particular season is not always strictly tropical. He suggested variations in tropical cyclone development processes noting that an individual tropical distrubance can develop by means of both tropical (latent heat) and extra-tropical (horizontal temperature gradients) energy sources. Spiegler (1971) and Ferguson (1973) identified systems from the 1970 North Atlantic hurricane season that had this dual nature. They made a case that some of the storms should be classified as tropical cyclones since they were warm-core[3] during part of their lifecycle. Indeed, Simpson et al. (1968) felt it necessary to adopt a new thinking to better identify and characterize hurricane activity. A study of hurricane *Diana* of 1984 by Bosart and Bartlo (1991) provides a detailed account of hurricane

---

[2]Mathematically, vorticity is given as $\zeta = k \cdot \nabla \times V$, where k is the unit vertical vector and $\nabla \times V$ is called the *curl* of the horizontal wind velocity V.

[3]Baroclinic storms are typically cold-core.

origin and development in a baroclinic environment. The early stages of formation in this storm were dominated by classical extra-tropical cyclone processes. A pre-existing middle-latitude front with attendant cyclonic vorticity and a strong anticyclone to the north provided fluxes of sensible and latent heat to the atmosphere in a region to the east of Florida. The fluxes triggered widespread convection in the form of showers and thundershowers, transforming the system into a warm-core tropical cyclone. They speculate that the vigorous convection collapsed the upper-level dome of colder air aloft. This lowered the center of mass in the column of air above the fledgling storm thereby increasing its kinetic energy through cyclonic rotation.

Other classifications of North Atlantic hurricanes have also been considered. Hebert and Frank (1973) suggest a dichotomy of hurricane types along the lines of initiation mechanisms. They separated hurricanes of baroclinic origin from hurricanes of tropical origin and noted that seasons during the late 1960s and early 1970s were dominated by hurricanes of baroclinic origin. Since many of the tropical-originating hurricanes develop from waves that move off the west coast of Africa, Avila and Clark (1989) suggest distinguishing African hurricanes from non-African hurricanes. Our work (Hess and Elsner 1994a and Elsner et al. 1996a) is inspired by these earlier efforts. This chapter provides a comprehensive climatology of baroclinically-enhanced hurricanes.

6.2 Abundance

A total of 196 baroclinically-enhanced hurricanes have occurred over the North Atlantic during the period 1886–1996. This represents 35.4% of all hurricane activity and compares with a total of 358 tropical-only North Atlantic hurricanes. Variations in the frequency of baroclinically-enhanced hurricanes are considered next.

6.2.1 Interannual Variability

Annual variation in the frequency of baroclinically-enhanced hurricanes is shown in Figure 6.1 and in Table 6.1. Seasonal numbers range from a maximum of seven in 1969 to zero in 26 of the 111 seasons. The seasonal average is 1.8 baroclinically-enhanced hurricanes with a standard deviation of 1.7 hurricanes. The average is two less than the average number of tropical-only hurricanes. The frequency of baroclinically-enhanced hurricanes is rising. The increase in the number of years since 1959 with three or more baroclinically-enhanced hurricanes in a single season is remarkable. Prior to 1959 there are only 8 years with three or more baroclinically-enhanced hurricanes while 25 of the years over the period 1959–1996 have at least three storms. Moreover, the average number of baroclinically-enhanced hurricanes in a season is one in the period 1886–1958, while the average is greater than three over the years since 1958.

Interestingly, years of abundant baroclinically-enhanced hurricanes tend to be years with fewer tropical-only hurricanes. For instance, over the 10 years with five or more baroclinically-enhanced hurricanes, the average number of tropical-only hurricanes is

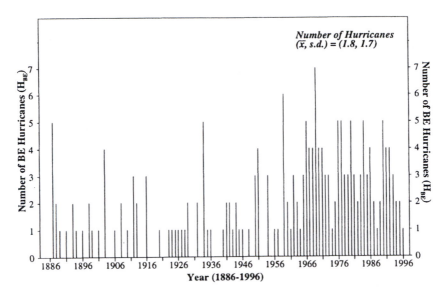

Fig. 6.1: Time series of the annual frequency of North Atlantic baroclinically-enhanced hurricanes over the period 1886–1996. The average ($\overline{x}$) and standard deviation ($s.d.$) are given in the upper-right corner.

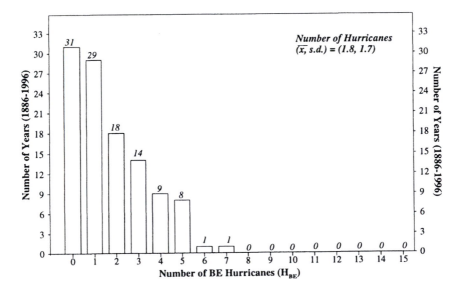

Fig. 6.2: Distribution of the annual abundance of Atlantic basin baroclinically-enhanced hurricanes over the period 1886–1996. Only 2 years have more than five baroclinically-enhanced hurricanes.

Table 6.1: Annual number of North Atlantic baroclinically-enhanced hurricanes over the period 1886–1996. Y denotes the last digit of the year. Decade totals are also given. Note the abundance of hurricanes since the 1950s.

Y	188	189	190	191	192	193	194	195	196	197	198	199
0	.	0	0	1	1	0	1	3	2	4	5	5
1	.	1	1	0	0	0	2	4	1	4	3	4
2	.	0	0	3	0	2	2	0	3	3	2	4
3	.	2	4	2	1	0	1	0	2	3	3	3
4	.	1	0	0	1	5	2	3	1	1	5	2
5	.	0	0	0	1	1	1	0	3	2	3	2
6	0	1	1	3	1	1	1	1	5	5	4	1
7	5	0	0	0	1	0	0	1	4	5	2	.
8	2	2	2	0	1	0	1	0	4	3	1	.
9	1	1	0	0	2	0	0	6	7	3	2	.
	8	8	8	9	9	9	11	17	32	33	30	21

2.2; this compares to an average of 3.2 tropical-only hurricanes averaged over all years. Likewise for the 13 years with seven or more tropical-only hurricanes, the average number of baroclinically-enhanced hurricanes is 1.1 compared with an overall average of 1.8. The Spearman rank correlation coefficient[4] indicates a weak inverse relationship between these two components of hurricane activity. The correlation value is -0.202 with a p-value of 0.039. Climate conditions that favor the formation of one type of hurricane tend to be the same conditions that inhibit the formation of the other type. This is not surprising considering hurricanes provide a portion of the upward transport of heat and momentum in the atmosphere important in maintaining the general circulation against friction as well as in limiting the buildup of energy in the tropics. However, tropical cyclones of the North Atlantic account for only 10–15% of global tropical storm activity, so the overall importance of this single tropical-cyclone basin is limited as a decrease in activity over one basin may be offset by increases over other basins. In fact, there is evidence of a negative correlation between hurricane activity over the North Atlantic and activity over the eastern North Pacific (see Chapter 10).

The annual distribution of baroclinically-enhanced hurricanes is shown in Figure 6.2. Because of the small number of baroclinically-enhanced hurricanes in any single season, the distribution is positively skewed with more than half the years experiencing fewer than two baroclincally-enhanced hurricanes. More years have zero baroclinic hurricanes than any other number. Only 30% of the years have 3 or more storms. The 1969 season had a record total of seven baroclinic hurricanes. It also had a record number of hurricanes.

---

[4]If two variables X and Y are ranked (from lowest to highest), then the Spearman's formula for rank correlation is given by $r_{rank} = 1 - 6 \sum D^2/[N(N^2 - 1)]$, where D is the difference between ranks of corresponding values of X and Y, and N is the number of paired values in the data.

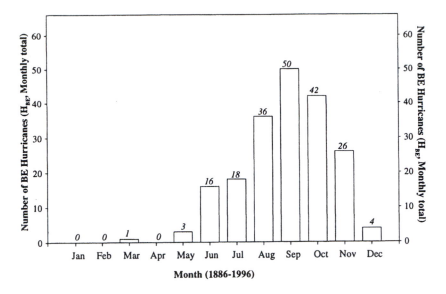

Fig. 6.3: Monthly distribution of North Atlantic baroclinically-enhanced hurricanes over the period 1886–1996. Although September is the peak month, it is followed closely by activity during October. November also sees a relatively large number of baroclinically-enhanced hurricanes compared to tropical-only hurricanes.

6.2.2 Seasonal Variability

The monthly distribution of baroclinically-enhanced hurricanes is shown in Figure 6.3. The distribution is skewed to the right. It peaks in September but is considerably flatter and broader than the distribution of tropical-only hurricanes. In fact, over the 111-year period from 1886–1996 exactly 50% (98 out of 196) of all baroclinically-enhanced hurricanes developed outside the peak months of August through October. This compares to only 8% of tropical-only hurricanes occurring during off-peak months. October is more active than August, and November is more active than June or July. June sees about as many baroclinically-enhanced hurricanes as July. The warm late-season waters of the western Caribbean Sea and Gulf of Mexico provide fuel to the frequent baroclinic disturbances from middle latitudes making November a relatively important month for baroclinically-enhanced hurricane development.

6.3 Season Length

Figure 6.4 shows the distribution of begin and end dates for baroclinically-enhanced hurricanes. The distributions are flatter than those for tropical-only hurricanes. The baroclinically-enhanced hurricane season begins in early June and continues through November with the most active period occurring from the middle of August through the early part of November. The most common begin and end dates are September 11th

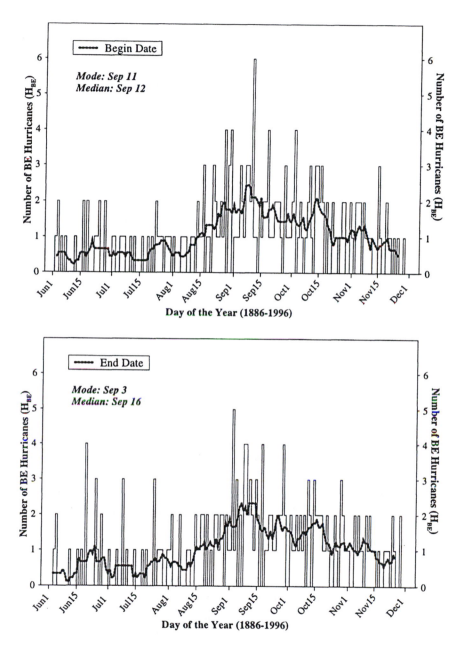

Fig. 6.4: Distributions of begin (top) and end (bottom) dates for baroclinically-enhanced hurricanes over the period 1886–1996 (Jun–Nov only). The dotted line is a 9-day running average. Note the increase in activity during the second week of October.

and September 3rd, respectively. The median begin and end dates are September 12th and September 16th, respectively.[5]

Length of the baroclinically-enhanced hurricane season for each year is shown in Figure 6.5. The large variability is indicative of the relatively few hurricanes in any one season (particularly before the middle 1960s) and the fact that baroclinically-enhanced hurricanes are relatively more common outside the active months of August through October. For years with only one baroclincally-enhanced hurricane, the season length is equal to the lifespan of the lone hurricane. Of the 33 years with three or more baroclinically-enhanced hurricanes, 16 (or 55%) had seasons longer than 3 months. The cumulative distribution of the onset and termination of the baroclinically-enhanced hurricane season is also shown in Figure 6.5. The onset is characterized by a relatively flat distribution with the exception of a period in late August when season onset is more likely. In half the years the onset of the baroclinically-enhanced season occurs by August 16th; this is coincident with the median date for the start of the tropical-only season. The cumulative frequency of termination dates suggest two regimes, one from the middle of June through August and a second one from September through November, with the exception of a nearly 2-week period in early October when season termination is rare. The median end date of the baroclinically-enhanced hurricane season is October 15th.

6.4   Intensity and Duration

The average maximum intensity of baroclinically-enhanced hurricanes for each season and the annual distribution of years by average maximum intensity are shown in Figure 6.6. The average maximum intensity refers to an average over all hurricanes in the season. Baroclinically-enhanced hurricanes are weaker than tropical-only hurricanes. The overall average maximum intensity is 82.5 kt, which is approximately 15 kt weaker than the average maximum intensity of tropical-only hurricanes. Most years have maximum intensities between 85 and 90 kt. This includes the known bias in the hurricane record prior to the 1900s, when maximum intensities were generally reported as 85 kt (see Chapter 3). Compared to tropical-only hurricane intensities, there is a distinct preference for much weaker baroclinically-enhanced hurricanes. This is because baroclinically-enhanced hurricanes tend to originate farther north and are consequently of shorter duration than the low-latitude-developing tropical-only hurricanes.

Duration of North Atlantic baroclinically-enhanced hurricanes is examined by considering the number of days at hurricane intensity for each hurricane averaged over all hurricanes in a season. The time series of annual-averaged duration is shown in Figure 6.7 (top). The long-term average duration is 3.1 days per storm with a standard deviation of 1.8 days. Despite the increase in abundance of baroclinically-enhanced hurricanes after 1958 there is no corresponding increase in duration. In fact, the annual-average duration over the period is relatively constant. Although the North Atlantic

---

[5] An upswing in baroclinic activity occurs around the second week of October.

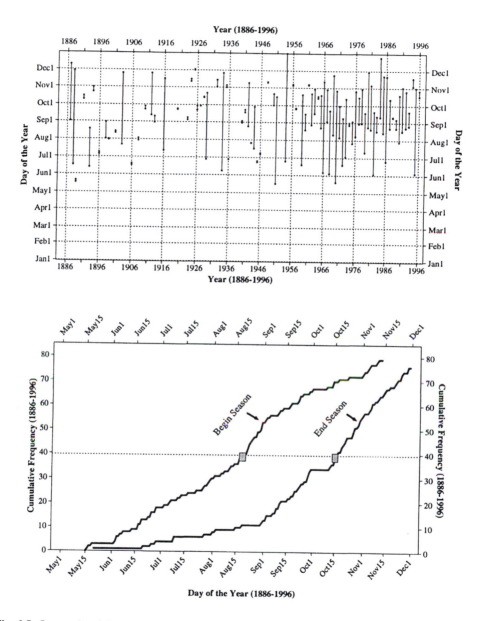

Fig. 6.5: Start and end dates of the baroclinically-enhanced hurricane season for each year in the period 1886–1996 (top) and the cumulative distribution of season onset and termination dates (bottom). The horizontal line in the cumulative distribution graph identifies the median dates.

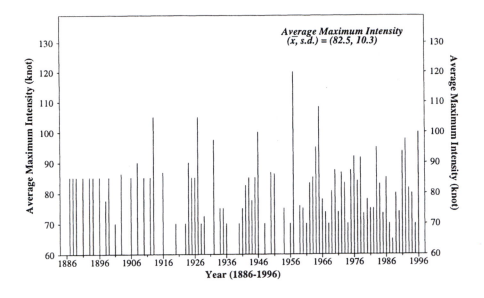

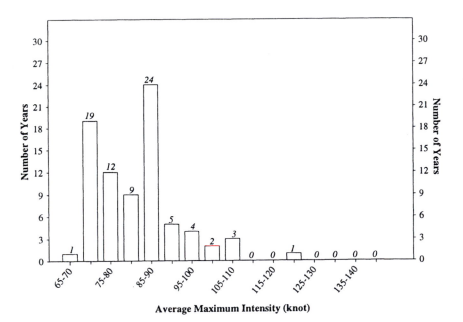

Fig. 6.6: Annual average maximum intensity (kt) of baroclinically-enhanced hurricanes for each year over the period 1886–1996 (top) and the distribution of years by average maximum intensity in 5 knot intervals (bottom).

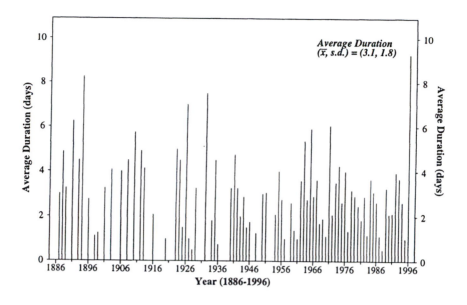

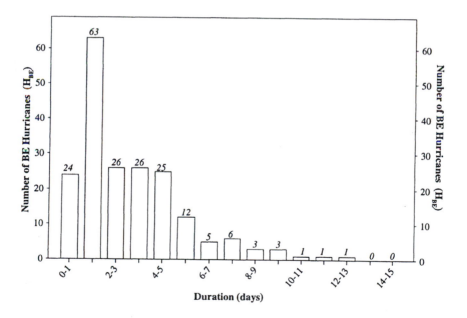

Fig. 6.7: Annual average duration of North Atlantic baroclinically-enhanced hurricanes for each year over the period 1886–1996 (top) and the distribution of baroclinically-enhanced hurricanes by duration in 1-day intervals (bottom).

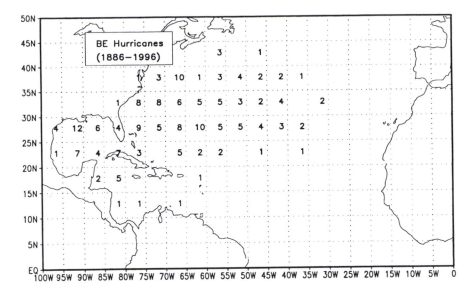

Fig. 6.8: Points of origin for baroclinically-enhanced North Atlantic hurricanes over the period 1886–1996 in 5° latitude-longitude grid boxes. Preferred regions of baroclinically-enhanced hurricane genesis include the northern Gulf of Mexico and the western North Atlantic.

basin is more favorable for baroclinically-enhanced hurricanes compared with the past, the later years are not more favorable for supporting longer duration storms. Care is advised in interpreting these statistics as years with few hurricanes strongly influence the averages. The distribution of baroclinically-enhanced hurricanes by duration has a large peak at 1 to 2 days with a broader peak through 5 days (Figure 6.7 bottom). Only 16% of baroclinically-enhanced hurricanes last longer than five days. This is in stark contrast to tropical-only hurricanes, which have an average seasonal duration of 5.6 days.

6.5 Origin

Genesis regions of baroclinically-enhanced hurricanes are examined by considering points of origin. Points of origin for baroclinically-enhanced hurricanes are shown alongside points of origin for tropical-only hurricanes in Chapter 5. Baroclinically-enhanced hurricanes tend to form at higher latitudes compared with tropical-only hurricanes particularly over the western and central North Atlantic. Over the western Caribbean and Gulf of Mexico there is only a weak dependence of latitude in separating the two hurricane types. This is because baroclinic factors are a contributing factor to the origin and development of tropical cyclones in the tropics near the continental landmasses. The number of baroclinically-enhanced origins in 5° degree latitude-longitude boxes is shown in Figure 6.8. Noticeably lacking are origins in the

Table 6.2: Development time. Values are the average time (in days) it takes for tropical-only (TO) and baroclinically-enhanced hurricanes (BE) to reach hurricane intensity. The categories include full development from 25 to 65 kt and early development from 25 to 45 kt.

Wind Speed Changes	Statistics	TO	BE
From 25 kt to 65 kt	N	69	29
(full development)	$\bar{x}$ (day)	3.4	4.0
	s.d. (day)	1.8	2.0
From 25 kt to 45 kt	N	69	28
(early development)	$\bar{x}$ (day)	2.3	2.8
	s.d. (day)	1.4	1.7

central and eastern North Atlantic south of 20°N. Genesis regions include the Gulf of Mexico and portions of the western North Atlantic. The large number of baroclinically-enhanced origins over the western Atlantic north of 25°N is in contrast to tropical-only hurricanes, which only rarely form in this region. The average latitude of baroclinic formation is 29°N. This is 10° father north than the average latitude of tropical-only hurricane formation. The most common latitude belt runs from 25–30°N. Tropical cyclone intensification at subtropical latitudes is a distinguishing feature of baroclinic-type hurricanes. The average longitude of formation is 69°W. This is a few degrees to the west of the average longitude of tropical-only hurricane formation. Baroclinically-enhanced hurricanes are likely to form in the semi-permanent baroclinic zone along the east coast of the United States.

The time it takes for an initial closed cyclonic circulation to develop into a hurricane varies widely between storms. The variation depends on many physical factors, and initial development is difficult to predict. It is interesting to compare the time it takes for baroclinically-enhanced and tropical-only hurricanes to develop. Middle-latitude baroclinic processes generally inhibit tropical cyclone development and intensification. Thus it might be anticipated that baroclinically-enhanced hurricanes will take longer to develop. Yet cyclone intensification might be expected to take place more rapidly for disturbances at higher latitudes due to a greater Coriolis force. Table 6.2 indicates that baroclinic development is slower. Under the assumption of equal population variances, the null hypothesis that the differences are due to chance is tested against the alternative that baroclinic development is slower using the student's t test. The null hypothesis is rejected at the 90% significance level (p-value of 0.076) for cyclones developing from 25 kt. The difference is significant for development to 45 kt, but not for intensification time after, indicating that it is the early stage of development that is slower for baroclinically-enhanced hurricanes. It should be noted that baroclinically-enhanced hurricanes are less likely to pose a serious threat to the U.S. coastline, in general. The exception is Florida. Here baroclinically-enhanced storms appear to be responsible for

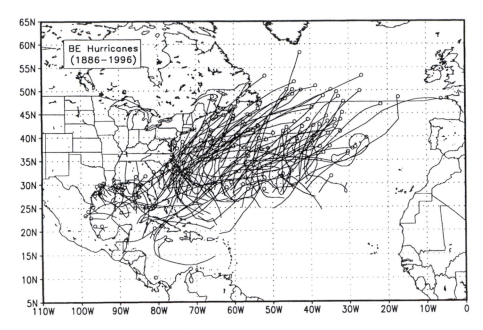

Fig. 6.9: Tracks of North Atlantic baroclinically-enhanced hurricanes over the period 1886–1996. Note the tendency of these hurricanes to track northeastward and away from the United States.

a greater percentage of hurricane-induced tornadoes. Additionally, the larger radius to maximum winds in baroclinic systems over the Gulf of Mexico enhance the potential for sustained gale-force winds creating greater wind-driven storm surges along the west coast of the peninsula.

6.6   Tracks

Baroclinically-enhanced hurricanes originating at higher latitudes tend to track north or northeastward and away from land areas (except Bermuda). This is especially true for development that occurs over the western North Atlantic north of the Greater Antilles. Here the characteristics of baroclinically-enhanced hurricane tracks are examined in detail including a look at their seasonal variations.

6.6.1   General Characteristics

Figure 6.9 shows the paths of all baroclinically-enhanced hurricanes over the North Atlantic over the period 1886–1996. In general, the tracks indicate little recurvature. Baroclinically-enhanced hurricanes tend to originate at higher latitudes often subsequent to recurvature of the incipient tropical storm. High-latitude hurricanes avoid an initial westward motion altogether. The exception is over the Gulf of Mexico, where baroclinically-enhanced hurricanes tend to move toward the west or northwest. This is particularly evident in the western Gulf. Texas and Louisiana are vulnerable to land-falling baroclinically-enhanced hurricanes. Over the western subtropical North Atlantic baroclinically-enhanced hurricanes track northeastward and away from the coast. Most of the hurricanes dissipate before reaching 50°N or 30°W. Because of the higher latitude of formation and the tendency to move northeastward toward colder waters, paths of baroclinically-enhanced hurricanes are shorter than tracks of tropical-only hurricanes. Baroclinically-enhanced hurricanes pose less of a threat to the northeast U.S. coastline.

6.6.2   Seasonal Variability

Similar to tropical-only hurricanes, there is considerable variation in the tracks of baroclinically-enhanced hurricanes over a season (Figure 6.10). June and July tracks are confined to the Gulf of Mexico and western North Atlantic. Texas and Florida are vulnerable to early-season baroclinically-enhanced landfalls. By August hurricane tracks are more numerous over the western and central subtropical Atlantic. In September there is a noticeable shift in activity away from the eastern U.S. seaboard into the central North Atlantic. Nearly all these storms become hurricanes north of 25°N latitude. The Gulf of Mexico remains a source region for baroclinic development during September. In October, and more so in November, baroclinically-enhanced hurricanes tend to track over a fairly narrow zone extending from the Bahamas northeast across the North Atlantic basin. The progression of baroclinically-enhanced activity farther offshore between summer and fall is particularly pronounced over the Bahamas. It coincides with a southward shift in rainfall over the region during this time of year as the baroclinic zone penetrates deeper into the tropics (Bosart and Schwartz 1979). Fronts that stall along the middle Atlantic states during August and September will progress farther south over Florida and the Bahamas during October and November. As explained, these fronts provide a mechanism for the development of baroclinically-enhanced hurricanes.

6.7   Dissipation

Areas where baroclinically-enhanced hurricanes dissipate are highlighted in Figure 6.11. Most hurricanes dissipate along the Gulf coast, middle Atlantic seaboard and at high latitudes over the western and central North Atlantic. The relative abundance of baroclinically-enhanced dissipation points along and near the coast of the middle Atlantic states is over a region where tropical-only hurricane dissipation is relatively less common. Moreover, hurricanes of baroclinic origin tend to dissipate over the open waters of the North Atlantic at a latitude farther south. Mid-season tropical-only hurricanes reach high latitudes at a time when the middle latitude westerlies are weakest. In contrast, vigorous high-latitude westerlies during early summer and fall will limit the northward extent of baroclinically-enhanced hurricane activity. However, the av-

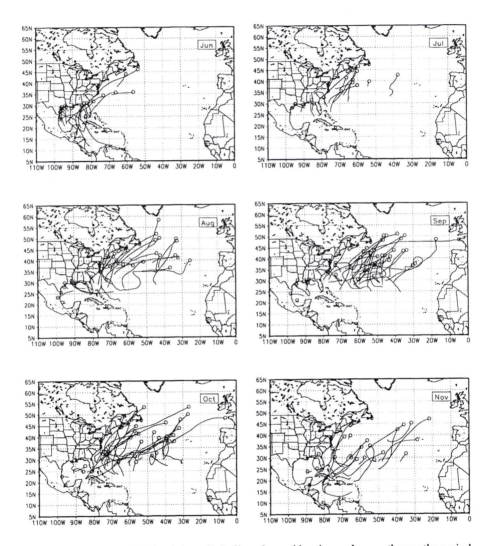

Fig. 6.10: Tracks of North Atlantic baroclinically-enhanced hurricanes by month over the period 1886–1996. Early season baroclinically-enhanced hurricanes have a tendency to make landfall along the Gulf coast or southeast United States. Middle- and late-season hurricanes have a strong inclination to avoid the U.S. coast altogether.

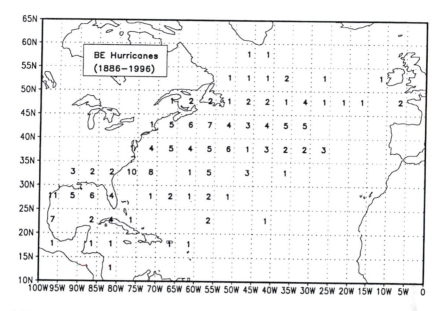

Fig. 6.11: Number of dissipation points for baroclinically-enhanced hurricanes in 5° latitude-longitude grid boxes over the period 1886–1996. Preferred regions of baroclinically-enhanced hurricane dissipation include the middle Atlantic seaboard and high latitudes over the western North Atlantic.

erage latitude of dissipation for all baroclinically-enhanced hurricanes is 36°N, which is about 3° farther north than the average dissipation of tropical-only hurricanes. This is because many tropical-only hurricanes make landfall in Mexico and the southern United States. The average longitude of dissipation is 62°W.

6.8 Baroclinically Initiated Hurricanes

A subset of baroclinically-enhanced hurricanes are hurricanes that *originate* from middle-latitude baroclinic disturbances. As mentioned, this class of North Atlantic hurricanes was suggested by Hebert and Frank (1973). Baroclinically-initiated hurricanes can originate in several distinct ways. Observations indicate that the most recurrent sources of baroclinically-initiated hurricanes are decaying troughs or baroclinic zones (fronts), upper-level cold lows, extra-tropical cyclones (often occluded), and mesoscale convective complexes (MCSs) with an associated mesoscale convective vortex (MCV).[6] Baroclinically-initiated hurricanes are more likely to originate in nonuniform airmasses, tend to be situated close to the middle-latitude jet stream, and generally develop over somewhat cooler waters compared with tropical-only hurricanes.

---

[6]Simulations of idealized convective systems indicate that the development of a MCV from a squall line is a natural consequence of the finite extent of the convective line as horizontal vorticity is tilted into the vertical (Davis and Weisman 1994).

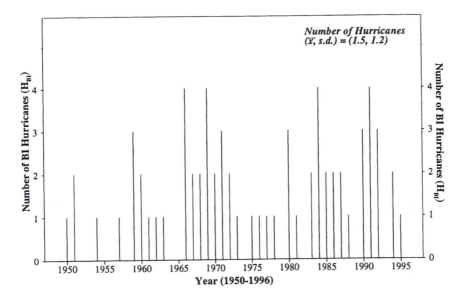

Fig. 6.12: Time series of the annual frequency of North Atlantic baroclinically-initiated hurricanes over the period 1950–1996. The average ($\overline{x}$) and standard deviation ($s.d.$) are given in the upper-right corner.

The separation of baroclinically-initiated hurricanes used here is taken from Kimberlain (1996). His work examined the relevant literature and synoptic maps to determine hurricane type for baroclinically-enhanced hurricanes over the period 1950 through 1996. The criterion was whether it could be determined that the initiation mechanism of hurricane development was of baroclinic origin or not. There is no attempt to use objective criteria to identify baroclinically-initiated hurricanes prior to 1950 as is done with tropical-only hurricanes. The climatology presented here is based on data over the period 1950 through 1996.

6.8.1   Abundance, Intensity, and Duration

A total of 67 baroclinically-initiated hurricanes have occurred over the 47-year period from 1950 through 1996. Annual variation is shown in Figure 6.12. The average number of baroclinically-initiated hurricanes in a season is 1.5 with a standard deviation of 1.2. The maximum number in a single season is four, however many seasons have no baroclinically-initiated hurricanes. The annual distribution is shown in Figure 6.13. Most seasons have less than 3 baroclinically-initiated hurricanes. Fourteen years had one baroclinically-initiated hurricane and 11 seasons had two. The monthly distribution is shown in Figure 6.14. The peak months are August through October. Compared with total activity baroclinically-initiated hurricanes are relatively more likely in the off-peak months of June, July, and November. A unique feature of baroclinically-

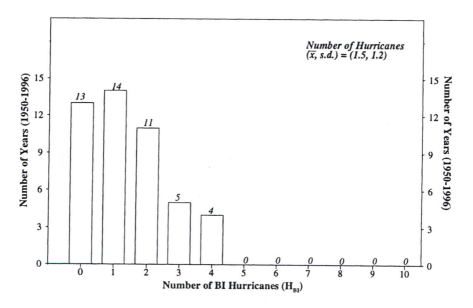

Fig. 6.13: Distribution of the annual frequency of North Atlantic baroclinically-initiated hurricanes over the period 1950–1996. More than half the years have either none or one baroclinically-initiated hurricane.

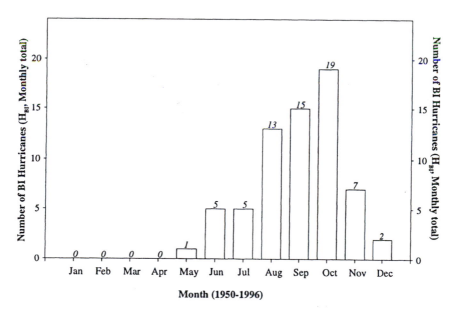

Fig. 6.14: Monthly distribution of North Atlantic baroclinically-initiated hurricanes over the period 1886–1996. In contrast to tropical-only hurricanes, baroclinically-initiated hurricanes are most frequent during October.

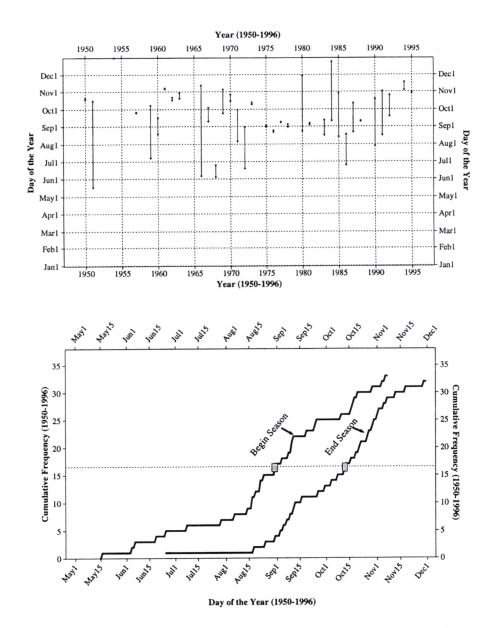

Fig. 6.15: Start and end dates of the baroclincally-initiated hurricane season for each year in the period 1950–1996 (top). Cumulative distribution of season onset and termination dates (bottom). The horizontal line in the cumulative distribution graph identifies the median dates.

initiated hurricanes is the maximum in frequency during October. October is a month in which overall hurricane activity is waning. Less than one third of the baroclinically-enhanced hurricanes are baroclinically-initiated, except during October when nearly half of all baroclinic activity is baroclinically-initiated.

Begin and end dates of baroclinically-initiated hurricane seasons over the period 1950–1996 are given in Figure 6.15. The cumulative distributions of begin and end dates for baroclinically-initiated hurricane seasons are also provided. The probability of the season beginning is roughly constant from the middle of May through early August, but increases thereafter. The median onset date is August 31, which is considerably later than for the season as a whole. The median termination date of the baroclinically-initiated season is October 12, which is a few days earlier than for the entire hurricane season.

Baroclinically-initiated hurricanes tend to form closer to the continent. Consequently they are the weakest of the North Atlantic hurricanes. Figure 6.16 shows the frequency and distribution of annual average maximum intensity. The overall average maximum intensity is 79.0 kt. This compares to 92.4 kt for all hurricanes. No trend in average maximum intensity is noted. In most years maximum intensities are between 70 and 75 kt. Greater than half the years in the record have average maximum intensities less than 85 kt. A secondary peak in the distribution occurs in the range of intensities between 90 and 95 kt. Only three years have an average maximum intensity exceeding 95 kt.

The distribution of baroclinically-initiated hurricane by duration is shown in Figure 6.17. Nearly two-thirds of all baroclinically-initiated hurricanes have a lifespan of 3 days or less, with the most common duration being 1 or 2 days. Hurricane *Ginger* of September 1971 (see Figure 6.18) is the longest-lived baroclinically-initiated hurricane. *Ginger* remained a hurricane for nearly 20 days, twice as long as the second longest baroclinically-initiated hurricane. Interestingly, *Ginger* is the longest lasting of any hurricane on record for the North Atlantic basin.

6.8.2   Origin, Tracks, and Dissipation

Baroclinically-initiated hurricanes form in areas that are generally not conducive to the development of tropical-only hurricanes (Figure 6.19). Points of origin are clustered in the northern Gulf of Mexico as well as in the western and central subtropical North Atlantic, generally north of 25°N latitude. Only one baroclinically-initiated hurricane originated in the Caribbean Sea, while two formed north of 40°N latitude. Baroclinically-initiated hurricanes are most likely to form off the southeast coast of the United States. This is the area where middle latitude fronts often stall. Also during the early and late weeks of the hurricane season, an occluded extra-tropical cyclone sometimes get trapped in the *eastern* North Atlantic under a middle-latitude blocking high pressure ridge. If the surface low drifts over warm ocean waters, it may develop into a weak hurricane as *Ivan* did in 1980.

The origin of baroclinically-initiated storms is concentrated in the latitude belt ex-

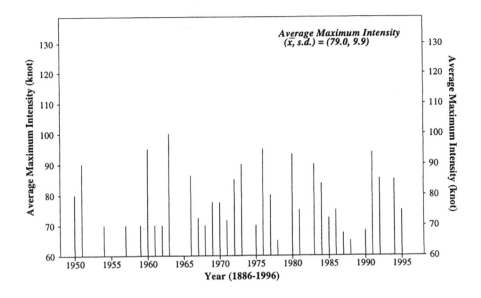

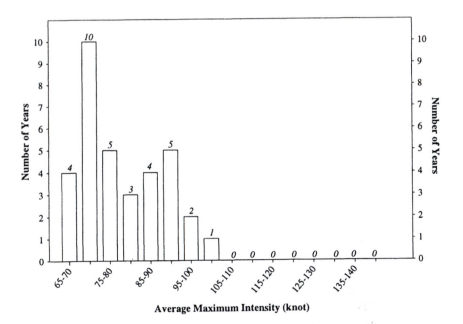

Fig. 6.16: Annual average maximum intensity (kt) of baroclinically-initiated hurricanes for each season over the period 1886–1996 (top) and the distribution of years by average maximum intensity in 5 kt intervals (bottom).

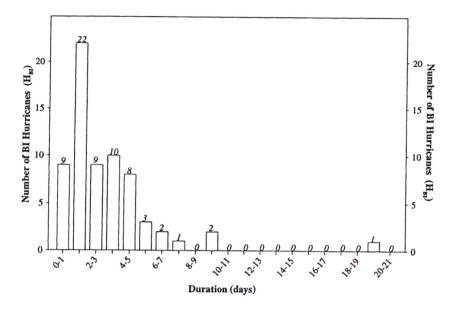

Fig. 6.17: Distribution of North Atlantic baroclinically-initiated hurricanes by duration in 1-day intervals. Hurricane *Ginger* of 1971 remained a hurricane for nearly 20 days. Note that most of baroclinically-initiated hurricanes have a lifespan of 3 days or less, and the most common duration is 1 or 2 days.

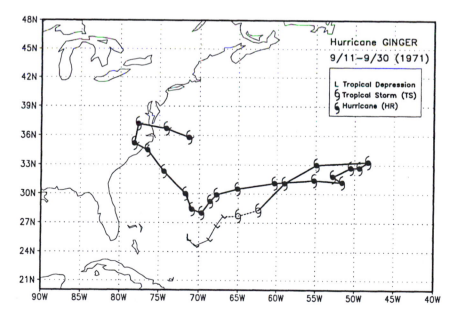

Fig. 6.18: Stages of development and intensification for hurricane *Ginger* in 1971. *Ginger* was a hurricane for almost 20 consecutive days.

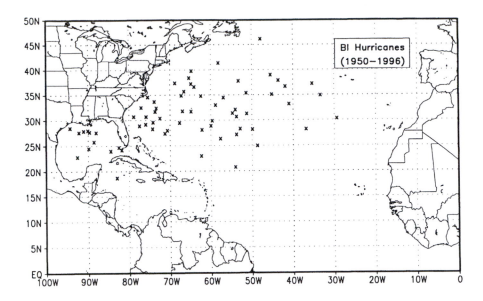

Fig. 6.19: Points of origin for baroclinically-initiated hurricanes over the period 1950–1996. Note the relatively high latitude of genesis for most baroclinically-initiated hurricanes.

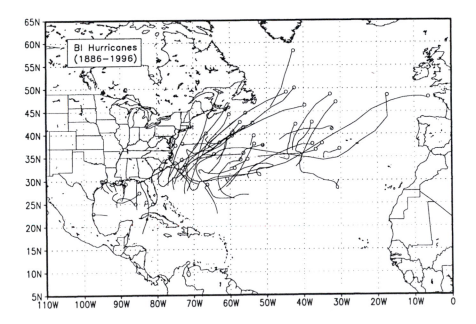

Fig. 6.20: Tracks of North Atlantic baroclinically-initiated hurricanes over the period 1950–1996. Note the high latitude of formation and direct track to the northeast.

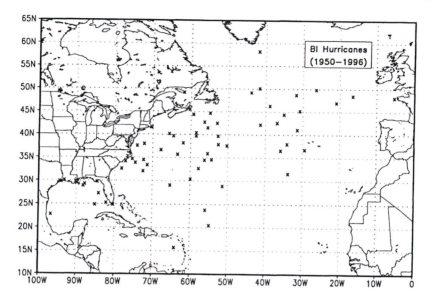

Fig. 6.21: Dissipation points for North Atlantic baroclinically-initiated hurricanes over the period 1950–1996.

tending from 27 to 29°N. The average latitude of formation is 30°N. This compares to an average latitude of 29°N for baroclinically-enhanced hurricanes and an average latitude of 19°N for tropical-only hurricanes. Baroclinically-initiated hurricanes are most likely to develop between 65 and 70°W, but there is considerable variation in longitude. The average longitude of formation is 66°W, which compares with 69°W for baroclinically-enhanced hurricanes. This is indicative of the fact that the western subtropical North Atlantic is more favorable for baroclinically-initiated development whereas the Gulf of Mexico is more favorable for baroclinically-enhanced hurricane development. As mentioned, the North American continent provides a source region for pre-hurricane baroclinic disturbances.

The track of a baroclinically-initiated hurricane is often in the direction of movement of the incipient disturbance. Tracks are short and direct. Figure 6.20 shows the tracks of baroclinic hurricanes over the period 1950 through 1996. Because of their relatively high latitude of formation, baroclinically-initiated hurricane paths are oriented to the north and northeast particularly over the western North Atlantic. Baroclinically-initiated hurricanes over the northern Gulf of Mexico tend to have at least a bit of a westerly component, which takes them into the Gulf coastal states. At higher latitudes, baroclinically-initiated hurricanes track with a larger eastward component compared to the tracks of tropical-only hurricanes. This indicates the presence of significant high-latitude westerlies concurrent with, and likely related to, the development of baroclinically-initiated hurricanes. A vigorous westerly current provides both a source region and a steering mechanism for these types of hurricanes.

Points of dissipation for all baroclinically-initiated hurricanes over the period are plotted in Figure 6.21. Dissipation includes storms that transition to extra-tropical cyclones. Although most of the points of origin for baroclinic hurricanes are at a latitude considerably farther north compared to points of origin for tropical-only hurricanes, points of dissipation tend to be farther to the south. Dissipation tends to be relatively close to origin. Dissipation is most common over the open waters of the western and central North Atlantic and over the north-central Gulf of Mexico. Dissipation or extra-tropical transition tends to occur in two areas. One area is located to the east of 45°W longitude and the other is located to the west of 50°W longitude. The region of fewer dissipations coincides with the average location of the middle Atlantic trough separating the two subtropical high pressure zones.

7

Major Hurricanes

When winds blow at speeds of 50 m s^{-1} ($\sim$ 100 kt) or higher the storm is called a major hurricane. The central pressure inside a major hurricane is typically less than 965 mb. The hurricane-driven storm surge can be expected to be as high as 3 meters or more. Damage potential ranges from extensive to catastrophic. Major hurricanes pose a significant threat to life and property in the United States, Mexico, Bermuda, and the nations of the Caribbean. Historically they are referred to as "great" hurricanes. Today they are called major or *intense* hurricanes. Most major North Atlantic hurricanes originate as tropical-only hurricanes. The probability of major hurricanes along the U.S. coastline shows large decadal variability in location. Florida is particularly vulnerable to the threat of major hurricanes. Periods of little or no major hurricane activity are interspersed with periods of frequent storms.

An average of two major hurricanes per year visit the North Atlantic basin. August and September see the vast majority of the intense hurricane activity. The average duration of a storm at major hurricane intensity is 2.5 days. A survey of major hurricanes of the North Atlantic is provided by Landsea (1993). The occurrence of major hurricanes making landfall in the United States is examined in Hebert and Taylor (1979a, and 1979b) and Hebert et al. (1990 and 1992). This chapter reviews various climatological aspects of major hurricanes. It includes a consideration of intense tropical-only and baroclinically-enhanced hurricanes. We make use of all of the major hurricanes listed in the best-track data set back to 1886. The most reliable hurricane data on intensity extends only back to about 1944. The bias correction (see Chapter 3) of Landsea (1993) is not considered.

7.1 Maximum Intensity

The highest wind speed recorded in the best-track data for each hurricane is defined as the hurricane's maximum intensity. The frequencies of maximum intensity for North Atlantic hurricanes are shown in Table 7.1. Of the 554 North Atlantic hurricanes over the period 1886 through 1996, 212 (or 38%) are category three or higher on the Saffir/Simpson scale. The average number of major hurricanes per season is 1.9. The av-

Table 7.1: North Atlantic hurricane frequencies. Values include total number (N), annual average ($\bar{x}$), and percentage (%) of minor (category 1, 2) and major (category 3, 4, 5) hurricanes by hurricane type. TO refers to tropical-only hurricanes, BE to baroclinically-enhanced hurricanes, and BI to baroclinically-initiated hurricanes.

Category	$\rightarrow$	All	TO	BE	BI
Minor Hurricanes ($33 \text{ m s}^{-1} \leq V < 50 \text{ m s}^{-1}$)					
Number of Cases	N	342	177	165	57
Annual Average	$\bar{x}$	3.1	1.6	1.5	0.5
Percentage	(%)	61.7	49.4	84.2	85.1
Intense Hurricanes ($V \geq 50 \text{ m s}^{-1}$)					
Number of Cases	N	212	181	31	10
Annual Average	$\bar{x}$	1.9	1.6	0.3	0.1
Percentage	(%)	38.3	50.6	15.8	14.9
Total	N	554	358	196	67

erage number of minor hurricanes is 3.1. Category five hurricanes are most severe, but the least common. On average the North Atlantic experiences 1 category five hurricane every 3 years. Category five hurricane account for less than 5% of all Atlantic hurricanes. Approximately half of all tropical-only hurricanes become major hurricanes. This compares to 16% of all baroclinically-enhanced and 15% of all baroclinically-initiated hurricanes. Only 6 baroclinically-initiated have reached major hurricane status since 1950. Nearly 90% of all major North Atlantic hurricanes are tropical-only. In comparison there is about one major baroclinically-enhanced hurricane every three years.

Distributions of hurricanes by maximum intensity for tropical-only and baroclinically-enhanced types are shown in Figure 7.1. For tropical-only hurricanes significant peaks are noted at 85 and 105 kt. Otherwise the distribution is flat. Considering the bias toward 85 kt hurricanes in the record before 1901, it is likely that a majority of tropical-only hurricanes become intense. The distribution of baroclinically-enhanced hurricanes by maximum intensity is positively skewed with a peak at 70 kt. Overall the two distributions are quite different. The distributions are most different at highest intensities. Baroclinically-enhanced hurricanes only rarely obtain wind speeds in excess of 115 kt. In contrast, 36% of the major tropical-only hurricanes have maximum winds exceeding this value. More than a quarter of these hurricanes have winds in excess of 135 kt. In contrast, 55% of all baroclinically-enhanced hurricanes have wind speeds less than 85 kt. This compares to 15% for tropical-only hurricanes.

7.2 Abundance

As mentioned, the expected annual number of major North Atlantic hurricanes is 1.9. However the frequency of major hurricanes varies on different time scales. Time vari-

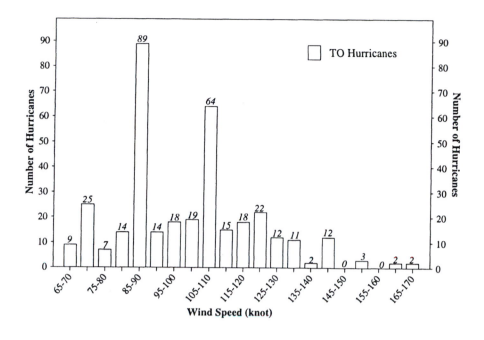

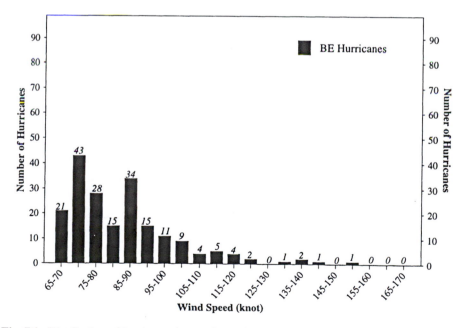

Fig. 7.1: Distribution of hurricanes by maximum intensity (wind speed) for tropical-only (top) and baroclinically-enhanced (bottom) North Atlantic hurricanes over the period 1886–1996. The large peaks are likely a result of biases in the best-track data before the turn of the century.

ations are examined on the annual, biennial, and semidecadal time scales. Seasonal variations are considered by looking at monthly and daily activity.

7.2.1 Interannual Variability

Multi-year variations in the number of major hurricanes are shown in Figure 7.2. Semidecadal abundance ranges from a low of two during the first pentad of the century to a high of 20 fifty years later in the period 1951–1955. The average number of intense hurricanes in these five-year intervals is 9.4, with a standard deviation of 4.9. Except for the increased activity during the late 1940s and early 1950s, the semidecadal abundance of major hurricanes is fairly constant throughout the 111-year record. The early years of 1906–1910 show a total of 12 intense hurricanes; the 6th most active pentad of this series. Though it is likely that not all early hurricanes were classified correctly as to their precise intensities, and some hurricanes were missed completely, the level of major hurricane activity during the early part of the record is consistent with the level of intense hurricane activity during the later decades of the 1970s and 1980s.

The biennial variation in the series ranges from a minimum of 0 to a maximum of 11. The two most active periods are 1949–1950 and 1996–1996. The most active biennial period before 1900 is 1893–1894, which featured five major hurricanes. The average biennial frequency of major hurricanes is 3.9 with a standard deviation of 2.8. Except for the heightened activity during the 1950s and the somewhat fewer great hurricanes before 1900, the frequency of intense hurricanes over the period is relatively constant over the 111-year period. An increase in major hurricane activity occurs during the middle 1990s.

7.2.2 Annual and Seasonal Variability

Annual variations in major hurricane frequencies are shown in Table 7.2 and in Figure 7.3. The heightened activity from the late 1940s into the early 1960s is clearly evident as is the return to more frequent major hurricane activity during the middle 1990s. Fewer major hurricanes before the turn of the century is also obvious. There are no major hurricanes in the best-track record from 1888 through 1892. The 1950 season tops the list with eight major hurricanes followed by the 1961 season with seven. The longest consecutive span of years with three or fewer intense hurricanes is the 25-year period from 1970 to 1994. Earlier, when overall there were fewer major hurricanes, at least some years had above average activity. After 1914 no two consecutive years occur without at least one intense North Atlantic hurricane.

The annual distribution of intense hurricanes (Figure 7.4) is positively skewed with an average of 1.9 major hurricanes and a standard deviation of 1.8. The mode of the distribution is one major hurricane. There are 5 years with exactly four, 5 years with exactly five, and 5 years with exactly six intense hurricanes. Approximately 23% of the years in the record are without a major hurricane. This is the same percentage of years with exactly two major hurricanes. Slightly more than 15% of the years have four

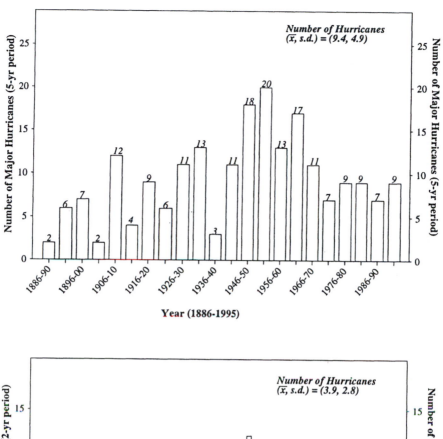

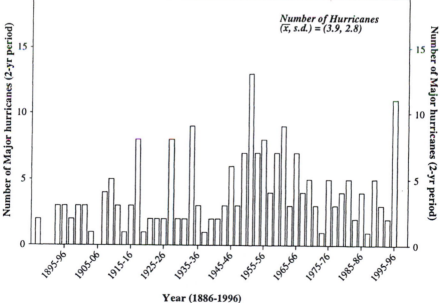

Fig. 7.2: Time series of the frequency of North Atlantic major hurricanes in 5-year intervals (top) and 2-year intervals (bottom). The average ($\bar{x}$) and standard deviation ($s.d.$) are given in the upper-right corner of each panel.

Table 7.2: Annual number of North Atlantic major hurricanes over the period 1886–1996. Y denotes the last digit of the year. Decade totals are also given. Note the relative abundance of major hurricanes during the 1950s and 1960s.

Y	188	189	190	191	192	193	194	195	196	197	198	199
0	.	0	2	3	0	1	0	8	2	2	2	1
1	.	0	1	0	2	1	2	5	7	1	3	2
2	.	0	0	1	1	4	1	3	1	0	1	1
3	.	3	1	0	1	5	2	4	2	1	1	1
4	.	2	0	0	2	0	3	2	6	2	1	0
5	.	1	0	3	0	3	3	6	1	3	3	5
6	0	2	4	6	6	1	1	2	3	2	0	6
7	2	0	0	2	2	0	2	2	1	1	1	.
8	0	0	1	0	1	1	4	5	0	2	3	.
9	0	3	4	1	1	1	3	2	5	2	2	.
	2	11	13	16	16	17	21	39	28	16	17	16

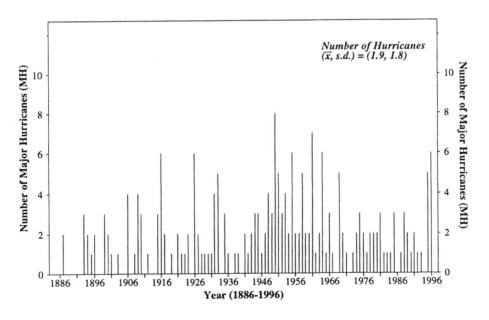

Fig. 7.3: Time series of the annual frequency of North Atlantic major hurricanes over the period 1886–1996. The average ($\bar{x}$) and standard deviation ($s.d.$) are given in the upper-right corner.

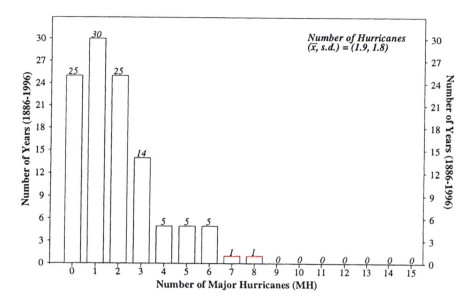

Fig. 7.4: Distribution of the annual frequency of North Atlantic major hurricanes over the period 1886–1996. The most common occurrence is one major hurricane each year.

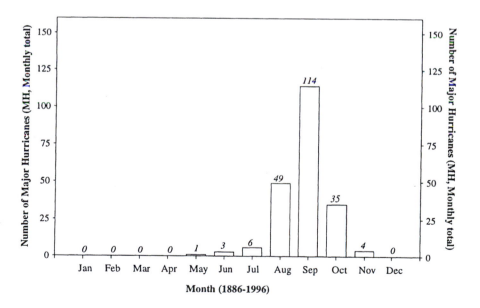

Fig. 7.5: Monthly distribution of North Atlantic major hurricanes over the period 1886–1996. The major hurricane season is narrowly concentrated around September with very few occurrences outside the months of August through October.

or more. It should be noted that two years (1950 and 1961) had more than six intense hurricanes.

Seasonal variations of major hurricane activity are shown in the monthly distribution (Figure 7.5). It should be kept in mind that a hurricane reaching major hurricane intensity in a particular month is considered a major hurricane for that month regardless of whether it remained at major hurricane intensity into the next month. Moreover, fifteen ($\approx 8\%$) of all major hurricanes in the North Atlantic reach major hurricane intensity more than once in their lifespan. Here we consider only the first occurrence for each major hurricane. As shown, major hurricane activity is rather narrowly focused on the month of September. The distribution is similar to the monthly distribution of tropical-only hurricanes (Chapter 5). Nearly 54% of all major hurricanes of the North Atlantic occur during September with less than 7% taking place outside the three peak months of August through October. August is more active than October, and July is more active than November. The occurrence of major hurricanes in November is quite rare. One major hurricane occurred before June 1st.

7.3   Season Length

Figure 7.6 shows the distribution of begin and end dates for all major hurricanes of the North Atlantic over the period 1886–1996. It includes the begin and end dates for major tropical-only hurricanes. Though major hurricanes can occur during June and July, it is not until August that they become likely. Clearly, the major hurricane season runs principally from August into the first part of October. By late October the season is essentially over. The season of major hurricanes runs coincident with the Cape Verde season (see Chapter 5). The majority of intense hurricanes are tropical-only. These storms often begin as African easterly waves deep in the tropics near the Cape Verde Islands. Indeed, the distribution of tropical-only major hurricanes is nearly identical to the distribution of all major hurricanes. The slight differences suggest that major baroclinically-enhanced hurricanes occur only during the height of the hurricane season. The median begin date of major hurricane occurrences is September 10th. Major hurricanes are most likely to form during the first week of September. Secondary peaks in activity are noted during late September into early October, and during the middle of October. An upswing in activity is noted during middle to late August. These local maximums might suggest a quasi-biweekly oscillation in intraseasonal major hurricane activity characterized by alternating active and break periods.

The length of each major hurricane season is shown in Figure 7.7 (top). In years with only a single major storm the season length is the number of days that particular hurricane remained a major hurricane. Season lengths vary considerably from year to year but are typically less than 2 months. As expected, years with more activity are characterized by longer seasons. The three major hurricanes of 1966 were spread over a record four-month period. During 1951 all 5 major hurricanes occurred before the middle of September. The cumulative distribution of season begin and end dates

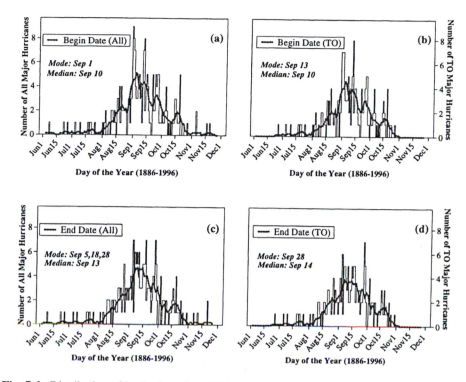

Fig. 7.6: Distribution of begin dates for all (a) and tropical-only (b) major hurricanes and the distribution of end dates for all (c) and tropical-only (d) major hurricanes. The solid line is a 9-day running average used to smooth the extremes in the daily variability.

are shown in Figure 7.7 (bottom). The major hurricane season is substantially shorter than the hurricane season as a whole. Major hurricane season begins (as defined by the median start dates) approximately 2 weeks after the start of the regular hurricane season and ends (as defined by the median end dates) approximately 20 days before the end of regular season. Half of all years in which major hurricanes occur, have their first major storm after the beginning of September and their last before the start of October.

7.4   Origin and Dissipation

Breeding and dissipation grounds for major hurricanes are indicated in Figure 7.8. The top panel shows the points at which North Atlantic hurricanes first obtain major hurricane intensity. Most major hurricanes reach category three west of 50°W longitude and south of 30°N latitude. One formed north of 40°N latitude and one east of 30°W longitude. There is a tendency for hurricanes to strengthen over the northwestern Caribbean Sea, the southeastern Gulf of Mexico, and the southwestern North Atlantic north of the Bahamas. A lack of major hurricane origins is conspicuous to the north of the Greater Antilles. The eastern and central Caribbean see few major hurricane developments.

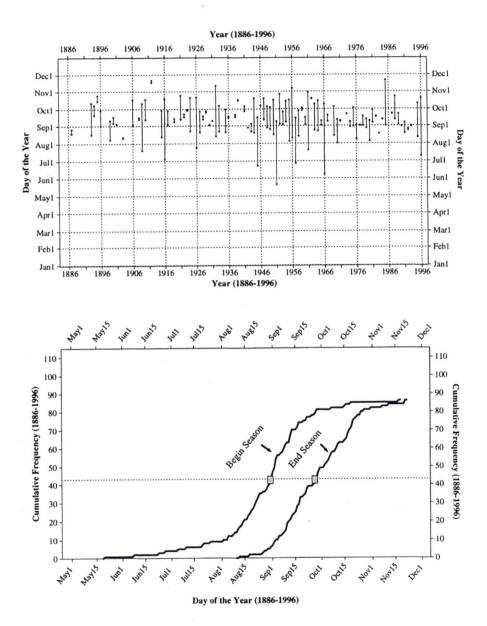

Fig. 7.7: Start and end dates of the North Atlantic major hurricane season for each year in the period 1886–1996 (top) and the cumulative distribution of season onset and termination dates (bottom). The horizontal line in the cumulative distribution graph identifies the median dates.

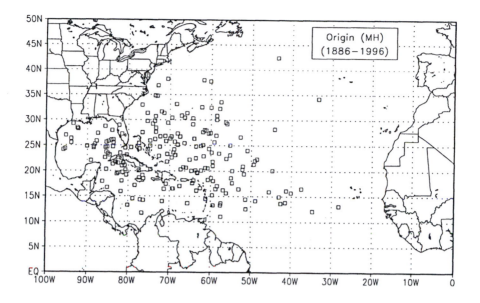

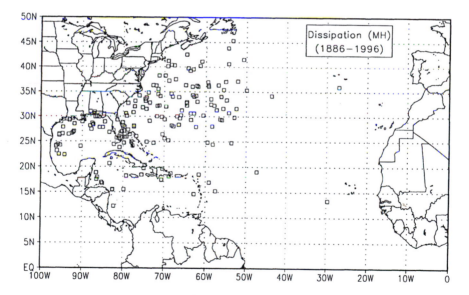

Fig. 7.8: Points of origin (top) and dissipation (bottom) for North Atlantic major hurricanes over the period 1886–1996. Major hurricane genesis is relatively common over the southeastern Gulf of Mexico, but relatively uncommon over most of the Caribbean Sea. Southeastern Florida is particularly vulnerable to major hurricane threats. The plotting symbol slightly distorts the precise location.

The latitudinal distribution of major hurricane origins is similar to the distribution of all hurricanes indicating that major hurricane activity is not restricted to the lowest latitudes. The average latitude of formation is 23°N. On the other hand, the longitudinal distribution of major hurricane origins is shifted considerably farther west compared to the distribution of all hurricanes. The shift amounts to about 3° in average longitude, and is a mainly a consequence of fewer major hurricanes forming east of 50°W. The coastal regions from Texas to Florida appear particularly vulnerable to hurricane intensification to major hurricane strength. Rapid intensification of storms along the immediate coastline poses a serious threat to life and property.

Dissipation points are clustered along the coastlines south of 30°N, but tend to occur over the open ocean to the north of this latitude. Most major hurricanes weaken to the west of 50°W longitude. Dissipation is common around the western, northern, and eastern borders of the Caribbean Sea. The majority of major North Atlantic hurricanes (66%) weaken between 23 and 35°N with an average latitude of dissipation of 29°N, which is 6° north of the average latitude of formation. The average longitude of dissipation is approximately 4° to the west of the average longitude of formation.

7.5 Duration

The average durations of North Atlantic major hurricanes for each year are shown in Figure 7.9 (top). Average duration is obtained by averaging the lifespan (at major hurricane intensity) of each major hurricane during the season. The overall average duration is 2.6 days with a standard deviation of 1.7 days. Spikes in the timeseries generally correspond to years that have a single, long-duration major hurricane. There is no significant trend in seasonally averaged lifespans. The last three decades, however, indicate somewhat shorter duration intense hurricanes corresponding to a period of fewer powerful storms. Major hurricanes lasted longer during the 1950s and early 1960s. The distribution of individual major hurricanes by duration is shown in Figure 7.9 (bottom). The distribution is markedly skewed. Most major hurricanes (72%) survive 3 days or less. The peak occurs at 1 to 2 days. At durations exceeding 3 days the distribution is flat through one week. Ten North Atlantic hurricanes survived at major hurricane intensity for longer than a week, one for more than two weeks.

7.6 Tracks

Major North Atlantic hurricane tracks are shown in Figure 7.10 by category. Tracks are shown only at major hurricane intensity. Reintensifying storms (8% of all major hurricanes) are not included. As anticipated, major hurricane tracks are largely confined to the western half of the North Atlantic between 15 and 40°N latitude. In fact, it is quite rare to observe a major hurricane east of 55°W longitude. Major hurricanes track northwestward into the Gulf of Mexico or curve northward near 70°N longitude between Bermuda and the U.S. east coast. The absence of major hurricane activity

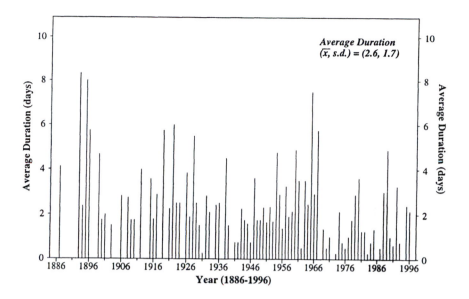

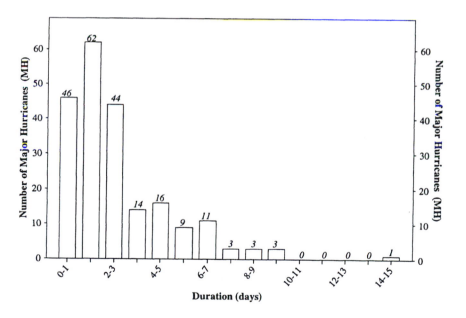

Fig. 7.9: Average duration of North Atlantic major hurricanes for each season over the period 1886–1996 (top). Distribution of major hurricanes by duration in 1-day intervals (bottom). Most hurricanes retain major hurricane intensity for 1 to 2 days. Ten storms over the period have lasted as major hurricanes for more than a week.

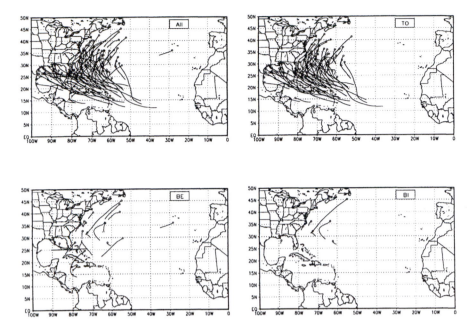

Fig. 7.10: Tracks of North Atlantic major hurricanes over the period 1886–1996 by classification. TO are tropical-only, BE are baroclinically-enhanced, and BI is baroclinically-initiated hurricanes. Only the portion of the hurricane track in which the hurricane was initially at category three or higher is shown. Note the lack of major hurricane activity over the extreme northeastern Gulf of Mexico.

over the northeastern Gulf of Mexico is conspicuous. The peninsula of Florida inhibits northwesterly tracking storms from entering this region as intense hurricanes.

Considering the entire track of hurricanes that become intense, Evans and McKinley (1997) note that a majority of hurricanes reach maximum intensity within 24 hours of recurvature with a preference for intensification to lead recurvature. This confirms the observation made in Simpson and Riehl (1980) that hurricanes tend to reach their greatest intensity near the time of recurvature. Major baroclinically-enhanced hurricanes tend to track over the western North Atlantic away from the coastline. Hurricane *Andrew* in 1992 was a notable exception. Most major baroclinically-enhanced hurricanes have a significant northward component to their forward motion. This is expected as these storms tend to intensify in association with a progressing middle-latitude baroclinic system. The atmospheric flow ahead of the system pushes the hurricane northward.

7.7   Major U.S. Hurricanes

Winds in excess of 50 m s^{-1} and storm surges exceeding 3 meters characterize major landfalling hurricanes. The Saffir/Simpson scale (Table 2.2) is used to warn coastal res-

Table 7.3: Intense U.S. hurricanes. The list is for the period 1900–1996. The symbol † indicates the hurricane was moving in excess of 30 mph, ‡ indicates winds and tides did not justify category three, and # indicates that the hurricane was classified category four because of extremely high tides. Categories and pressures refer to values at landfall. Note 1 mb = 0.0295 inches of mercury. Source is the National Hurricane Center (NHC).

Rank	Hurricane or Location	Year	Category	Millibars	Inches
1	Florida Keys	1935	5	892	26.35
2	*Camille*	1969	5	909	26.84
3	*Andrew*	1992	4	922	27.23
4	Florida Keys and TXs	1919	4	927	27.37
5	Lake Okeechobee, FL	1928	4	929	27.43
6	*Donna*	1960	4	930	27.46
7	Galveston, TX	1900	4	931	27.49
7	Grand Isle, LA	1909	4	931	27.49
7	New Orleans	1915	4	931	27.49
7	*Carla*	1961	4	931	27.49
11	*Hugo*	1989	4	934	27.58
12	Miami, MS and AL	1926	4	935	27.61
13	*Hazel*	1954	4†	938	27.70
14	$FLse$	1947	4	940	27.76
15	TXn	1932	4	941	27.79
16	*Gloria*	1985	3†‡	942	27.82
16	*Opal*	1995	3‡	942	27.82
18	*Audrey*	1957	4#	945	27.91
18	Galveston, TX	1915	4#	945	27.91
18	*Celia*	1970	3	945	27.91
18	*Allen*	1980	3	945	27.91
22	New England	1938	3†	946	27.94
22	*Frederic*	1979	3	946	27.94
24	Northeast	1944	3†	947	27.97
24	Carolinas	1906	3	947	27.97

idents and emergency management of their potential destructiveness. Major hurricanes rank as category three or higher indicating the possibility for extensive to catastrophic damage. Florida is the most vulnerable U.S. state, but major hurricanes can occur anywhere from Texas to Maine. Significant inter-decadal variability of major U.S. hurricanes is noted. A hurricane that moves across the coastline of the United States as a category three storm is called a major U.S. hurricane. Here we examine some statistics of major U.S. hurricanes.

Table 7.3 ranks the twenty-five most intense U.S. hurricanes on record by minimum central pressure. The intensity is estimated at landfall. Leading the list is the devastating 1935 hurricane that struck the Florida Keys. The storm knocked out road and rail access to the Keys and resulted in a major loss of life. The central pressure at landfall was estimated at 892 mb. This pressure is only 4 mb higher than the lowest pressure ever recorded for any North Atlantic hurricane. Both hurricanes *Camille* in 1969 and *Andrew* in 1992 had central pressures at landfall estimated below 925 mb. Hurricane *Allen* in 1980 reached category five intensity three times through the Caribbean and Gulf of Mexico. *Allen* went through several periods of rapid intensification. The lowest pressure reported in *Allen* was 899 mb (26.55 in) on August 7, 1980 off the northeastern tip of the Yucatan Peninsula. At landfall along the southern Texas coast, *Allen* had a central pressure estimated at 945 mb. Central pressures in *Hugo* dropped from 970 mb to 918 mb in 24 hr when it was east of the Lesser Antilles. Hurricane *Opal* deepened from 965 mb to 916 mb in 16 hr over the central Gulf of Mexico. Each decade of the 20th century is represented in the list of the most intense U.S. hurricanes.

7.7.1 Decadal Variability

The intensities of storms at landfall are generally not available until the late 19th century. The best-track data provide Saffir/Simpson scale at landfall for hurricanes beginning in 1899. Here storms back to 1891 are considered by estimating the category based on the once-per-day wind speed estimates. Following Hebert and Taylor (1979b), the tracks of major hurricanes striking the United States in 10-year intervals beginning with 1891 are shown in Figures 7.11 through 7.16. Included, for each track, is the year and the Saffir/Simpson scale at landfall. Multiple landfalls from a single storm have multiple category listings. Names of the hurricanes are also provided. Of all U.S. hurricanes, 39% are category three or higher at landfall, and 12% are category four or higher. Only two category five hurricanes have made landfall in the United States over the past 106 years. Ten of the 11 decades had at least one category four or higher hurricane make landfall. The exception is the 1970s. As noted in Hebert and Taylor (1979b), there is a tendency for intense hurricanes to cluster in particular areas during certain decades. This pattern is evident in these figures.

In the last decade of the 19th century, U.S. major hurricanes clustered along the southeast coast and Florida. The exception was 1900, which saw a category four hurricane strike Galveston, Texas. The Galveston hurricane caused 10,000 deaths, the worst loss of life of any U.S. hurricane on record (see Chapter 17). This hurricane foreshadowed two decades of activity concentrated along the northwestern Gulf coast and southern Florida. The exception was a category three storm that punished North and South Carolina in 1906. The first two decades of the 20th century saw a total of 13 major U.S. hurricanes. The 1919 hurricane make landfall over Key West, Florida and Corpus Christi, Texas as a category four storm resulting in a significant loss of life. During the 1920s the focus of major hurricane activity shifted to Florida and the northeastern Gulf coast. Florida was particularly hard hit with four major storms in nine years. Three of the storms hit southeastern Florida. Major U.S. hurricanes during the 1930s were scattered from Texas to New England. A 1933 hurricane struck southern Texas as a category three storm. A 1935 storm crossed the Florida Keys as a category five hurricane and a 1938 category three hurricane battered the northeast states of New York, Connecticut, Rhode Island, and Massachusetts.

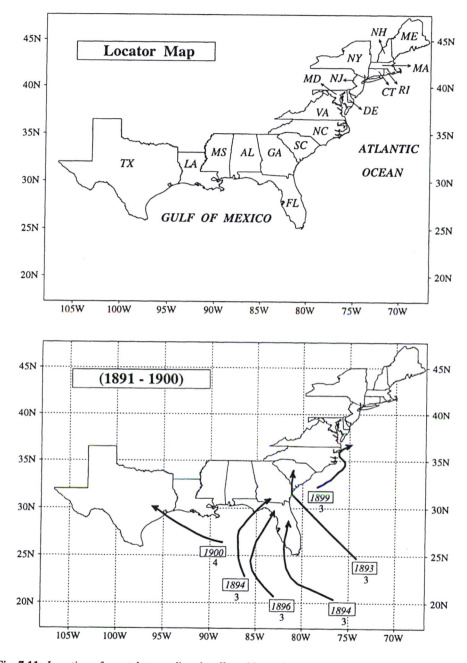

Fig. 7.11: Location of coastal states directly affected by major hurricane landfalls (top). Tracks of major U.S. hurricanes over the period 1891–1900 (bottom). The entire track is not necessarily shown. The number under the year is the Saffir/Simpson category at landfall. Florida was particularly hard hit during the last decade of the 19th century.

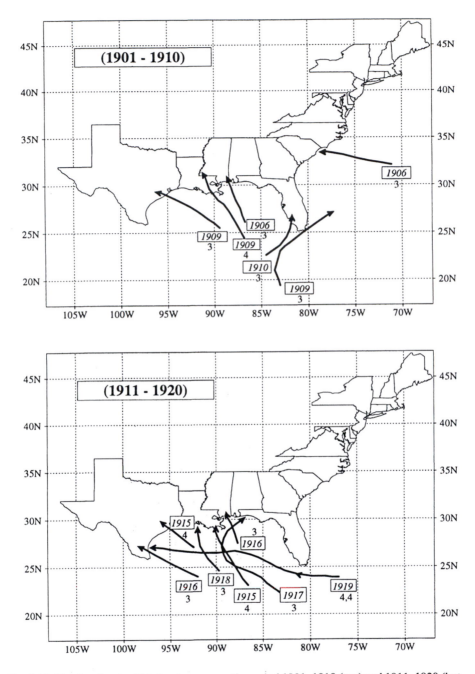

Fig. 7.12: Tracks of major U.S. hurricanes over the period 1901–1910 (top) and 1911–1920 (bottom). The entire track is not necessarily shown. The number under the year is the Saffir/Simpson category at landfall. The northwestern Gulf coast was particularly hard hit.

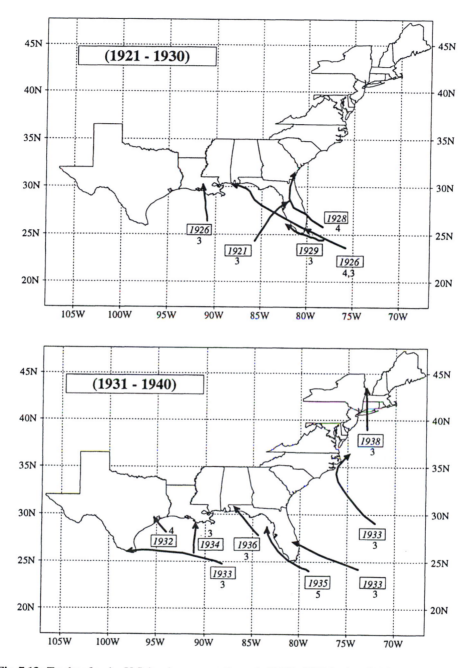

Fig. 7.13: Tracks of major U.S. hurricanes over the period 1921–1930 (top) and 1931–1940 (bottom). The entire track is not necessarily shown. The number under the year is the Saffir/Simpson category at landfall. Florida was again the focus for much of the major hurricane activity. Florida was hit by three major hurricanes in 4 years, two of which hit the southern part of the state.

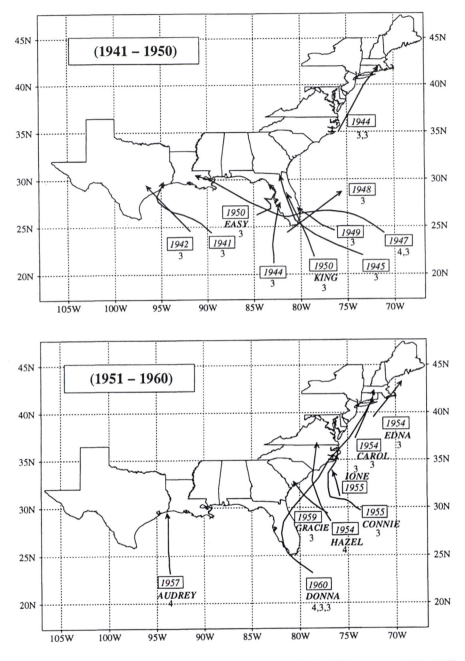

Fig. 7.14: Tracks of major U.S. hurricanes over the period 1941–1950 (top) and 1951–1960 (bottom). The entire track is not necessarily shown. The number under the year is the Saffir/Simpson category at landfall. Major hurricane activity along the U.S. coastline peaked during two decades.

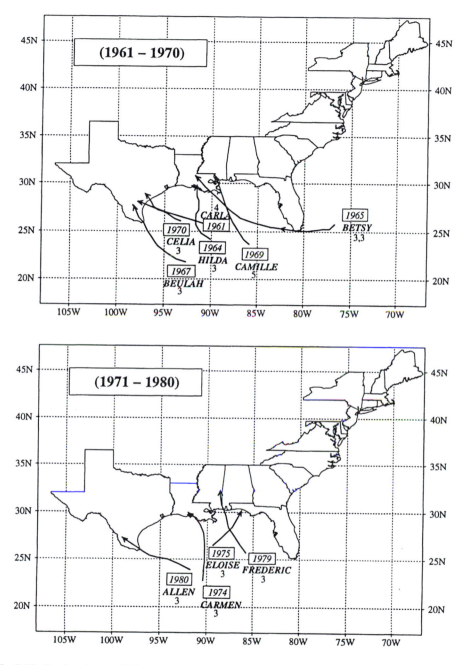

Fig. 7.15: Tracks of major U.S. hurricanes over the period 1961–1970 (top) and 1971–1980 (bottom). The entire track is not necessarily shown. The number under the year is the Saffir/Simpson category at landfall. Major hurricane activity was again focused along the northwestern Gulf coast.

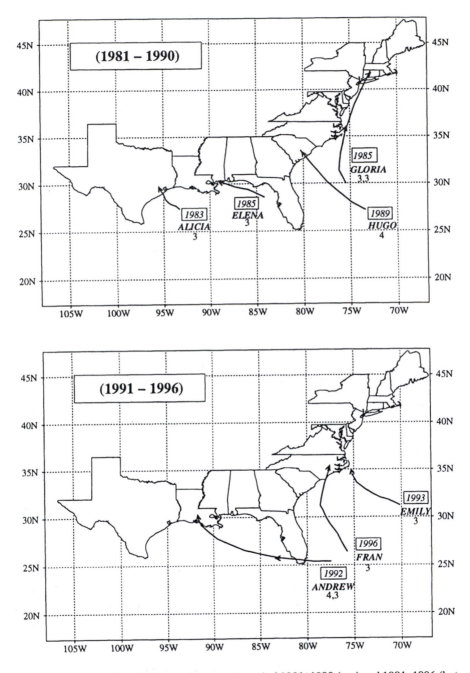

Fig. 7.16: Tracks of major U.S. hurricanes over the period 1981–1990 (top) and 1991–1996 (bottom). The entire track is not necessarily shown. The number under the year is the Saffir/Simpson category at landfall. Major hurricane activity along the U.S. coastline has diminished.

Table 7.4: Decadal frequencies of major U.S. hurricane strikes. A single, major hurricane strik-ing two different regions gets counted in both regions. For example, the 1992 hurricane *Andrew* gets counted as a major hurricane for both Florida and Louisiana. *LA–AL* represents the states of *LA, MS*, and *AL*, *GA–NC* represents the states of *GA, SC*, and *NC*, and *VA–ME* represents the states of *VA, MD, NJ, NY, CT, RI, MA, NH*, and *ME*.

Years	TX	LA–AL	FL	GA–NC	VA–ME
1891–1900	1	0	3	2	0
1901–1910	1	2	2	1	0
1911–1920	3	3	2	0	0
1921–1930	0	2	4	0	0
1931–1940	2	1	3	1	1
1941–1950	2	1	7	1	1
1951–1960	2	1	1	5	3
1961–1970	3	3	1	0	0
1971–1980	1	2	1	0	0
1981–1990	1	1	1	2	1
1991–1996	0	1	1	2	0
1891–1996	16	17	26	14	6

The 1940s were marked by an increase in the number of major U.S. hurricanes that continued through the 1960s. This was also a time when major hurricanes were more common over the North Atlantic basin as a whole. Florida was particularly hard hit again. Seven major storms smashed the peninsula in seven years beginning in 1944. The two major Florida hurricanes in 1950 (*Easy* and *King*) ended a 31-year period, beginning in 1919, when Florida averaged one major hurricane strike every other year. This level of activity is in marked contrast to the more recent decades of few major hurricanes in Florida. Texas, North Carolina, and New York were also hit hard during the 1940s. In the 1950s, U.S. major hurricane activity shifted nearly exclusively to the east coast. During 1954 and 1955 seasons alone, five major hurricanes hit the United States between North Carolina and Massachusetts. The exception was hurricane *Audrey* which struck near the Texas-Louisiana border in 1957.

The 1960s and 1970s saw major hurricane activity return to the northwestern Gulf coast. Most activity occurred west of Florida. The 1960s saw hurricane *Carla* strike Texas as a category four hurricane and hurricane *Camille* strike Louisiana as the only other category five hurricane the 20th century. Similar to the 1930s, major hurricane activity since 1981 has been scattered with no preferred location. Major storms struck from Texas to New England during the 1980s. Hurricane *Hugo* was the only category four landfall during this decade. Hurricane *Andrew* devastated parts of southeastern Florida as a category four hurricane during the otherwise tranquil early and middle 1990s. Hurricane *Fran* in 1996 was the last major hurricane to strike the United States in the 111-year record.

Table 7.4 shows the frequency of major hurricane landfalls by region in 10-year increments. Florida and the Gulf coastal states of Texas, Louisiana, Mississippi, and Alabama experience the greatest threat from major hurricanes. On average Florida gets hit by at least two major hurricanes a decade, while Texas averages about three major hits every 2 years. However, there is large inter-decadal variability. In fact, Florida has not seen a multiple-hit decade since the 1950s. The stretch of coast from Louisiana to Alabama also averages about three major hurricane landfalls every 2 years. The northeast coastal states from Virginia to Maine average one major hurricane every two decades. However, during the 1950s this region was hit three times in 7 years, twice in 1954 alone.

An inverse relationship between the occurrence of major hurricanes in Florida and along the east coast from Georgia to Maine is apparent. Decades that feature above average activity in Florida tend to feature below average activity along the east coast. Cumulatively there were 21 and 7 major hurricane strikes to Florida and the east coast, respectively over the period 1891–1950. In contrast, over the period from 1951–1996, there were 5 strikes to Florida and 13 strikes to the east coast. A notable difference between major hurricane activity in Texas and Florida is the degree of interdecadal variability. The interdecadal standard deviation in major hurricane frequency is 1.9 for Florida and 1.0 for Texas. This suggests that Florida is more vulnerable to hurricane disasters as Florida hurricane activity is more sensitive to swings in the climate compared to Texas hurricane activity. Changes in large-scale climate factors tend to influence the frequency of significant Florida hurricanes to a greater extent than their influence on major Texas hurricanes.

7.7.2 Annual and Seasonal Variability

Annual and seasonal frequencies of major U.S. hurricanes since 1891 are shown in Figure 7.17. A major hurricane that makes two landfalls is counted as one major U.S. hurricane. The average annual frequency is 0.63 or about five major hurricanes every 8 years. Roughly half (54) of the 106 years have no major landfalling hurricane. The years 1909, 1933, and 1954 each had three major hurricanes. There is a slight, statistically insignificant, negative trend in major U.S. hurricanes due to the absence of multiple strike years in the more recent decades. In fact, with the exception of 1985, no year since 1956 has had more than one major hurricane hit the United States in a season.

Major U.S. hurricanes are most likely during the peak of the hurricane season. Interestingly, the probability increases from early August and reaches a maximum during the second two weeks of September. In fact, two-thirds of all major U.S. hurricanes occur during the 2-week period at the end of September. This peak is shifted a few weeks later compared to the peak in frequency of all North Atlantic hurricanes (see Chapter 4). Outside this peak 2-week period most major hurricane landfalls occur during late August and early September. The likelihood of a strike in October is considerably lower, but the second two weeks are as likely as the first two weeks of the month. Late-season major hurricanes out of the Caribbean sometimes pose a threat to Florida. No

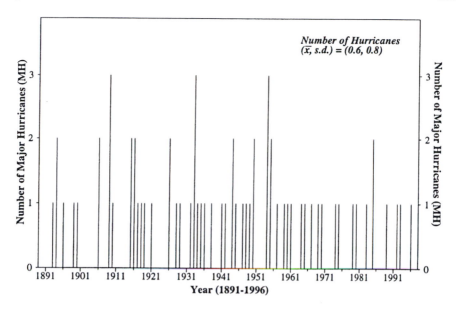

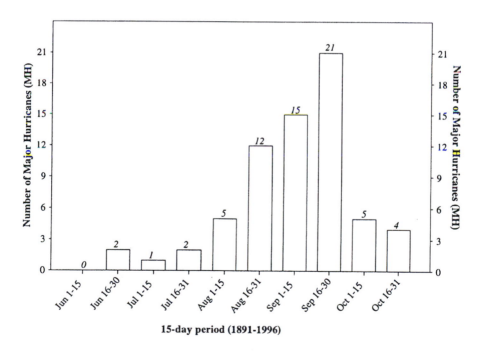

Fig. 7.17: Annual number of major U.S. hurricanes (top) and the seasonal distribution of major landfalling hurricanes in 15-day intervals (bottom) over the period 1891–1996. Fewer multiple-hit years are noted after 1955. A major U.S. hurricane is unlikely before August.

major hurricanes have hit the United States before June 16th. The latest hurricane in a season to make landfall in the United States as a category three or higher occurred nearly Tampa Bay, Florida on October 25, 1921.

7.7.3 Tracks by Region

The complete tracks of hurricanes that have hit the United States as major hurricanes over the period 1886 through 1996 by region are shown in Figure 7.18. As expected, major hurricanes affecting the Texas coast originate southeast of the coastline [Figure 7.18 (a)]. About half of the major hurricanes originate in the Gulf of Mexico while the other half begin in the Caribbean Sea or the western North Atlantic. Most of the hurricanes start south of 25°N and east of 90°W. On average, the storm heading (direction of approach) near landfall is perpendicular to the coastline. Figure 7.18 (b) shows major hurricanes of the northern Gulf coast from Louisiana to Alabama tend to originate in the western Caribbean Sea and the southern Gulf of Mexico. Similar to major hurricanes of Texas most of the storms develop south of 25°N and east of 90°W. A few of the hurricanes originate in the western North Atlantic. Again, the storm heading at landfall is generally perpendicular to the coastline.

Major hurricanes of western Florida have two rather distinct areas of formation [see Figure 7.18 (c)]. One is located to the west of 80°W over the eastern Gulf of Mexico and western Caribbean Sea. These storms approach from the south and turn eastward into Florida from the west. Another area of formation is over the western and central North Atlantic. These hurricanes sometimes strike southeastern Florida before crossing the peninsula to menace the western coastline of Florida. The heading at landfall is quite variable from storm to storm. A major hurricane approaching southwestern Florida from the south has the potential to cause significant problems as evacuation routes through the peninsula are in the direction of storm motion. Over eastern Florida, the threat of major hurricanes comes from the east, the south, and the southeast [Figure 7.18 (d)]. Northeastern Florida has not been hit by a major hurricane in the past 100 years. However, hurricane *Dora* was a strong category two or possibly a weak category three storm when it made landfall near St. Augustine early in September, 1964 (Barnes 1998).

Major hurricanes of the southeastern U.S. coastline typically form south of 22°N and east of 65°W as seen from Figure 7.18 (e). In particular, the region northeast of Puerto Rico is a favored breeding ground for these hurricanes. Unlike hurricanes that sometimes threaten Florida, the storms striking the southeast coast originate several thousand kilometers from the shoreline. This allows greater time for preparation. The general heading of hurricanes at landfall is nearly perpendicular to the coast of South Carolina, but is more tangential to the coastline of North Carolina. This is partly due to the orientation of the coastlines and partly due to recurvature of storms farther north. As in northeastern Florida, no major hurricane has struck the Georgia coastline in the past 100 years. However, according to Ludlum (1963), it is likely that the southeast U.S. coastline between St. Augustine, *FL*, and Savannah, *GA*, was hit be intense

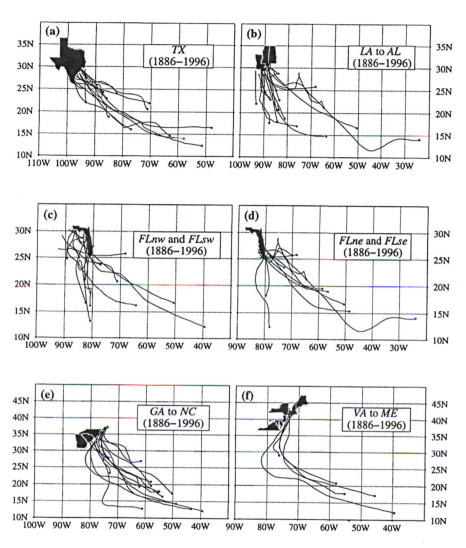

Fig. 7.18: Tracks of hurricanes that have resulted in major U.S. landfalls in Texas (a), from Louisiana to Alabama (b), in western Florida (*FLnw, FLsw*) (c), in eastern Florida (*FLse, FLne*) (d), from Georgia to North Carolina (e), and from Virginia to Maine (f). The solid circle indicates the origin of hurricane-force winds. The majority of major landfalling storms originate as hurricanes south of 20°N latitude. This is true even for the northeastern coastline. The northern Gulf coast is threatened by storms that intensify rapidly over the western Caribbean Sea and southern Gulf of Mexico. Western Florida is vulnerable to storms approaching from a variety of different directions. Note that axis scaling is different for each panel.

hurricanes in 1824 and 1837. Major hurricanes of the northeastern U.S. coast also tend to originate or track northeast of Puerto Rico [Figure 7.18 (f)]. The tendency is for these hurricanes to recurve to the north or northeast between 25 and 30°N latitude, then continue northward until landfall in southern New England. These hurricanes obtain a significant forward motion before striking the coast. Long Island, New York is particularly vulnerable to intense storms as the coastline is nearly perpendicular to the direction of hurricane movement. According to Foster and Boose (1992), major hurricanes struck New England in 1635, 1788, 1815, and 1869.

8

U.S. Hurricanes

Landfalling hurricanes cause significant social and economic hardships. Understanding when and where they occur is important for planning and mitigation. This chapter examines the statistics of U.S. landfalling hurricanes. Landfall is observed when all or part of the hurricane eye wall (see Chapter 1) crosses the shoreline. Landfalling hurricanes have the greatest potential to do damage and inflict casualties. The impacts are most acute along the immediate coastline and barrier islands. Over the open ocean, small islands are particularly vulnerable to a disaster even when the eye of the hurricane stays out to sea. Major hurricanes that skirt the coast can cause considerable damage to property even if the center remains out over the ocean. Storm surge and wind are responsible for much of the damage along the immediate coastline. Flooding rains are a significant concern inland.

The frequency of U.S. hurricanes is uniform over the past 150 years. Regionally, however, there are differences. Hurricane hit rates in south Florida appear to fluctuate on various time scales, whereas they are relatively stable along the Gulf coast. Hurricane strikes along the Texas coastline tend to occur early in the hurricane season, while strikes along the Florida coastline tend to occur in September and October. Most U.S. hurricanes are from tropical-only hurricanes and most strike as category one storms. We examine these facts in more detail here.

Christopher Columbus was fortunate in avoiding tropical cyclones on his maiden voyage to the New World.[1] In 1495, the town of Isabella, on the north coast of Hispaniola, became the first European settlement destroyed by a hurricane. During August of 1508, the Spanish explorer Ponce de León experienced two tropical cyclones in less than 14 days. The first storm forced his ship onto the rocks in the port of Yuna, Hispaniola, while the second washed his vessel ashore along the southwest coast of Puerto Rico (Hughes 1987). Since this early time, there have been numerous accounts of hurricane impacts in the Caribbean and North America from travelers and settlers. From these records it is possible to extract useful statistical information on early hurricane frequencies. We begin with a description of the data sources.

---

[1]He did, however, do battle with several of them on later trips.

8.1   Data Sources

According to Ludlum (1963), William C. Redfield was the first to comprehensively consider the occurrence, frequency, and spatial distribution of U.S. tropical cyclones.[2] His careful blending of data from ship logs, newspaper accounts and interviews with ship crews allowed him to detail the occurrence of tropical cyclones in early America. Subsequently, Andres Poey y Aguirre of Havana compiled a chronological list of 400 hurricanes that was first published in the *Journal of the Royal Geographical Society* of London in 1855. Other chronicles of U.S. hurricanes include Blodget (1857), Lapham (1872), and Loomis (1876). According to E. B. Garriot in his 1900 study, *West Indian Hurricanes*, nearly 400 hurricanes affected the Caribbean region and Florida over the period 1493 to 1870 (Williams and Duedall 1997).

In 1929, Joseph Henry of the U.S. Weather Bureau published a comprehensive list of American hurricanes (Henry 1929) based on F. G. Tingley's annual hurricane maps. I. R. Tannehill used Tingley's charts in his 1938 hurricane book. Tannehill's book includes updates to the track maps through later editions until 1950. Gordon E. Dunn and Banner I. Miller revised and expanded the growing list of U.S. hurricanes in their 1960 book *Atlantic Hurricanes*. David M. Ludlum's treatise *Early American Hurricanes: 1492–1870*, published in 1963 by the American Meteorological Society as part of a series "The History of American Weather," remains the authoritative chronology of the earliest U.S. hurricanes. The work is devoted to accounts of tropical cyclones in America back to the time of Columbus. The historical documentation of North Atlantic tropical cyclones begins with records kept by 15th century Europeans during explorations of the New World.

More recently, Fernández-Partágas made important contributions to our historical knowledge of hurricanes. His research focused on uncovering and interpreting written accounts of tropical cyclones over the North Atlantic from ship logs, newspapers, and other sources. The work updates and adds to the tracks of hurricanes over the period 1851 through 1900 (see Fernández-Partágas and Diaz (1996c). For instance, the *New York Times* reports of damage and casualties often contain enough detailed to reconstruct the location and intensity of a hurricane at landfall. Arguably this work provides justification to extend the U.S. hurricane record back to 1851. For the years that overlap the best-track record (1886–1900) we refer to the updates provided in Fernández-Partágas and Diaz (1995a, 1995b, 1996a, 1996b).

A concatenation of the Fernández-Partágas and Diaz data set and the best-track data set serves as the primary source for most of the tables and graphs in this chapter. Because the greater uncertainty associated with the accounts of hurricanes in Ludlum (1963) these data are used only as a secondary source. As Ludlum (1963) notes, the judgment of whether a particular tropical cyclone reached hurricane intensity is largely subjective since instruments to measure wind speeds were not in widespread use until the 1870s. He notes that uprooted trees, major structural damage to buildings,

---
[2]Redfield's work appeared in the *American Journal of Science and the Arts* in 1831.

Table 8.1: Data sources for U.S. hurricanes. The range of years over which the sources are consulted is provided.

Number	Range of Years	Author(s)	Usage
1	1901–1996	Neumann et al. (best-track)	Primary
2	1851–1900	Fernández-Partágas and Diaz	Primary
3	1492–1850	Ludlum	Secondary

toppled ship masts or boats on beam-ends are considered sufficient evidence of hurri-cane force winds. During early times most coastal areas lacked economic development so reports of damage were scarce. Table 8.1 lists the data sets that were consulted for use in this chapter. The best-track data are used over the period 1901–1996 and the Fernández-Partágas and Diaz data are used over the earlier period of 1851–1900.

8.2 Landfalls

Several notes are needed concerning hurricane landfalls. First, landfall occurs when all or part of the hurricane eye wall passes directly over the coast or over the adjacent barrier islands. Since the eye wall extends outwards a radial distance of 50 km or more from the hurricane center, landfall can occur even if the exact center of lowest pressure remains offshore. Moreover, by the above definition, landfall never occurs at a single location. Thus, when grouping landfalls by state it is common for the hurricane to be counted as striking more than a single state at once. For counting strike frequencies, a hurricane that makes landfall near the border of two states is counted as a single U.S. hurricane, though it gets counted as a strike for both state totals. Furthermore, a hurricane that crosses a coastline and then moves on to strike another part of the coast is counted as a single U.S. hurricane with two strikes. Storms that retain hurricane intensity as they move inland to affect states other than the coastal state are counted as hits for those states.

Second, it is important to realize that hurricane size (see Chapter 1) can vary con-siderably between individual storms. The size dictates the spatial extent of the damage. A large hurricane can leave a damage swath 200 km wide as it moves directly onshore. A hurricane that moves tangent to the coastline can produce an even larger corridor of damage. The patterns of wind, rainfall, and storm surge are often asymmetric about the hurricane eye. Wind and storm surge are generally higher in the right side of the hurricane as viewed toward the direction of motion. Because the circular motion of the hurricane is counterclockwise (cyclonic), the right side is where the forward speed of motion enhances the rotational speed.

Third, the date of hurricane landfall for each state is interpolated from the track charts of Neumann et al. (1993) and Fernández-Partágas and Diaz (1995a, 1995b, 1996a, 1996b) with a new day starting at 12 am Eastern Standard Time (EST). There is

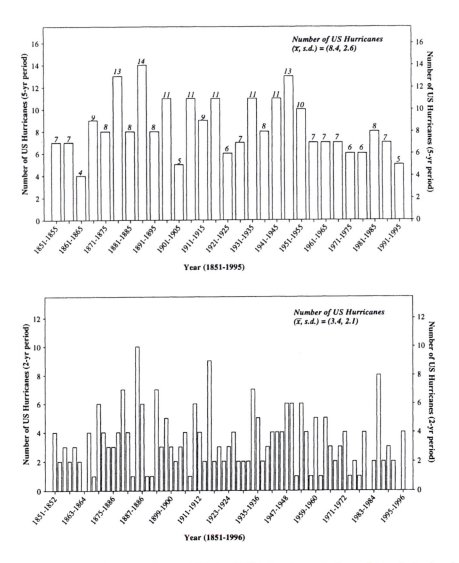

Fig. 8.1: Time series of the abundance of U.S. landfalling hurricanes in 5-year intervals (top) and 2-year intervals (bottom) over the period 1851–1996. The average number ($\overline{x}$) of U.S. hurricanes making landfall and the standard deviation ($s.d.$) are given in the upper-right corner of the graphs.

some uncertainty as to the precise date for some hurricanes, but particularly for those where landfall occurs between observation times. In general, U.S. landfalls include any hurricane that crosses the coast between Port Isabel, Texas (near Brownsville) and Eastport, Maine.

8.3   Frequencies

Here we consider the frequency of North Atlantic hurricanes hitting the U.S. coastline during the period 1851–1996. A single hurricane striking the U.S. coast more than once (as *Andrew* did in Florida and Louisiana in 1992) is counted as a single U.S. landfalling hurricane (U.S. hurricane). The total number of U.S. hurricanes over this period is 246. The frequency of U.S. hurricanes in 5-year intervals beginning with 1851–1855 is shown in Figure 8.1. The frequency of U.S. hurricanes is remarkably constant over the period. This observation combined with the fact that there is no trend in the intensity of hurricanes argues against the claim that climate change is causing the observed rises in hurricane related property losses (see Chapter 18). The average number of landfalling hurricanes in these 29 pentads is 8.4 with a standard deviation of 2.6. The variability ranges from a minimum of four landfalling hurricanes from 1861–1865 to a maximum of 14 from 1886–1890. It should be noted that there are no above-average pentads since 1955.

The biennial frequency shows similar characteristics with no important overall trend, but a slight tendency for fewer landfalling hurricanes in the more recent decades. The maximum 2-year total is 10 landfalling hurricanes during 1885 and 1886. The most recent period of heightened landfall activity is 1985–1986. Notice that the active 1995 and 1996 hurricane seasons produced only an average number of U.S. landfalling storms. Simpson and Riehl (1980) note the decline in U.S. hurricanes starting around 1950 and suggest that it might be associated with the intrusion of middle-latitude westerlies into the tropics of the western North Atlantic Ocean and Caribbean Sea. Indeed, in Chapter 6, we saw that baroclinic hurricane activity is on the upswing as of late. From the vantage of a longer record, the frequency of U.S. hurricanes appears to be rather stable.

The annual time series and distribution of U.S. hurricanes are shown in Figure 8.2. The average is 1.7 hurricanes per year (approximately five U.S. hurricanes in 3 years) with a standard deviation of 1.4 and a maximum of seven in 1886 (see also Table 8.2). There were 6 U.S. hurricanes in 1916 and 6 in 1985. There were 5 hurricanes in 1893 and 5 in 1933. In 30 of the 146 years (20.5%) there are no U.S. hurricane landfalls. The most recent year of no landfalls is 1994. The longest streak of years without a U.S. hurricane is three from 1862–1864. The longest streak of years with at least one landfall is 13 from 1938–1950. The most common strings of successive years with landfalls is two and three. As expected, the most likely number of hurricanes is one followed closely by two. Three hurricanes hit the United States in a single year 26 times with ten of these occurring in the period 1944 through 1959.

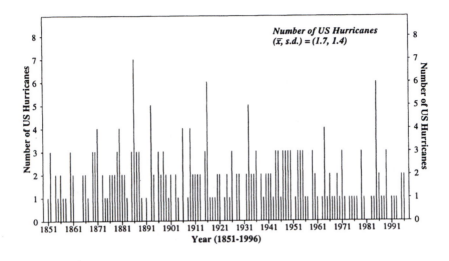

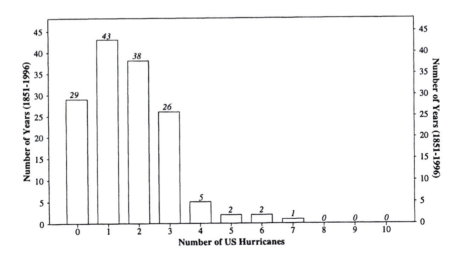

Fig. 8.2: Time series of the annual abundance of U.S. landfalling hurricanes (top) and the distribution of annual abundance (bottom) over the period 1851–1996. Notice that there are more years with two U.S. hurricanes than years with no strikes. The average ($\bar{x}$) and standard deviation (s.d.) are given in the upper-right corner of the graph.

Table 8.2: U.S. landfalling hurricanes. Values are annual numbers over the period 1851–1996. A single hurricane that makes landfall more than once is counted only once. The total number of hurricanes making landfall in the United States over the 146-year period is 246. The greatest number in a single year is seven in 1886.

Year	180	181	182	183	184	185	186	187	188	189
0	.	.	.	.	.	.	3	3	4	0
1	.	.	.	.	.	1	2	4	2	1
2	.	.	.	.	.	3	0	0	2	0
3	.	.	.	.	.	0	0	2	1	5
4	.	.	.	.	.	2	0	1	0	2
5	.	.	.	.	.	1	2	1	3	0
6	.	.	.	.	.	2	2	2	7	3
7	.	.	.	.	.	1	1	2	3	2
8	.	.	.	.	.	1	0	2	3	3
9	.	.	.	.	.	0	3	3	1	2

Year	190	191	192	193	194	195	196	197	198	199
0	1	2	2	0	2	3	2	1	1	0
1	2	2	2	0	2	0	1	3	0	1
2	0	2	0	2	2	1	0	1	0	1
3	2	2	1	5	1	3	1	0	1	1
4	1	0	2	2	3	3	4	1	1	0
5	0	3	1	2	3	3	1	1	6	2
6	4	6	3	3	1	1	2	1	2	2
7	0	1	0	0	3	1	1	1	1	.
8	1	1	2	2	3	0	1	0	1	.
9	4	1	2	1	3	3	2	3	3	.

The observed frequency of U.S. landfalling hurricanes by month and their associated probabilities are given in Table 8.3. As seen, September is the most active month having a slightly better than even chance of at least one landfalling hurricane each year. However, the probabilities of two or more U.S. hurricanes in a month are nearly equal for August and September at approximately 10%. This is the same probability of observing at least one hurricane in June. The probability of at least one U.S. hurricane in July is only slightly greater than the probability for June. The probability of at least one October hurricane landfall is nearly 30%, but the chance of two U.S. hurricanes in October is only about 2%.

The monthly distribution of U.S. hurricanes is shown in Figure 8.3. September is the peak month followed in order by August, October, and July. The three peak months account for 83% of all U.S. landfalling hurricanes. Both June and July are more active than November as late-season hurricanes tend to track on a more northeasterly path away from the United States. The earliest hurricane to hit the United States was *Alma* which struck northwest Florida on June 9, 1966, while the latest strike occurred on November 30, 1925 near Tampa, Florida (Hebert and Taylor 1979a).

Table 8.3: U.S. landfalling hurricanes by month. Values are frequencies and probabilities based on data over the period 1851–1996. The whole numbers are the number of years with that many hurricanes.

Number of U.S. Hurricanes	Jun	Jul	Aug	Sep	Oct	Nov
0	132	128	95	67	103	141
1	13	16	40	64	40	5
2	0	1	11	13	3	0
3	1	1	0	2	0	0
Observed Probability	Jun	Jul	Aug	Sep	Oct	Nov
At least 1 hurricane	0.10	0.12	0.35	0.54	0.29	0.03
2 or more hurricanes	0.01	0.01	0.08	0.10	0.02	0.00
3 or more hurricanes	0.01	0.01	0.00	0.01	0.00	0.00

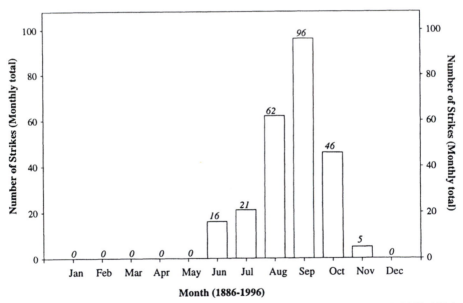

Month (1886-1996)

Fig. 8.3: Monthly distribution of U.S. landfalling hurricanes over the period 1851–1996. September is the peak month followed in order by August and October. Landfalling hurricanes are rare in November. Together, June and July account for 15% of all U.S. hurricanes. August has 3 times as many landfalls as July. September is the peak month followed by August and October. U.S. hurricanes are rare in November. Landfalls during June and July tend to occur along the Gulf coast, especially over Texas and Louisiana. Florida sees a majority of its hurricane landfalls after August.

Table 8.4: U.S. versus North Atlantic hurricane activity. Values are given for years of above-average North Atlantic hurricane activity in the three categories of all, tropical-only (TO), and baroclinically-enhanced (BE) and compared with the number of U.S. landfalling hurricanes (U.S.) over the period 1886–1996. Average values ($\bar{x}$) for each column are also given.

Year	All	U.S.	Year	TO	U.S.	Year	BE	U.S.
1887	10	3	1886	8	7	1887	5	3
1893	10	5	1893	8	5	1934	5	2
1916	11	6	1916	8	5	1959	6	3
1933	10	5	1933	10	5	1966	6	1
1950	11	3	1950	8	3	1969	7	2
1955	9	3	1955	9	3	1976	5	1
1969	12	2	1995	9	2	1977	5	1
1980	9	1	1996	8	2	1980	5	1
1995	11	2	.	.	.	1984	5	1
1996	9	2	.	.	.	1990	5	0
$\bar{x}$	10.2	3.2	$\bar{x}$	8.5	4.0	$\bar{x}$	5.4	1.5

8.4   Relation to Basin-Wide Activity

Since the number of U.S. hurricanes is bounded from above by the total number of hurricanes, it is reasonable to expect that the abundance of U.S. landfalls is correlated to total basin-wide activity. In general, if total activity is high, then the number of U.S. hurricanes is also above average. This is verified by noting that the average number of landfalling hurricanes is 3.2, with a standard deviation of 1.6 in the 10 years with nine or more North Atlantic hurricanes. This compares to the overall average of 1.7 landfalling hurricanes per year. This relationship between basin-wide activity and landfalls is even stronger for tropical-only hurricanes. For the 8 years with eight or more tropical-only hurricanes the average number of U.S. landfalling hurricanes is 4.1, with a standard deviation of 1.9. In contrast, years in which baroclinically-enhanced hurricane activity is above normal, the number of U.S. hurricanes is near the overall average. Considering the 10 years with five or more baroclinically-enhanced hurricanes, the average number of U.S. hurricanes is 1.5 and the standard deviation is 1.0. Thus the expected relationship between heightened activity and U.S. hurricanes is partially explained through variations in tropical-only activity. Caution is urged in interpreting these statistics since they are influenced by a few years with many landfalls.

Table 8.4 lists the years of high activity for all, tropical-only, and baroclinically-enhanced hurricanes and the corresponding total number of U.S. landfalling hurricanes. Half of the eight most active tropical-only seasons have five or more U.S hurricanes, while none of the 10 most active baroclinically-enhanced years have more than three U.S. hurricanes. Again, caution should be used in generalizing from this table, particularly in light of the observation that the ratio of U.S. hurricanes to basin-wide activity

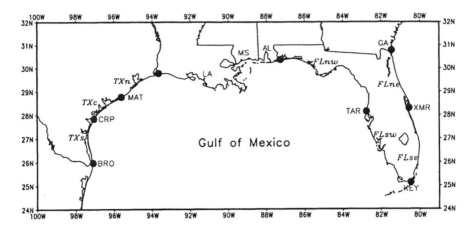

Fig. 8.4: Map of the U.S. Gulf coast from Texas to Florida. The dots represent the regional borders of the coastline for Texas and Florida. BRO is Brownsville, CRP is Corpus Christi, MAT is Matagorda Bay, TAR is Tarpon Springs, KEY is Key Largo, and XMR is Cape Canaveral. The coastal states are labeled with their two-letter abbreviations. TXs refers to the south Texas coast, TXc to the central Texas coast, and TXn to the north Texas coast. $FLnw$ refers to the northwest Florida coast, $FLsw$ to the southwest Florida coast, $FLse$ to the southeast Florida coast, and $FLne$ to the northeast Florida coast.

seems to be lower for the more recent years, which is consistent with the likelihood that some hurricanes at great distance from land were not observed during the earliest years of the record. This fact is also consistent with a greater number of the more recent North Atlantic hurricanes being baroclinically-enhanced.

8.5   Decadal Variability

The probability of a hurricane varies from state to state. Here we consider landfalls by state in 10-year intervals. Because of the extended length of coastlines in Florida and Texas, these states are subdivided into regions. For Texas, the southern region extends from Brownsville to Corpus Christi, the central region from Corpus Christi to Matagorda Bay, and the northern region from Matagorda Bay to the Louisiana border (see Figure 8.4). For Florida, the northwest region extends from the Alabama border to Tarpon Springs, the southwest region extends from Tarpon Springs to Key Largo at the southern tip of the peninsula, the southeast region from Key Largo to Cape Canaveral, and the northeast region from Cape Canaveral to the Georgia border. Southwestern Florida includes the Florida Keys. Since a hurricane landfall is sometimes felt in more than one state, state totals cannot be added to obtain a national total. Similarly regional totals in Texas and Florida do not add to state totals. Other things being equal, hurricane landfall frequency depends on the length of coastline considered.

8.5.1 Texas

The frequency and distribution of hurricanes along the Texas coast over the period 1851–1996 are considered here. Table 8.5 gives the number of landfalls in 10-year intervals beginning with 1851–1860 for the north (TXn), central (TXc), and southern (TXs) Texas coasts. Decadal activity ranges from a minimum of one in the first decade to a maximum of seven in two consecutive decades starting in 1931. However, activity over the 10-year period beginning in 1931 is concentrated along the southern Texas coast while the activity over the next 10 years is focused in the north. Overall the north coast is slightly more vulnerable to hurricane strikes than the south or central regions. North Texas was hit by 5 hurricanes during the period 1941–1950 and 4 hurricanes during the 1980s.

Figure 8.5 shows the time series of hurricane landfalls in Texas by region. Each dot represents one landfall. An open circle above the dot indicates a year with two hurricane strikes. As seen, the time interval between successive hits (recurrence interval) varies considerably due to a clustering of events. This is especially true for the northern and southern coastlines. The early 1940s were active along the north coast. The south coast was hit by six hurricanes in 11 years over the period from 1909 through 1919. As defined here, the central Texas coastline has fewer hurricane strikes than either the north coast or the south coast. North Texas was hit twice in a single year during 1886 and again during 1989. On average multiple-hit years are less frequent in Texas than in Florida or North Carolina.

The rate of hurricane landfalls is better illustrated in Figure 8.6, which shows the cumulative frequency of Texas landfalls by region over the 146-year period. The horizontal axis is the year and the ordinate is the cumulative frequency of strikes beginning with the 1851 season. On average the hit rate (as indicated by the slope of the frequency curve) along the north Texas coast is fairly uniform at approximately two hits per decade. Active and break periods are clearly indicated. The hit rate was higher over the years beginning in 1940. The break in activity during the 1960s and 1970s is evident. The hit rate for the central Texas coast is also uniform at about four hits every 30 years. The absence of central Texas hurricanes from the last decade of the 19th century through the first two decades of the 20th century represents a departure from the average. Another break in activity is noted at the end of the record. Hurricane *Fern* in 1971 was the last hurricane to make landfall along the central Texas coastline.

The cumulative distribution of landfalls along the south Texas coast is considerably less uniform. The hit rate varies substantially over the 146-year period. Four active and five break periods are apparent. During active periods, hit rates are close to six hits per decade, while during break periods landfalls are absent. The active period beginning in 1909 is evident as is the period in the early 1930s. An extended break of several decades with no landfalls in south Texas is noted beginning in 1937. It is likely that active and break periods are related to low-frequency changes in climate. If this is true, then hurricane activity in north Texas is apparently insensitive to the same climate factors affecting the south.

Table 8.5: Texas hurricanes. Values are the number of hurricane landfalls by region in 10-year intervals over the period 1851–1996. Regional totals are also given. A dot indicates no hurricanes. North Texas (TXn) was hit quite frequently during the 1940s and again in the 1980s.

Year	TXn	TXc	TXs	10-Year Total
1851–60	.	1	.	1
1861–70	1	1	.	2
1871–80	1	1	2	4
1881–90	2	2	2	6
1891–00	2	1	.	3
1901–10	1	.	2	3
1911–20	1	.	4	5
1921–30	.	2	.	2
1931–40	2	1	4	7
1941–50	5	2	.	7
1951–60	2	.	.	2
1961–70	1	1	2	4
1971–80	.	1	1	2
1981–90	4	.	.	4
1991–96	.	.	.	.
1851–1996	22	13	17	52

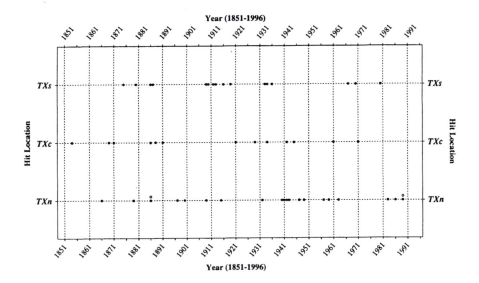

Fig. 8.5: Time series of hurricane landfalls in Texas by region over the period 1851–1996. Each dot represents one landfall. Circles above the dot indicate multiple-hit years.

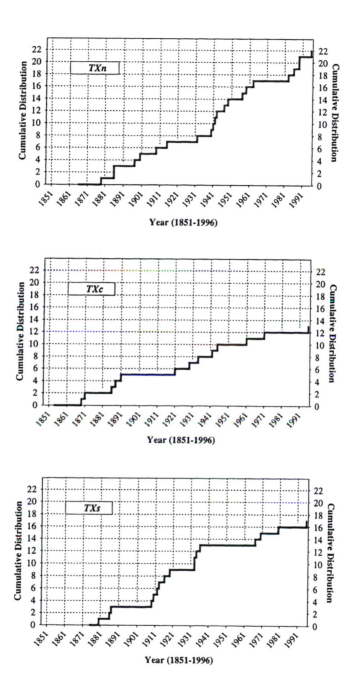

Fig. 8.6: Cumulative frequency of hurricane landfalls in Texas by region over the period 1851–1996. Larger slopes correspond to more frequent strikes. Hit rates are indicated by the slope of the line. Hit rates along the south Texas coastline are quite variable over the period.

8.5.2   Florida

The state of Florida, with 2171 kilometers of coastline,[3] leads the nation in frequency of hurricane strikes. Low-lying terrain and the fact that no location is more than 113 km from the coast makes the entire state vulnerable to a hurricane catastrophe. The early Spanish explorers, including Ponce de León, Panfilo de Narváez, and Tristán de Luna, were the first to chronicle the impact of hurricanes on the state. A 1528 expedition, led by Narváez in the Tampa Bay area, was shipwrecked along the northwestern coast of Florida in a what was likely hurricane. Spanish fleets were destroyed frequently by hurricanes during the 1540s and 1550s. The Tampa Bay hurricane of 1559 killed at least 600 onboard 13 ships. Historical accounts of Florida hurricanes are given in Henry et al. (1994) and in Barnes (1998).

Table 8.6 gives the number of landfalls in 4 regions of Florida in 10-year intervals beginning with 1851. The coastal shape makes multiple-hit hurricanes more likely in Florida than anywhere else. For example, hurricane *Allison* in 1995 struck the southeastern coast of Florida then re-intensified over the northeastern Gulf of Mexico before striking northwestern Florida. Decadal hurricane landfall activity as a whole ranges from a low of three in the 1860s and 1980s to a high of 17 during the 1940s. Hurricane strikes were also quite numerous during the 1920s, a decade that featured only 2 Texas hurricanes. With the exception of the 1960s, the decades since the 1940s have had fewer than 6 landfalls per ten-year period.

Northeastern Florida gets hit the least often. The region was hit by 3 hurricanes during the 1920s and only 4 hurricanes since then through 1996. Northwestern Florida was hit by 8 hurricanes during the 1880s. Southeastern Florida was hit by 8 storms in the 1940s. Figure 8.7 shows the time series of hurricane landfalls in Florida by region. The time interval between successive strikes is not uniform, particularly for the southwestern and southeastern Florida. This clustering of strikes suggests that successive landfalls are not independent. Overall hurricane activity across Florida was high during the 1920s, 1930s, and 1940s but was tranquil from the late 1960s through the 1980s. A terrible hurricane struck the Florida Keys in 1935. The storm was the first category five U.S. hurricane of the 20th century. The eye passed over both Long Key and Lower Metecumbe Key during the evening of September 2nd.

Hurricane hit rates in Florida are inferred from the cumulative distributions shown in Figure 8.8. As a generalization, the hit rate in northwestern and northeastern Florida is uniform, particularly after 1886. The rate is nearly three strikes per decade in the northwest and about one per decade in the northeast. The hit rate for northeast Florida is substantially lower due to a short, concave coastline. The hit rate in northwestern Florida was higher in the period from 1871 through 1890. In contrast to the north, hurricane landfalls in the south show nonuniform cumulative distributions indicative of a clustering in landfall occurrences. In particular, southeastern Florida hurricanes

---

[3]The Atlantic coastline of Florida measures 933 km and the Gulf coastline of Florida measures 1239 km. Measurements are made with a unit measure of 30 minutes of latitude on charts as near the scale of 1:1,200,000 as possible based on data provided by the U.S. National Oceanic and Atmospheric Administration.

Table 8.6: Florida hurricanes. Values are the number of hurricane landfalls by region in 10-year intervals over the period 1851–1996. A dot indicates no hurricanes. The southern coastline of Florida ($FLse$ and $FLsw$) was particularly hard hit during the 1940s.

Year	$FLnw$	$FLne$	$FLsw$	$FLse$	10-Year Total
1851–60	3	.	2	.	5
1861–70	.	.	3	.	3
1871–80	7	.	3	2	12
1881–90	8	.	.	1	9
1891–00	4	.	2	.	6
1901–10	1	.	1	4	6
1911–20	4	.	2	.	6
1921–30	4	3	5	4	16
1931–40	3	.	2	4	9
1941–50	3	1	6	8	17
1951–60	2	1	1	.	4
1961–70	3	2	3	3	10
1971–80	2	1	.	1	4
1981–90	2	.	1	.	3
1991–96	2	.	1	2	5
1851–1996	48	8	32	29	117

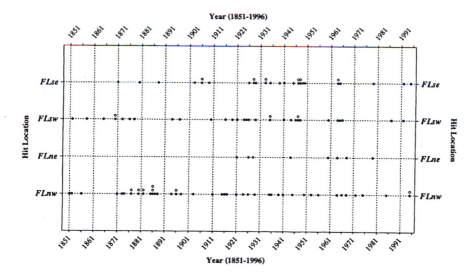

Fig. 8.7: Time series of hurricane landfalls in Florida by region over the period 1851–1996. Each dot represents one landfall. Circles above the dot indicate multiple-hit years.

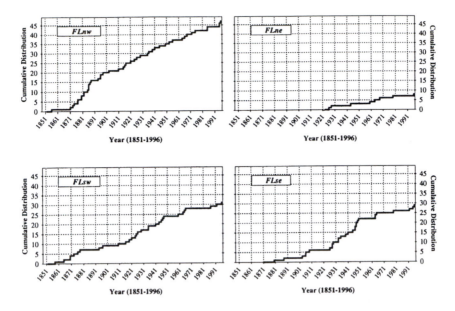

Fig. 8.8: Cumulative frequency of hurricane landfalls in Florida by region over the period 1851–1996. Hit rates are indicated by the slope of the line. Larger slopes correspond to more frequent strikes. Hit rates along the southern Florida coastlines are quite variable.

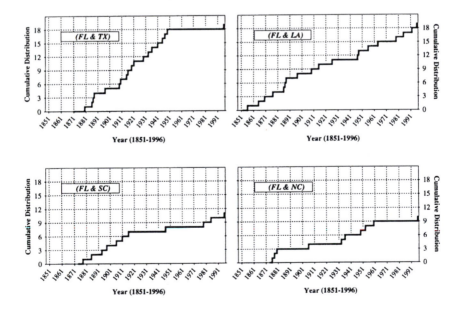

Fig. 8.9: Cumulative distribution of years with at least one hurricane strike in Florida and at least one strike in Texas, Louisiana, North and South Carolina. Note that dual-state hurricane years appear at a steady rate for Florida and Louisiana, but not for Florida and North Carolina. Florida and Texas have not been hit by a hurricane in the same year since 1941.

show active and break periods lasting a decade or more. The most pronounced active period in this part of the state occurs from the middle 1920s through the middle 1940s. The period from the middle 1960s through the 1980s represents a prolonged break of few strikes. In southwestern Florida hurricane landfalls appear to come in waves of activity at all time scales suggestive of scaling behavior (fractals). The interdecadal variance of hurricane landfalls over northwestern Florida is smaller than the variance of landfalls over southern Florida. From an insurance perspective (see Chapter 19), the coastline of southern Florida is a more risky investment even though overall the hit rates are similar. This is because the active and break regimes make it difficult to assign a single hit rate to the coasts of southern Florida when estimating return periods (see Chapter 11). The different hit rates in different epochs suggest that the frequencies of south Florida hurricanes might be more sensitive to changes in climate than the frequency of hurricanes in north Florida.

It is interesting to consider years in which two different states that do not share a common border get hit by at least one hurricane each. A dual-state hit can occur from a single hurricane or from separate hurricanes occurring in the same season. In both situations the years are counted as dual-state hurricane years. Like the September 1919 storm, some of the strongest hurricanes have struck both Florida and Texas. Figure 8.9 gives the cumulative distribution of dual-state years for Florida versus Texas, Louisiana, South Carolina, and North Carolina. Considering Florida and Texas, the first dual-state hurricane season was 1871. This occurrence was followed by a steady rate of approximately two dual-state (FL & TX) years per decade until 1940. From 1941 through 1996 there are no seasons with hurricanes in both Florida and Texas. Instead the period is marked by alternating years of Florida and Texas hurricanes. The frequency of dual-state hurricane seasons is more regular for Florida and Louisiana as indicated by an overall straight line in the cumulative distribution. The line corresponds to about 2 dual-state years per decade. Dual-state years for Florida and South Carolina occur episodically. The active period from 1874 to 1916 was replaced by an inactive stretch through the remainder of the record. Florida and North Carolina show a similar cumulative distribution. Four dual-state seasons occurred in seven years between 1874 and 1880. After which years of Florida hurricanes alternated with years of North Carolina storms. Several dual-state seasons occurred from the early 1930s through the 1950s.

8.5.3   Louisiana to Maryland

The frequency of hurricane landfalls along the Gulf and southeast U.S. coasts excluding Texas and Florida are given in Table 8.7. For reference, Figure 8.10 is a map of the U.S. east coast from South Carolina to Maine. Among the coastal states, Louisiana gets hit the most often followed by North and South Carolina. With the exception of Virginia and Maryland, most states get hit by at least one hurricane each decade. Particularly active decades include the 1850s and 1970s in Louisiana and the 1870s and 1950s in North Carolina. Georgia and South Carolina were hit quite frequently

Fig. 8.10: Map of the U.S. east coast from South Carolina to Maine. The coastal states are labeled with their two-letter abbreviations.

during the 1890s. The 1920s saw no hurricanes in North Carolina. This was a decade which featured relatively few storms in Texas but frequent hurricanes in Florida.

Time series of hurricane landfalls for each state are shown in Figure 8.11. The temporal distribution of strikes for each state is roughly uniform. The exceptions are North and South Carolina. Both states tend to get hit in bunches. For instance, North Carolina was hit by six hurricanes in the 3-year period 1952–1954 followed by no strikes in the next 3 years. Most states see one or two landfalls per decade, but get hit hard during an active decade. Louisiana was hit a record 5 times over the period 1851–1860. The sum of landfalls over these states is fairly constant on the decadal scale with the exception of the 1860s and again during the 1960s when landfalls were rather infrequent.

Hurricane hit rates for these states are shown in Figure 8.12. Average hit rates as indicated by the slope of the distribution vary for each state, but are generally uniform in time with the noted exception of North and South Carolina. Louisiana in particular shows a strikingly linear distribution at a rate of three hits per decade. Hit rates are lower for Mississippi, Alabama, Georgia, and South Carolina. North Carolina's hit rate is less uniform with occasional periods of heightened hurricane activity. A partial

Table 8.7: Hurricanes in states from Louisiana to Maryland excluding Florida. Values are the number of hurricane landfalls in 10-year intervals. A dot indicates no hurricanes. State totals are also given. North Carolina was particularly hard hit during the 1950s.

Year	LA	MS	AL	GA	SC	NC	VA	MD
1851–60	5	4	1	1	.	1	.	.
1861–70	2	.	1	.	.	2	.	.
1871–80	2	.	.	1	2	5	.	.
1881–90	4	2	1	1	2	3	.	.
1891–00	3	1	.	3	4	2	.	.
1901–10	2	2	1	.	2	3	.	.
1911–20	3	1	4	1	2	2	.	.
1921–30	2	.	1	1	1	.	.	.
1931–40	3	.	1	1	1	3	1	.
1941–50	2	1	1	1	1	3	.	.
1951–60	3	1	.	.	4	7	1	.
1961–70	3	1	.	.	.	.	.	.
1971–80	4	1	2	1	1	1	.	.
1981–90	4	1	1	.	2	3	1	.
1991–96	1	.	.	.	.	3	.	.
1851-1996	43	15	14	11	22	38	3	0

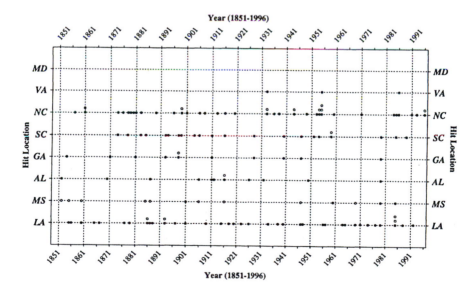

Fig. 8.11: Time series of hurricane landfalls in states from Louisiana to Maryland over the period 1851–1996. Each dot represents one landfall. Circles above the dot indicate multiple-hit years.

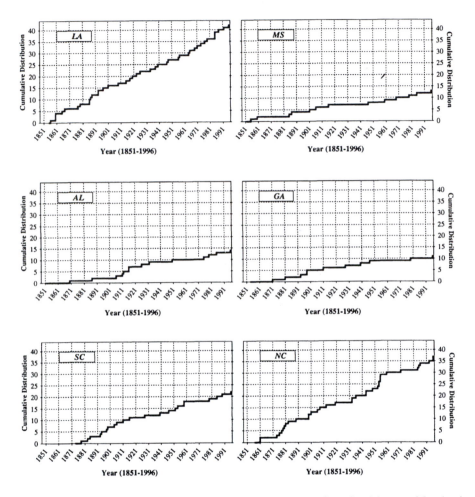

Fig. 8.12: Cumulative distribution of hurricane strikes in states from Louisiana to Maryland over the period 1851–1996. Hit rates are indicated by the slope of the line with larger slopes corresponding to more frequent hurricane strikes. Various periods show different hit rates. The hit rate for Louisiana is remarkably uniform over the 146-year period. North Carolina's hit rate varies considerably over the 146-year period. Hurricanes were most frequent in North Carolina during the middle 1870s into the early 1880s and again during the early 1950s.

Table 8.8: Average annual number of hits ($\bar{x}$) by state or region and the percentage of hits from tropical-only (TO) hurricanes. The number of cases (N) is the total number used in the averages and percentages. Averages are based on data over the period 1851–1996, while percentages are based on data over the period 1886–1996.

Strike (Hit) Location		Hits (1851–1996)		TO (1886–1996)	
		$\bar{x}$	N	(%)	n
Texas north	TXn	0.15	22	60.0	12
Texas central	TXc	0.09	13	60.0	6
Texas south	TXs	0.15	17	86.7	13
Louisiana	LA	0.29	42	66.7	22
Florida northwest	$FLnw$	0.32	47	82.4	28
Florida northeast	$FLne$	0.05	8	100	8
Florida southwest	$FLsw$	0.21	31	65.2	15
Florida southeast	$FLse$	0.20	29	81.5	22
North Carolina	NC	0.25	37	75.0	21
New York	NY	0.09	13	80.0	8

explanation centers on the dichotomy in North Atlantic hurricanes. The frequency of tropical-only and baroclinically-enhanced hurricanes is inversely related. A year with many tropical-only hurricanes tends to be a year with fewer baroclinically-enhanced hurricanes (see Chapter 6). Thus there is some compensation in North Atlantic hurricane activity on the interannual time scale. Years in which climatological conditions favor tropical-only hurricanes tend to be years in which conditions are less favorable for the development of baroclinically-enhanced hurricanes. Since North Carolina (as well as much of Florida) is affected primarily by tropical-only hurricanes (see Table 8.8), hit rates are more sensitive to changes in tropical climate. In contrast, Louisiana is hit by both tropical-only and baroclinically-enhanced hurricanes. Changes in climate affecting the frequency of tropical-only hurricanes do not substantially alter the overall hit rate as there is a tendency for an increase in the frequency of baroclinically-enhanced hurricanes. This may well be the case in other parts of the Gulf coast including northern and central Texas as well as southwestern Florida.

8.5.4   Delaware to Maine

Hurricane landfalls for coastal states north of Maryland are tabulated by decade in Table 8.9. Since the coastlines of these states are relatively short, a hurricane tends to make landfall simultaneous over several states. Thus hurricane activity between states is not independent. Furthermore, as the states are located north of where most Atlantic hurricanes turn to the north, the frequency of strikes is less. The orientation of Long Island makes New York quite vulnerable to hurricane landfalls. Connecticut also receives its share of hurricanes. It is rare for a state to get hit more than once in 10 years. The 1950s were by far the most active decades on record for the northeast coastline.

Table 8.9: Hurricanes in states from Delaware to Maine. Values are the number of hurricane landfalls in 10-year intervals. State totals are also given. Hurricanes along the northeast U.S. coast were quite numerous during the 1950s.

Year	DE	NJ	NY	CT	RI	MA	NH	ME
1851–60	.	.	1	1	1	.	.	.
1861–70	.	1	2	1	1	.	.	1
1871–80	.	.	.	.	.	.	.	.
1881–90	.	.	.	.	.	.	.	.
1891–00	.	.	2	2	.	1	.	1
1901–10	.	1	1	1	.	.	.	.
1911–20	.	.	.	.	.	1	.	.
1921–30	.	.	.	.	.	.	.	.
1931–40	.	.	1	1	1	.	.	.
1941–50	.	.	1	1	1	1	.	.
1951–60	.	.	2	2	2	3	1	3
1961–70	.	.	.	.	.	.	.	1
1971–80	.	.	1	.	.	.	.	.
1981–90	.	.	1	1	.	.	.	1
1991–96	.	.	1	1	1	1	.	.
1851–1996	0	2	13	11	7	7	1	7

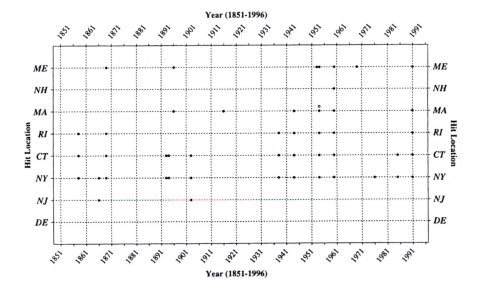

Fig. 8.13: Time series of hurricane landfalls in states from Delaware to Maine over the period 1851–1996. Each dot represents one landfall. Circles above the dot indicate multiple-hit years.

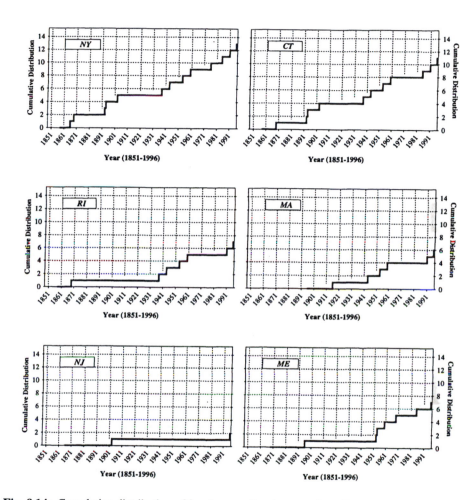

Fig. 8.14: Cumulative distribution of hurricane strikes in states from Delaware to Maine over the period 1851–1996. Hit rates are indicated by the slope of the line. Larger slopes indicate more frequent hurricanes. Various periods show different hit rates. The hit rate for New York fluctuates considerably including a period from 1904–1937 without a strike. New Jersey has only been hit twice over the 146-year period.

Table 8.10: U.S. hurricane strikes by region. Values are annual frequencies and probabilities. Florida experiences the most multiple-hit years of any of the regions. *LA–AL* denotes the coastal region from Louisiana to Alabama, *GA–NC* the coastal region from Georgia to North Carolina, and *VA–ME* the coastal region from Virginia to Maine.

Number of Hurricanes	*TX*	*LA–AL*	*FL*	*GA–NC*	*VA–ME*
0	102	99	83	99	126
1	39	40	37	34	18
2	4	5	24	12	1
3	0	2	2	1	1
4	1	0	0	0	0
Observed Probability	*TX*	*LA–AL*	*FL*	*GA–NC*	*VA–ME*
At least 1 strike	0.30	0.32	0.43	0.32	0.14
2 or more strikes	0.03	0.05	0.18	0.09	0.01
3 or more strikes	0.00	0.01	0.01	0.00	0.00

Time series of northeast landfalls by state are shown in Figure 8.13. Here we see the strong interrelationship of landfalls between states as a single hurricane often affects more than one state. Decadal hit rates (Figure 8.14) are quite low in comparison to the Gulf or southeast states. Landfalls appear to be somewhat nonuniformly distributed over time, although the infrequency of storms makes it difficult to make any firm conclusions. Low-frequency shifts in climate factors likely influence the probability of landfalls along this stretch of coastline.

8.6   Annual and Seasonal Frequencies

The expected occurrence of hurricane landfalls by region are shown in Table 8.10. The empirical annual probability of at least one hurricane in a season is greatest for Florida. The annual probability along the northern Gulf and southeast coasts are also quite high at 32%. The probability of exactly one strike in a season is larger for Texas than for Florida. Florida tends to have a higher number of multiple-hit years. This is indicative of a clustering of hurricanes on the intraannual time scale. A similar clustering was noted previously for portions of Florida on the interannual time scale. The probability of two or more Florida strikes in a given season is better than 18%. Thus, given a landfall in Florida during the year, there is a 41% chance that it will get hit again later in the same year. This compares with Texas which has an 11% chance of being hit again in the same year. The southeast coast is also susceptible to multiple-hit years. North Carolina was hit by three separate hurricanes in 1955 (*Connie, Diane,* and *Ione*). Hurricanes *Bertha* and *Fran* hit North Carolina in 1996. As expected, the northeast (VA to ME) is considerably less likely to see a hurricane. The annual probability of at least one hit somewhere along this stretch of coastline is 14%. The probability of two or more hurricanes in a season is 1%.

Table 8.11: Monthly frequencies and percentages of U.S. strikes by region. The total number of cases (N) is given along with the percentage (%) of all U.S. strikes for that region and month. *LA–AL* denotes the coastal region from Louisiana to Alabama, *GA–NC* the coastal region from Georgia to North Carolina, and *VA–ME* the coastal region from Virginia to Maine.

Month	Cases	TX	$LA–AL$	FL	$GA–NC$	$VA–ME$
Jun	N	9	3	6	0	0
	(%)	2.5	0.8	1.7	0.0	0.0
Jul	N	7	4	6	6	1
	(%)	1.9	1.1	1.7	1.7	0.3
Aug	N	17	18	17	22	15
	(%)	4.7	5.0	4.7	6.0	4.1
Sep	N	15	35	47	33	30
	(%)	4.1	9.6	12.9	9.0	8.3
Oct	N	4	12	36	9	5
	(%)	1.1	3.3	9.9	2.5	1.4
Nov	N	0	0	5	1	0
	(%)	0.0	0.0	1.4	0.3	0.0

Monthly frequencies and percentages of hurricane landfalls by region are given in Table 8.11. The percentages indicate the ratio of hurricane landfalls for the region to the total number of landfalls. For instance, the nine June hurricanes of Texas represent 2.5% of all U.S. strikes. Early season (June and July) hurricanes are most likely in Texas than in any of the other four regions. A hurricane strike along the northeast coast is unlikely before August. In August the frequency of hurricanes is nearly uniform over the coastal segments, with a slight preference for landfalls along the southeast coast. Percentages peak in September for each region except Texas. Most hurricane landfalls during this month occur over Florida. The preference for Florida hurricanes continues into October. If a hurricane hits the United States in October, it is more likely to be in Florida than anywhere else and it is 9 times more likely to be Florida than Texas. Florida gets hit by tropical-only and baroclinically-enhanced hurricanes. September and October are the most active months for Florida hurricanes. Strong tropical-only hurricanes threaten Florida from the southeast. Hurricanes originating over the western Caribbean during October tend to track in the direction of Florida. The hurricane that hit North Carolina on November 1, 1861 is the only hit outside of Florida during the month of November.

8.7 Landfalls by Category

Hurricanes are classified into five categories according to their potential for damage. The Saffir/Simpson scale (see Chapter 2) relies on maximum sustained wind speed, central pressure, and storm surge to estimate hurricane potential. As the wind speed (v) of a hurricane increases the energy of the wind increases by v^2. A doubling of wind

Table 8.12: Distribution of U.S. landfalling hurricanes by the Saffir/Simpson scale (SS). Winds are given in knots (kt). A disproportionate number of hurricanes are of category three.

SS Category	Winds (kt)	Frequency
5	≥ 135	2
4	115–130	14
3	100–110	48
2	85–95	34
1	65–80	60
Total	≥ 65	158

speed from 35 to 70 m s^{-1} represents a quadrupling of the wind energy. However, the power to do damage is energy over time sustained for a distance, which increases by v^3. Maximum sustained one-minute averaged winds in a hurricane are estimated in 5-knot intervals. Categories one and four encompass the largest range of wind speeds. Category five is not bounded from above.

Table 8.12 shows the distribution of U.S. hurricanes by category at landfall based on data from 1900 through 1996. For hurricanes that make landfall more than once, only the category of the first landfall is included. Thirty-eight percent are minimal category one hurricanes at landfall. The majority of hurricanes (59%) are either category one or two at landfall. Landfalling category four or five hurricanes comprise 10% of all landfalls. Only two category five hurricanes have made landfall in the United States during the 20th century. There is a disproportionate number of category three landfalls. Indeed, there are more category three strikes than category two strikes. Assuming no significant data bias, this may suggest hurricanes with wind speeds in the range of 90 kt are less stable. At this intensity a hurricane will tend to either weaken or strengthen.

8.8   Preferred Paths

Different hurricanes striking the same stretch of coastline tend to approach along similar paths. For example, Texas hurricanes typically approach through the northern Caribbean Sea into the Gulf of Mexico. In contrast, hurricanes of northeastern Florida prefer a path through the western Caribbean Sea. Here the paths of hurricanes approaching the coastline are examined based on data over the period 1886 through 1996. The Gulf and Atlantic coasts are divided into 16 separate latitude-longitude boxes of dimension 2.5°. The 16 boxes are numbered in Figure 8.15 (top). The frequencies of hurricanes appearing in each box over the 111-year period are given in Figure 8.15 (bottom). The boxes represent threat regions (threat boxes) for the corresponding coastal locations. Box 8, covering southern Florida, is visited most frequently by hurricanes. An earlier analysis of this kind by Hope and Neumann (1971) considered all tropical cyclones. Here only hurricanes are considered.

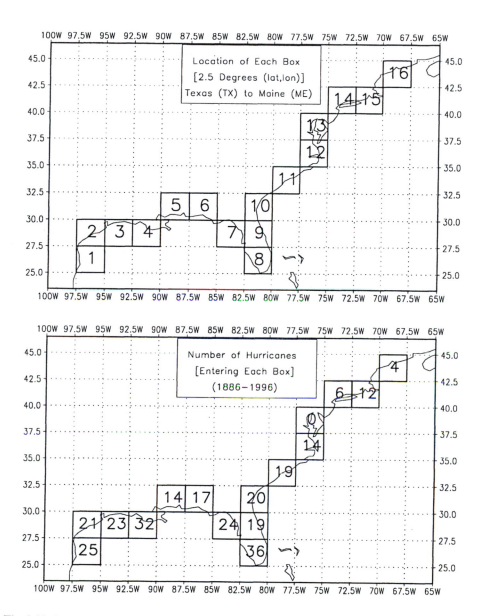

Fig. 8.15: Location of the threat boxes along the U.S. coastline (top). The number in the box corresponds to the number used in Figures 8.16 through 8.30. The number of hurricanes appearing in each box over the period 1886–1996 is written in the box (bottom).

As an example, Figure 8.16 shows the spatial distribution of empirical probabilities in 2.5° latitude by 2.5° longitude boxes for hurricanes entering threat boxes 1 (top) and 2 (bottom). A hurricane located over the central regions of the Gulf of Mexico in the area bounded by 90 and 92.5°W longitudes and by 20 and 22.5°N latitudes has a 33% chance of threatening the south Texas coastline. That is, one-third of all hurricanes observed in this region of the Gulf pass into threat box 1. A contiguous area of high percentage is interpreted as a preferred path for hurricanes approaching this part of the coastline. Total percentages in the boxes may exceed 100% as some storms threaten more than one coastal region. Hurricanes passing through adjacent threat boxes do not represent independent events.[4] The preferred path provides an overview of where hurricanes are likely to be before they approach the coast. The information does not necessarily translate into a good forecast of future hurricane tracks. For instance, the number of hurricanes appearing in many of the grids is small, resulting in low statistical confidence. High percentages for some boxes are merely a reflection of few hurricane visits to the region. Moreover, many hurricane landfalls result from tropical storm intensification close to the coast.

Boxes 1 and 2 include the south and central coasts of Texas, respectively. Hurricanes affecting this part of the Gulf coast tend to approach from the east and south through the northern Caribbean Sea. Tropical storms intensifying over the southern and central Gulf of Mexico also pose a hurricane threat to Texas. The Texas coast is most vulnerable to hurricanes that track over the Lesser Antilles and continue westward through the northern Caribbean. In particular, hurricanes passing near Jamaica have a relatively high chance of threatening the Texas coastline. As expected, probabilities are greatest for areas immediately adjacent to the threat box. Figure 8.17 on the facing page shows selected cities and population densities for coastal counties in threat boxes 1 and 2. Population data are from the U.S. Bureau of the Census. Major cities in these two regions include Brownsville, Corpus Christi, Houston, and Galveston. Population densities are highest for the coastal counties of Cameron and Galveston.

Figures 8.18 and 8.19 pertain to threat boxes 3 and 4, which includes the northern Texas coast and most of coastal Louisiana. A large percentage of hurricanes of the southern and western Gulf of Mexico move toward this part of the Gulf coast. In particular, hurricanes originating over the Bay of Campeche have a tendency to strike southeastern Louisiana, but not the northern Texas coast. Hurricanes visiting the western Gulf occur early in the season. October hurricanes in this region are relatively uncommon. In fact most hurricanes (68%) making landfall within 150 miles of Lake Charles, Louisiana (Calcasieu Parish) are either the first, second, or third tropical storm of the season (Roth 1998). Early season hurricanes tend to approach from origins in the western and southern Gulf of Mexico. These storms provide only a few days to prepare. Major cities in these regions include Port Arthur and New Orleans. Population densities are greatest in Jefferson County, Texas and Jefferson Parish, Louisiana.

Hurricanes menacing the central Gulf coast approach from the south and southeast

---

[4]The area of the boxes decreases with increasing latitude.

(Figure 8.20). Most threats come from the south and central Gulf of Mexico having originated over the Caribbean Sea or the southwestern North Atlantic. A percentage of hurricanes of the western Caribbean Sea eventually threaten the central Gulf coast. A percentage of hurricanes passing over the Bahamas make their way to the northwestern coastline of Florida. Large percentages in the grid boxes located south of 15°N over the tropical North Atlantic are due the rare occurrence of hurricanes in these regions. Principal cities of these regions include Mobile, Pensacola, and Panama City (Figure 8.21). Coastal population densities are fairly high from St. Tammany Parish in Louisiana to Okaloosa County in Florida.

Hurricane threats to western and southern Florida come from storms tracking through the Bahamas, the Florida Straits, and the western Caribbean Sea (Figure 8.22). South Florida is particularly vulnerable to hurricanes from the Caribbean Sea and the tropical Atlantic. Hurricane *Inez* in 1966 tracked straight across the Florida Keys from Key Largo to Key West. Hurricanes of the far western Gulf of Mexico are not a threat to western Florida. Hurricanes moving through the southeastern Gulf will threaten Ft. Myers and St. Petersburg. Population densities are quite high along much of the Florida peninsula south of Levy County. This region experienced some of the largest growth rates during the 1990s. Pinellas is Floridas most densely populated county. Major cities in these threat boxes include Tampa, St. Petersburg, Fort Myers, Miami, and Ft. Lauderdale (Figure 8.23). Much of the coastline is developed and heavily populated, particularly in the Tampa Bay area from Tarpon Springs to Sarasota, and in the Miami area from South Miami through North Palm Beach.

Hurricanes that threaten northeastern Florida and the coastline of Georgia have a tendency to remain north of the Caribbean Sea (Figure 8.24). Some of these hurricanes originate east of the Lesser Antilles. A relatively large percentage of hurricanes passing south of Puerto Rico eventually threaten the Georgia coast. Northeastern Florida and Georgia are also threatened by hurricanes that cross the Florida peninsula from the Gulf of Mexico. In fact, it is more likely that a hurricane in the northeastern Gulf of Mexico will menace Daytona Beach than one located over the eastern Bahamas. Major cities in these regions include Daytona Beach, Jacksonville, and Savannah (Figure 8.25). Population densities are highest for coastal counties in Florida and for Chatham County in Georgia.

The coasts of South Carolina, North Carolina, and Virginia are imperiled by hurricanes tracking northwestward, generally to the north of the Greater Antilles (Figure 8.26). Recurving storms that skirt coastal Florida are a potential threat to these regions. Hurricanes passing north of Puerto Rico pose a threat to North Carolina. Occasionally hurricanes from the eastern Gulf of Mexico move across Florida before swinging northward and striking the Carolinas. It is relatively rare for a hurricane developing over the Caribbean Sea to move north and threaten the North Carolina or Virginia coasts, although South Carolina is occasionally threatened by these hurricanes. Major cities in these threat boxes include Charleston, Wilmington, and Norfolk (Figure 8.27).

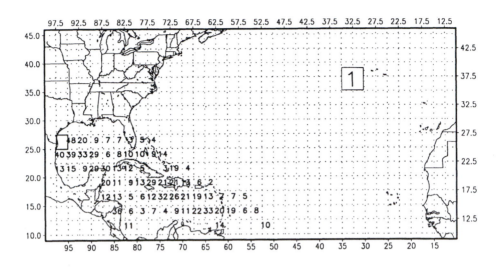

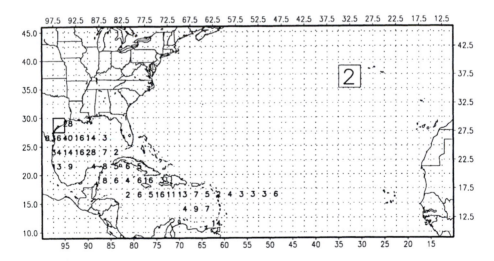

Fig. 8.16: Percent of all hurricanes passing through all $(2\frac{1}{2})°$ latitude by longitude boxes before passing through threat boxes 1 (top) and 2 (bottom). Box numbers are labeled in the upper-right portion of the plots. Boxes with percentages less than 1 are left blank. Highest percentages are typically located adjacent to the threat box.

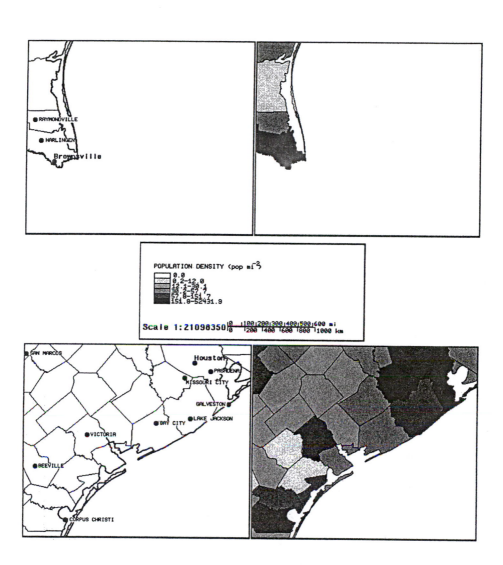

Fig. 8.17: Locations of cities and population density (1996 census) by counties for the areas delineated by threat boxes 1 (top) and 2 (bottom). Only the location of some cities are provided. County population densities are expressed in population per square mile. Population densities are shaded from light to dark using an approximate exponential scale as shown in the legend in order to capture the extremely broad range of densities along the coastline. All the location maps are drawn to the same scale as indicated in the legend.

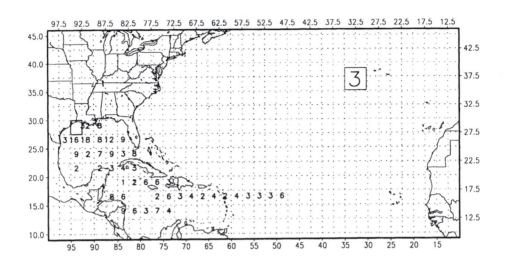

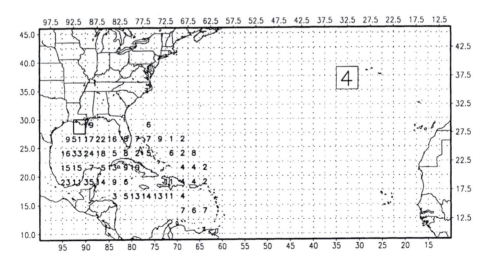

Fig. 8.18: Percent of all hurricanes passing through all $(2\frac{1}{2})°$ latitude by longitude boxes before passing through threat boxes 3 (top) and 4 (bottom). Box numbers are labeled in the upper-right portion of the plots. Boxes with percentages less than 1 are left blank. Highest percentages are typically located adjacent to the threat box.

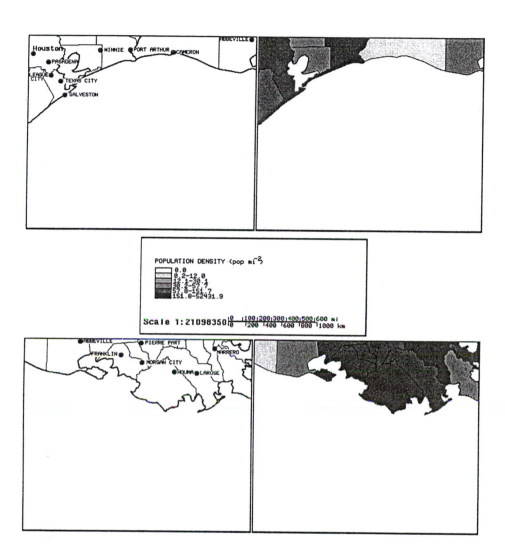

Fig. 8.19: Locations of cities and population density (1996 census) by counties for the areas delineated by threat boxes 3 (top) and 4 (bottom). Only the location of some cities are provided. County population densities are expressed in population per square mile. Population densities are shaded from light to dark using an approximate exponential scale as shown in the legend in order to capture the extremely broad range of densities along the coastline. All the location maps are drawn to the same scale as indicated in the legend.

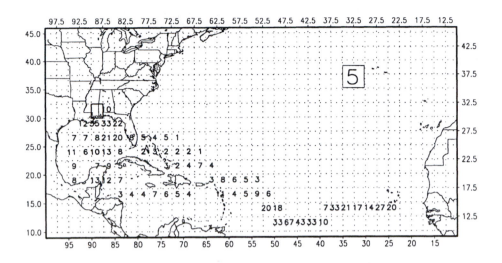

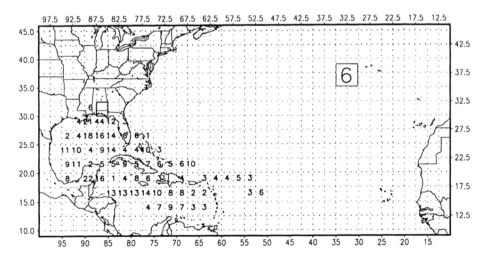

Fig. 8.20: Percent of all hurricanes passing through all $(2\frac{1}{2})°$ latitude by longitude boxes before passing through threat boxes 5 (top) and 6 (bottom). Box numbers are labeled in the upper-right portion of the plots. Boxes with percentages less than 1 are left blank. Highest percentages are typically located adjacent to the threat box.

Fig. 8.21: Locations of cities and population density (1996 census) by counties for the areas delineated by threat boxes 5 (top) and 6 (bottom). Only the location of some cities are provided. County population densities are expressed in population per square mile. Population densities are shaded from light to dark using an approximate exponential scale as shown in the legend in order to capture the extremely broad range of densities along the coastline. All the location maps are drawn to the same scale as indicated in the legend.

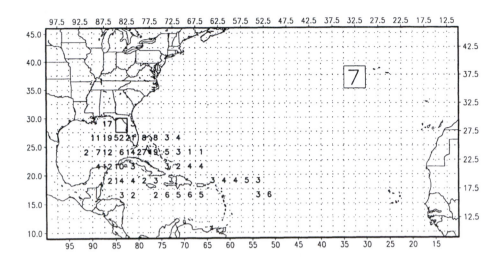

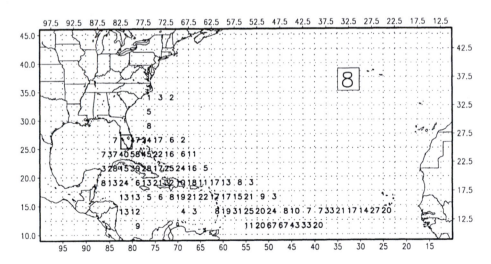

Fig. 8.22: Percent of all hurricanes passing through all $(2\frac{1}{2})°$ latitude by longitude boxes before passing through threat boxes 7 (top) and 8 (bottom). Box numbers are labeled in the upper-right portion of the plots. Boxes with percentages less than 1 are left blank. Highest percentages are typically located adjacent to the threat box.

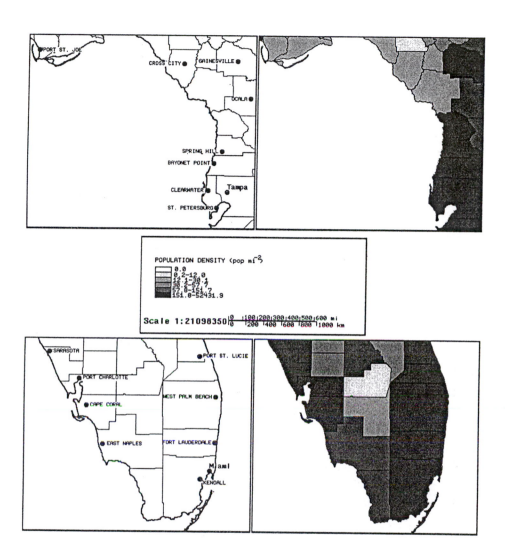

Fig. 8.23: Locations of cities and population density (1996 census) by counties for the areas delineated by threat boxes 7 (top) and 8 (bottom). Only the location of some cities are provided. County population densities are expressed in population per square mile. Population densities are shaded from light to dark using an approximate exponential scale as shown in the legend in order to capture the extremely broad range of densities along the coastline. All the location maps are drawn to the same scale as indicated in the legend.

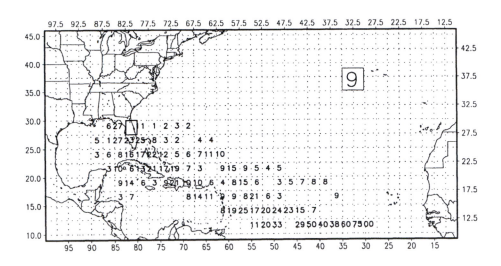

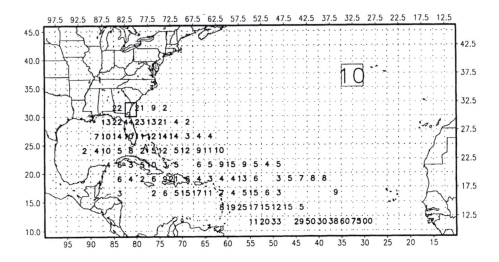

Fig. 8.24: Percent of all hurricanes passing through all $(2\frac{1}{2})°$ latitude by longitude boxes before passing through threat boxes 9 (top) and 10 (bottom). Box numbers are labeled in the upper-right portion of the plots. Boxes with percentages less than 1 are left blank. Highest percentages are typically located adjacent to the threat box. The 00 in the grid over the eastern tropical Atlantic denotes 100%.

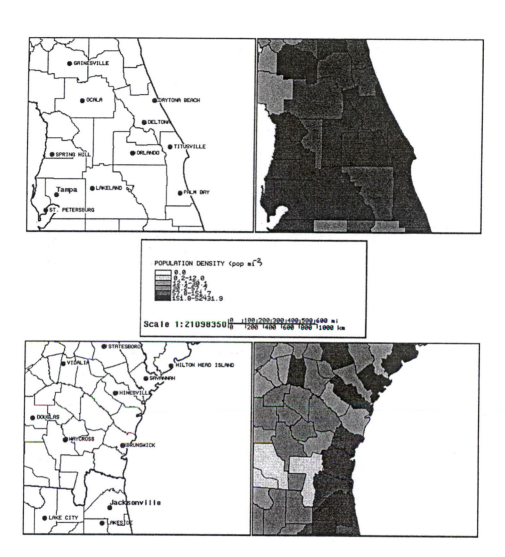

Fig. 8.25: Locations of cities and population density (1996 census) by counties for the areas delineated by threat boxes 9 (top) and 10 (bottom). Only the location of some cities are provided. County population densities are expressed in population per square mile. Population densities are shaded from light to dark using an approximate exponential scale as shown in the legend in order to capture the extremely broad range of densities along the coastline. All the location maps are drawn to the same scale as indicated in the legend.

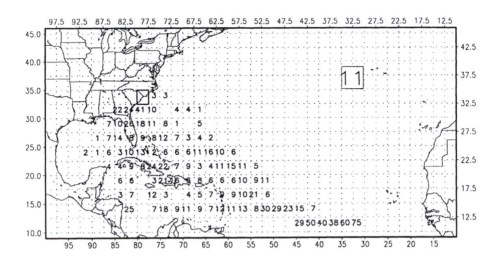

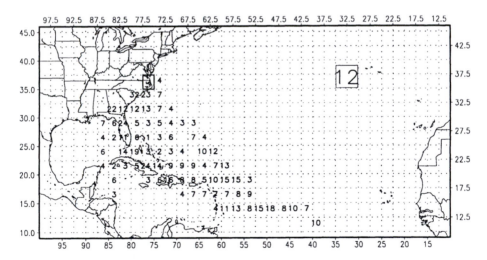

Fig. 8.26: Percent of all hurricanes passing through all $(2\frac{1}{2})°$ latitude by longitude boxes before passing through threat boxes 11 (top) and 12 (bottom). Box numbers are labeled in the upper-right portion of the plots. Boxes with percentages less than 1 are left blank. Highest percentages are typically located adjacent to the threat box.

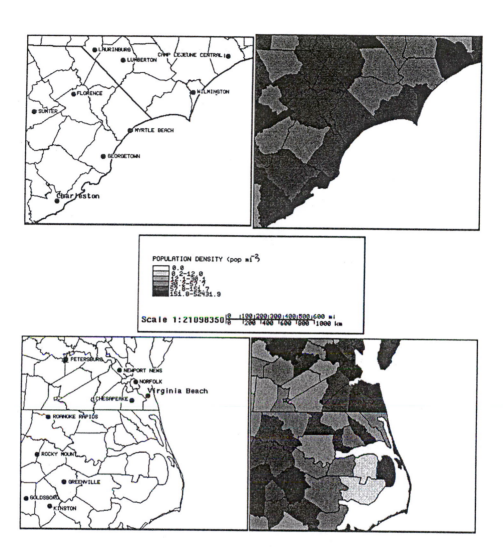

Fig. 8.27: Locations of cities and population density (1996 census) by counties for the areas delineated by threat boxes 11 (top) and 12 (bottom). Only the location of some cities are provided. County population densities are expressed in population per square mile. Population densities are shaded from light to dark using an approximate exponential scale as shown in the legend in order to capture the extremely broad range of densities along the coastline. All the location maps are drawn to the same scale as indicated in the legend. Coastal county population densities are greatest in Charleston County in South Carolina as well as in New Hanover and Onslow counties in North Carolina. Population density is quite high around the Norfolk area.

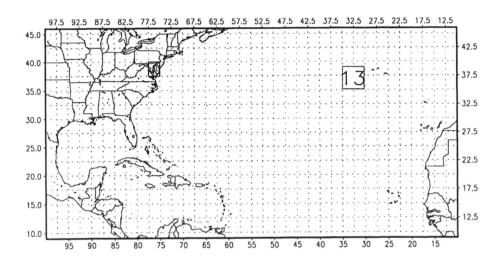

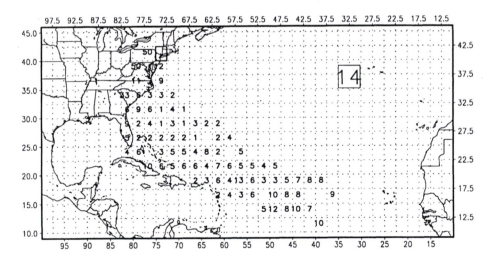

Fig. 8.28: Percent of all hurricanes passing through all $(2\frac{1}{2})°$ latitude by longitude boxes before passing through threat boxes 13 (top) and 14 (bottom). Box numbers are labeled in the upper-right portion of the plots. Boxes with percentages less than 1 are left blank. Highest percentages are typically located adjacent to the threat box.

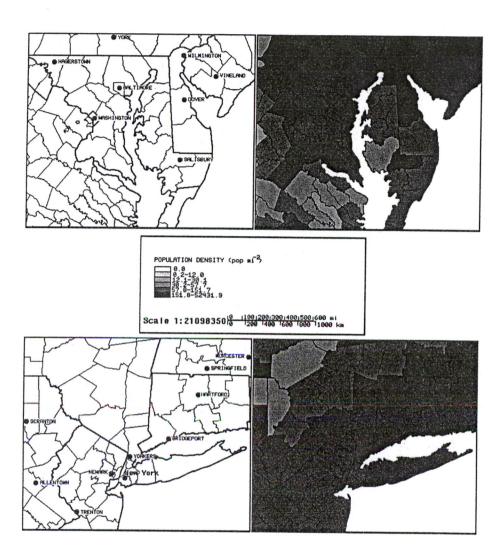

Fig. 8.29: Locations of cities and population density (1996 census) by counties for the areas delineated by threat boxes 13 (top) and 14 (bottom). Only the location of some cities are provided. County population densities are expressed in population per square mile. Population densities are shaded from light to dark using an approximate exponential scale as shown in the legend in order to capture the extremely broad range of densities along the coastline. All the location maps are drawn to the same scale as indicated in the legend.

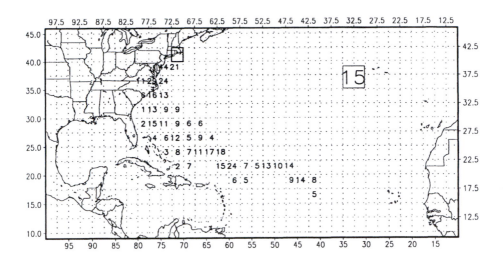

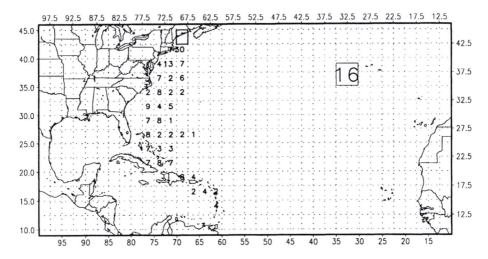

Fig. 8.30: Percent of all hurricanes passing through all $(2\frac{1}{2})°$ latitude by longitude boxes before passing through threat boxes 15 (top) and 16 (bottom). Box numbers are labeled in the upper-right portion of the plots. Boxes with percentages less than 1 are left blank. Highest percentages are typically located adjacent to the threat box.

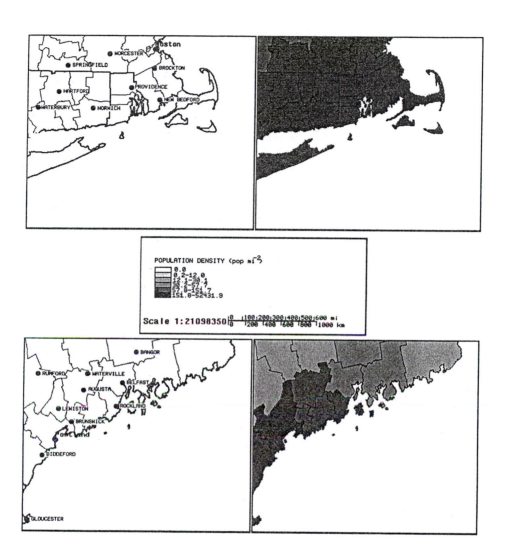

Fig. 8.31: Locations of cities and population density (1996 census) by counties for the areas delineated by threat boxes 15 (top) and 16 (bottom). Only the location of some cities are provided. County population densities are expressed in population per square mile. Population densities are shaded from light to dark using an approximate exponential scale as shown in the legend in order to capture the extremely broad range of densities along the coastline. All the location maps are drawn to the same scale as indicated in the legend.

Hurricanes affecting the southern New England coastline originate in the tropical or western North Atlantic (Figure 8.28). These hurricanes track northwestward before striking the coast. The percentages are small for any location indicating that most hurricanes of the western North Atlantic avoid this part of the coastline. Indeed, no tropical cyclone has entered threat box 13 as a hurricane over the period of record. Although the eye of hurricane *Donna* in September of 1960 never made landfall, Ocean City, Maryland reported winds in excess of 70 kt before the anemometer broke (Barnes 1998). Occasionally, hurricanes brushing the eastern Florida coastline make it north to affect Long Island. Population and population densities are extremely high along the coastline (Figure 8.29). This area includes the major cities of Washington, Baltimore, Trenton, New York, and Hartford.

Preferred paths of hurricanes influencing the New England coast and population statistics for the region are shown in Figures 8.30 and 8.31, respectively. This region includes eastern Long Island and Cape Cod. Both extend eastward into the North Atlantic making them susceptible to northward tracking hurricanes. Indeed this region is more likely to receive a hurricane strike from a tropical-only hurricane originating at low latitudes then it is from a baroclinically-enhanced hurricane developing in the western or central North Atlantic (see Table 8.8). Hurricane *Carol* of 1954 was an example of a recurving tropical-only hurricane that struck northern New England. Hurricanes tracking over the area located to the north and east of the Bahamas are more likely to strike the northern New England coastline than any other spot along the Atlantic seaboard. Hurricanes of the Gulf of Mexico and most of the Caribbean Sea rarely reach New England. Major cities in this region include Providence, Boston, and Portland. Population densities are high for Massachusetts but decrease northward into Maine.

9

Hurricanes of Puerto Rico, Jamaica, and Bermuda

This chapter surveys the hurricane climatology for regions of the West Indies and the western North Atlantic. Similarities in hurricane activity are noted between the regions. Yet each region displays unique characteristics. In particular, data are examined for the occurrence of hurricanes over and around Puerto Rico, Jamaica, and Bermuda. Hurricanes that influence Puerto Rico tend to arrive in bunches separated by long intervals of little activity. This suggests their frequencies may be related to long-period changes in climate. Higher surface pressures over the area might inhibit activity during some decades. While hurricanes of Puerto Rico are most likely during September, Jamaica sees a proportionally larger number of storms during October. This is related to the previously noted annual increase in the probability of storms forming over the western Caribbean after the seasonal peak in activity. Bermuda also sees late-October hurricanes, but annual hurricane threat frequencies for Bermuda are positively correlated with frequencies for Puerto Rico and negatively correlated with frequencies for Jamaica. This chapter looks at these facts in some detail.

The islands of the West Indies and Bermuda were pushed off the ocean floor by volcanoes hundreds of million years ago. The West Indies are a chain of islands that separate the Caribbean Sea from the Atlantic Ocean (Figure 9.1). The three main island groups include the Bahamas, the Greater Antilles, and the Lesser Antilles. Coral reefs surrounding many of the islands, including Bermuda,[1] offer some protection against the full force of hurricane-generated waves. The reefs themselves are influenced in both diversity and extent by hurricane-generated waves and surges. The Bahamas are an independent nation consisting of 3,000 small islands (20 inhabited) and coral reefs off the southeast coast of Florida. The capital is Nassau on New Providence Island. Grand Bahama Island is the northerly most island and includes the city of Freeport. Andros Island to the south of Grand Bahama is the largest. The Greater Antilles include the large islands of Cuba, Jamaica, Hispaniola, and Puerto Rico. Puerto Rico is the easternmost island of the Greater Antilles, lying to the east of the Mona Passage. Jamaica is located to the south of Cuba. The Lesser Antilles include the islands of Guadeloupe,

---

[1] Bermuda is the most northerly collection of coral islands in the world.

Fig. 9.1: Map of the Caribbean Sea and adjacent areas. The Greater Antilles include the islands of Cuba, Jamaica, Hispaniola (Haiti and the Dominican Republic), and Puerto Rico.

Martinique, St. Lucia, Barbados, and Grenada. The Lesser Antilles are subdivided by the Dominica Passage; the Leeward Islands located to the north and the Windward Islands to the south. Bermuda, located near 32°N latitude and 65°W longitude, is not a Caribbean island.

While this chapter focuses on hurricanes of Puerto Rico, Jamaica, and Bermuda other areas in and around the Caribbean are affected by North Atlantic hurricanes. Situated about 64 km (40 mi) from the eastern tip of Puerto Rico, the U.S. Virgin Islands are clustered in the Anegada Passage of the northwestern Lesser Antilles. The three main islands include St. Thomas, St. Croix, and St. John. Though it is rare for a hurricane to pass direct over such small islands, the region is frequently threatened by hurricanes. The island of St. Thomas was heavily damaged by hurricane *Marilyn* during September of 1995. The storm caused an estimated $2.1 billion in damage and 13 deaths. The Windward Islands are susceptible to hurricanes from July through August, while the Leewards and Greater Antilles are threatened from August through October (Caviedes 1991). Dominica was rocked hard by a hurricane in 1834 and by hurricane *David* in 1979. After devastating parts of the Lesser Antilles, *David* tracked northwestward and inundated the Dominican Republic with copious rainfall.

The Turks and Caicos islands, located to the southeast of the Bahamas, are frequently threatened by hurricanes. Venezuela and Colombia, located on the Caribbean coast of South America, occasionally feel the effects of hurricanes passing through the

Table 9.1: Data sources for Caribbean hurricanes. The range of years over which the sources are consulted is provided.

Range of Years	Author(s)	Usage
1901–1996	Neumann et al. (1993) (best-track)	Primary
1851–1900	Fernández-Partágas and Diaz (1995a,b, 1996a,b)	Primary
Before 1851	Tannehill (1950), Millás (1968), Tucker (1996)	Secondary

southern Caribbean Sea. The countries of Central America bordering the Caribbean Sea including Panama, Costa Rica, and especially Nicaragua, Honduras, and Belize are sometimes visited by North Atlantic hurricanes. Mexico is quite vulnerable to hurricanes, particularly along the Yucatan Peninsula. The three Cayman islands are located in the northwestern Caribbean Sea to the south of Cuba and to the northwest of Jamaica. The Cayman's were hard hit by a hurricane in 1932 and they felt the direct effect of powerful hurricane *Gilbert* in 1988.

Data on hurricanes of Puerto Rico, Jamaica, and Bermuda are available from the sources listed in Table 9.1. The primary sources include the Fernández-Partágas and Diaz (1995a, 1995b, 1996a, 1996b) and the Neumann et al. (1993) data sets. These data were used in previous chapters. Additional data for years before 1851 are taken from Millás (1968), Tannehill (1950), and Tucker (1996). The chronology of Millás is the most complete and comprehensive for early Caribbean hurricanes since the time of Columbus. Early dates are reset to the present-day Gregorian calendar in his chronology. Tucker's list is the most comprehensive for Bermuda, and extends back to the settlement of the island in 1609.

9.1 Hurricanes of Puerto Rico

Owing to its location in the tropical cyclone-prone North Atlantic, the commonwealth of Puerto Rico[2] is quite vulnerable to the ravages of hurricanes. Yet its relatively small size (covering about 8,768 km^2) and its east-to-west orientation make a direct hurricane landfall a relatively rare event. Often many years pass without a direct hit. The island commonwealth includes the smaller islands of Mona, Culebra, and Vieques. The interior of Puerto Rico is mountainous and very rugged, with the highest elevations in the south and east. Several hundred small rivers (streams) cut through the island. The interior mountains are the source of the streams, which tend to run north or south to the coasts. The streams on the south and east sides of the island are short and fall quickly to the sea, while those on the north and west sides are longer and slope more gently. None of the streams are large enough for navigation except by small boat. During heavy rainfall the streams become ragging torrents capable of widespread destruction. Flooding from heavy rains is a significant problem in Puerto Rico.

---

[2]*Puerto Rico*, which means *rich port* in Spanish, was the name for San Juan in early colonial times.

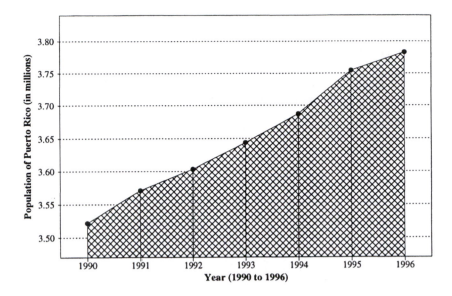

Fig. 9.2: Population estimates (from the U.S. Bureau of the Census) for Puerto Rico over the period 1990–1996. Population estimates are given in millions of people. Overall, Puerto Rico is more crowded (greater population density) than any state in the United States. Population trends are expected to continue.

The population of Puerto Rico continues to swell. The 1996 estimate puts the number of residents at just short of 3.8 million. This represents a 7% increase in population since 1990 (Figure 9.2). The increase occurs over a period during which time Puerto Rico was threatened by few hurricanes. Only hurricane *Hortense* in 1996 made direct landfall. *Hortense* was a category one storm. About 67% of all Puerto Ricans live in urban areas including San Juan, Ponce, Mayagüez, Carolina, and Bayamón. San Juan is located on the northeast coast (Figure 9.3) and is the capital of Puerto Rico. Both Carolina and Bayamón are within the San Juan metropolitan area. In 1985 the U.S. Federal Emergency Management Agency (FEMA) estimated that 47% of Puerto Ricans live in flood-prone areas. The population in Guaynabo, located in the hills to the south of San Juan, rose by 13% since 1990 and in 1996 exceeded 100,000 residents. Indeed, Puerto Rico has a population density exceeding 430 people per square kilometer, making it more crowded than any state. Additionally, more than one and half million tourists visit Puerto Rico annually, with many coming from the United States. Estimates put the economic impact of tourism at one and half billion annually although manufacturing is the most important industry. Increases in population at a time when major hurricane threats were low combined with a topography that concentrates the population in a few, very dense urban centers prone to flooding, suggests Puerto Rico will likely experience significant problems during a major hurricane.

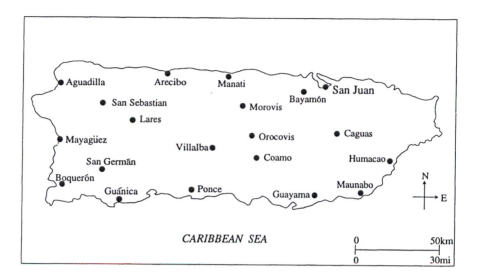

Fig. 9.3: Principal cities and towns of Puerto Rico. The San Juan metropolitan area includes the cities of Bayamón to the south and Carolina (not shown) to the east. Ponce and Mayagüez are major metropolitan areas in the south and west, respectively.

9.1.1   Landfalls

A chronological list of Puerto Rican hurricanes over the 146-period from 1851 through 1996 is shown in Table 9.2. Over this time 15 hurricanes made direct landfall for an average of about one hit every 10 years. The average is misleading in that hurricanes tend to strike in groups. For instance, in the short 9-year period from 1893 through 1901 there were four hurricanes. Another group of 4 landfalls occurred over the 7-year period from 1926 through 1932. More than half the hurricanes occurred in a combined 16-year stretch. There were no direct hits in Puerto Rico over the period 1955 through 1988. Six of the 15 hurricanes were major storms at landfall and 2 were category five. *San Felipe II*[3] in 1928 caused catastrophic damage to the island. After crossing Puerto Rico, *San Felipe II* tracked westward and devastated the Bahamas and southeastern Florida. A week after *San Ciriaco* demolished Puerto Rico as a category five hurricane on August 8, 1899, it brushed the eastern coast of Florida and then struck North Carolina's Outer Banks.

As seen from Table 9.2, August and September are the most common months for hurricanes in Puerto Rico. These two months represent the peak of the North Atlantic hurricane season. This is when hurricanes develop from tropical waves at low latitudes (south of 20°N). The waves frequently originate over Africa and pass near the islands of Cape Verde.[4] The waves are sometimes associated with the inter-tropical conver-

---

[3]Historically, Puerto Rican hurricanes were named for the saint day on which the hurricane occurred. The roman numeral indicates it was the second storm to hit the island on the particular day.

[4]Cape Verde is an independent African island country in the eastern North Atlantic.

Table 9.2: Hurricane landfalls in Puerto Rico. The track refers to the direction of movement and position of path across the island (see Figure 9.4). Track I is from southeast to northwest across the entire island, track II is east to west, track III is southeast to northwest across the northeast corner, and track IV is southeast to northwest across the southwest corner. The category is an estimate of the Saffir/Simpson scale at landfall. TO refers to the classification as a tropical-only hurricane. Hurricanes before 1886 are not been classified by type.

Year	Date	Track	Category	Type	Name
1852	9/5	IV	3	.	*San Lorenzo*
1867	10/29	II	3	.	*San Narcisco*
1876	9/13	II	2	.	*San Felipe I*
1893	8/16	III	1	TO	*San Roque*
1896	8/31	IV	3	TO	*San Ramon*
1899	8/8	I	5	TO	*San Ciriaco*
1901	7/7	IV	1	TO	*San Cirilo*
1916	8/22	II	2	TO	*San Hipolito*
1926	7/23	IV	1	TO	*San Liborio*
1928	9/13	I	5	TO	*San Felipe II*
1931	9/10	II	2	TO	*San Nicolas*
1932	9/26	II	3	TO	*San Ciprian*
1956	8/12	I	1	TO	*Santa Clara*, (*Betsy*)
1989	9/18	III	2	TO	*Hugo*
1996	9/9	IV	1	TO	*Hortense*

gence zone (ITCZ) over the central Atlantic. Hurricanes originating at low latitudes over the eastern and central North Atlantic generally take a westerly or northwesterly path toward the Caribbean Sea and Puerto Rico. Hurricanes that impact Puerto Rico are almost exclusively of the tropical-only variety. Indeed, all the landfalls since 1886 are tropical-only hurricanes. Two hurricanes struck Puerto Rico during the month of July and one hit during October.

Hurricanes follow 4 main tracks across Puerto Rico (Figure 9.4). Tracks I, III, and IV represent the typical direction of motion for hurricanes over this region of the Atlantic basin. The motion is to the northwest around the backside of the Bermuda high. Historically, tracks II and IV are the most common. Two hurricanes over the 146-year period followed track II across length of the island.[5] All else being equal, hurricanes following paths I or IV have the greatest potential for damage. In these cases much of Puerto Rico lies in the right semicircle with respect to the forward track of the hurricane. The two most intense hurricanes took a type-I track, moving in a southeast to northwest direction across the center of the island. Figure 9.5 shows the actual tracks of the last twelve hurricanes to hit Puerto Rico through 1996. All of the hurricanes originated at a latitude south and west of Puerto Rico—east of the Lesser Antilles. Ten of the twelve storms originated in a narrow 4° area between 14 and 18°N latitude. The

---

[5]Hurricane *Georges* in 1998 followed a track II path across Puerto Rico as a category two storm.

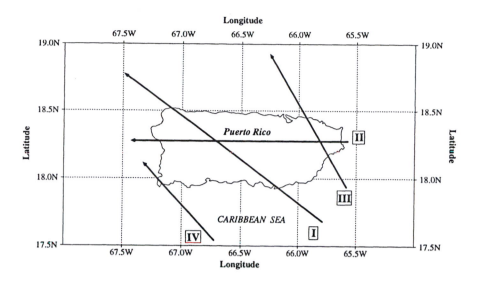

Fig. 9.4: Characteristic tracks of Puerto Rican hurricanes. The roman numerals correspond to the tracks used to describe the paths of the hurricanes listed in Table 9.2. Tracks I and IV pose a serious threat for widespread high winds as the strongest part of the hurricane crosses over much of the island. Track II has the potential to keep the eye wall over the island for several hours or more.

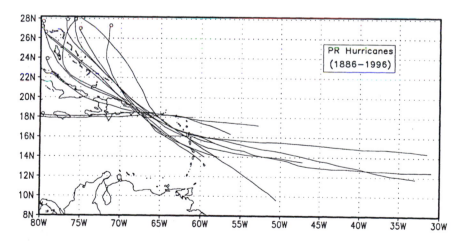

Fig. 9.5: Tracks of all hurricanes making direct landfall in Puerto Rico over the period 1886–1996. The tracks indicate the path at tropical storm strength or greater. The circle does not necessarily indicate the point of hurricane dissipation.

preferred direction of movement after striking Puerto Rico is northwestward through the Bahamas. Two hurricanes remained in the Caribbean Sea after slamming Puerto Rico. Both weakened to tropical-storm intensity before striking Jamaica.

A broader perspective on Puerto Rican hurricanes is obtained by combining data from the primary and secondary sources. Puerto Rican hurricanes in 50-year intervals back through the 16th century are shown in Figure 9.6 (top). The frequency of direct hits ranges from a minimum of two to a maximum of 13. Hurricanes were apparently more common during the 18th and first half of the 19th centuries compared to the 20th century. Some of the interdecadal variability is attributable to incomplete records during the 15th and 16th centuries. Differences in the interpretion of historical accounts between the various authors (Millás, Tannehill, and Fernández-Partágas and Diaz) as to whether a hurricane made a direct hit or a glancing blow contributes to the uncertainty of the earlier chronology. In any event, the longer list of storms allows for a finer examination of the seasonal variability of Puerto Rican hurricanes [Figure 9.6 (bottom)]. The intraseasonal distribution of hurricanes is symmetric with a peak during the first two weeks of September. Puerto Rico's hurricane season runs from the middle of July through the middle of October.

9.1.2 Threats

Although rarely hit directly, Puerto Rico feels the effect of many hurricanes. Here the climatology of hurricanes that have threatened the island is examined. A Puerto Rico hurricane threat is defined as a hurricane that has at least one 6-hourly observation within 500 km of the center of the island. This definition is arbitrary but practical. It is arbitrary since it is likely that some hurricanes that pass within the 500 km radius (particularly those to the far north of the island) never produce as much as a squall whereas others that remain farther away result in considerable wind and rain over the island. For instance, hurricane *Hazel* of 1954 moved northward out of the Caribbean Sea through the Greater Antilles. *Hazel* remained more than 700 km to the east of Puerto Rico, but a long spiral band of convection passed over the island causing more than 50 cm (20 in) of rainfall over many locations in the interior mountains.

The 500 km radius is practical because the best-track data set provides 6-hourly estimates of hurricane intensity at latitudes and longitudes so regional threats can easily be determined. In considering the risk of tropical cyclones to San Juan, Neumann (1987) used radii of 139 km and 278 km depending on the calculation. Since the interest here is hurricanes where the sample size is considerably smaller (the set of hurricanes is smaller than the set of tropical storms and hurricanes combined) a threat radius of 500 km is chosen. In any case, hurricane threats as defined here provide a useful way to consider the statistics of hurricanes over a part of the Caribbean region that includes Puerto Rico.

The time series of hurricane threats to Puerto Rico in 5- and 2-year intervals is shown in Figure 9.7. Hurricane threats in the individual pentads range from a minimum of one from 1971 through 1975 to a maximum of 10 from 1951 through 1955.

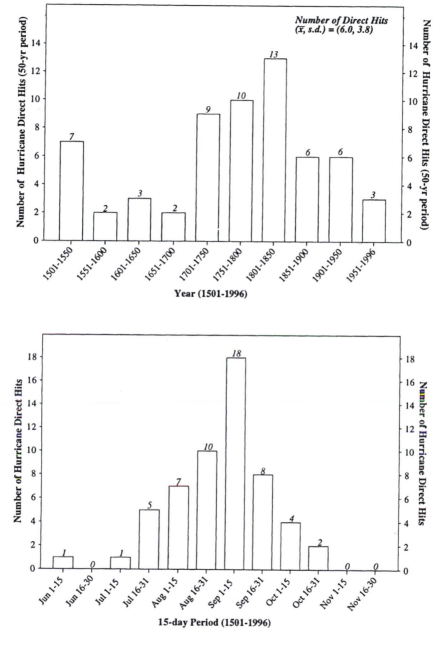

Fig. 9.6: Abundance of landfalling hurricanes over Puerto Rico in 50-year intervals (top) and the distribution of strikes in 15-day periods (bottom). The early 19th century was most active. Hurricane season peaks during the first 2 weeks of September. The average ($\bar{x}$) and standard deviation ($s.d.$) are given in the upper-right corner of the top figure.

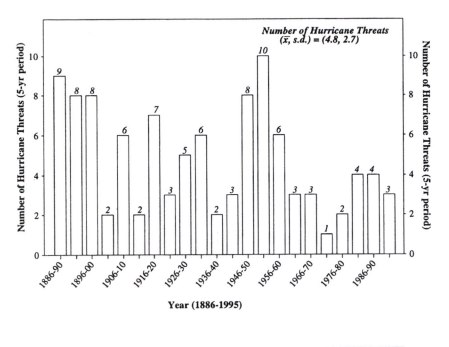

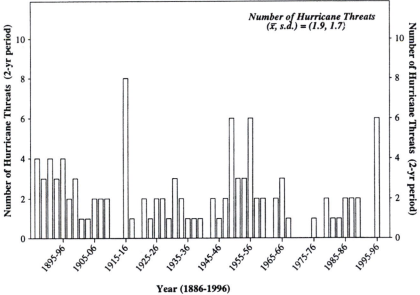

Fig. 9.7: Time series of the frequency of hurricane threats to Puerto Rico in 5-year intervals (top) and 2-year intervals (bottom) over the period 1886–1996. The 50-year average ($\bar{x}$) and standard deviation ($s.d.$) are given in the upper-right corner.

Table 9.3: Hurricane threats to Puerto Rico. Values are frequencies and probabilities based on data over the period 1886–1996. Whole numbers are the count of years with that many hurricanes.

Number of Threats	Number of Years	Observed Probability
0	47	0.42
1	34	0.31
2	20	0.18
3	8	0.07
4 or more	2	0.02
Amount	Number of Years	Observed Probability
At least 1 threat	64	0.58
2 or more threats	30	0.27
3 or more threats	10	0.09
4 or more threats	2	0.02

The average number of threats over 5 years is 4.8 or 1 threat per year. The late 19th century was a particularly active period for Puerto Rican hurricanes. Three consecutive pentads in this period had 8 or more hurricanes. Another period of heightened activity is noted during the late 1940s through the 1950s. Recall this is a period when major hurricane activity in the United States was high (Chapter 7). In the 2-year intervals, hurricane threats are most numerous during 1915–1916. The average is 1.9 threats with a standard deviation of 1.7. The last 2-year period in the record has 6 hurricane threats. This was a period of increased tropical-only hurricane activity over the North Atlantic basin (Chapter 12).

Periods of increased hurricane activity are characterized by consecutive years of multiple threats. Figure 9.8 gives the time series of annual hurricane threats to Puerto Rico over the period 1886 through 1996. The average annual number of threats to Puerto Rico is one with a standard deviation of approximately one. The most hurricane threats in 1 year is six in 1916. The longest streak of years with at least one threat is 11 (1886–1896). The longest string of consecutive years with no threats is 7 (1968–1974). A recent streak of no threats includes the 5-year period from 1990 through 1994. This period is characterized by few tropical-only hurricanes and little to no hurricane activity over the Caribbean Sea (see Chapter 12). Table 9.3 gives the observed frequency and empirical probabilities of annual hurricane threats. Each year Puerto Rico has a 58% chance of at least one threat and a 27% chance of two or more. Eight years in the record have exactly three Puerto Rican hurricane threats. It should be emphasized that these represent hurricane threat frequencies and not landfall probabilities, which are much smaller. Only 15 hurricanes made a direct landfall in Puerto Rico over the period 1851–1996.

The monthly distribution of hurricane threats is shown in Figure 9.9. Nearly 85% of all hurricane threats occur in August or September, with September seeing 50%

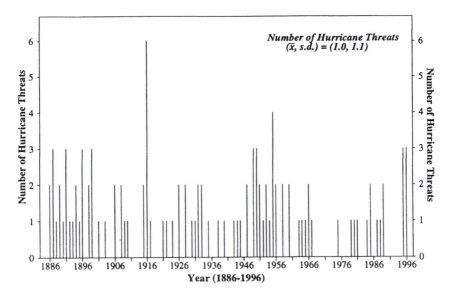

Fig. 9.8: Time series of the annual frequency of hurricane threats to Puerto Rico over the period 1886–1996. The average ($\overline{x}$) and standard deviation (*s.d.*) of the number of threats is given in the upper-right corner.

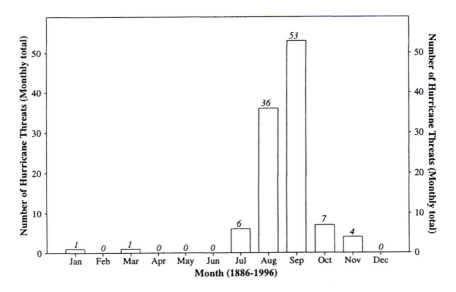

Fig. 9.9: Monthly distribution of hurricane threats to Puerto Rico over the period 1886–1996. Puerto Rico is only rarely threatened by hurricane landfalls outside the months of August and September.

more threats compared to August. This is consistent with the fact that virtually all threats come from tropical-only hurricanes. As was shown in Chapter 6, tropical-only hurricanes form predominantly during the middle of the hurricane season from easterly waves. Threats diminish markedly during October. In fact, October accounts for less than seven percent of all Puerto Rico hurricane threats. No hurricane has threatened Puerto Rico during May or June.

9.1.3   Rainfall

Storm surge accounts for much of the loss of life and property damage from hurricanes that strike the United States. The situation is different for Puerto Rico. Flooding from heavy rains is often the most devastating aspect of hurricanes. Heavy rainfall from tropical cyclones tends to concentrate over the central interior mountain range (Cordillera Central) resulting in rapid runoff of the water from the ground into streams and small rivers. The swollen rivers cause mudslides (particularly along denuded slopes) and widespread flooding. Minimal strength storms can result in horrific disasters from massive flooding. Hurricane *Hortense* made landfall as a category one hurricane with maximum sustained winds of 70 kt during September of 1996. Extremely heavy rainfall was reported in Caguas as well as in other towns over the eastern half of the island. Moderate to severe flooding occurred over most of the mountainous interior and in areas along the northern, eastern, and southern coasts. The diversity of weather phenomena responsible for non-tropical-cyclone rainfall is large and results in rainfall patterns that are less related to topography (Carter and Elsner 1996, Carter 1997).

Studies of rainfall patterns over Puerto Rico are possible because of a dense network of rain gauges covering the island. Here comparisons are made between rainfall patterns produced by tropical cyclones and those produced by other weather systems. Days during the hurricane season (June through November) over the period 1955–1993 are grouped into either tropical-cyclone or non-tropical-cyclone days based on whether a storm is threatening (at least one 6-hour observation within 500 km of the center of the island). Rainfall averages for both groups of days are computed at 80 rain gauges. An objective analysis (Barnes 1964) is used to grid the averages at $\left(\frac{1}{16}\right)^{\circ}$ latitude-longitude intervals using a variable radius of influence to include at least three gauges for the average at every grid point. Three passes with the analysis provides proper smoothing of the rainfall patterns.

Three maps of Puerto Rico are shown in Figure 9.10. The top map indicates the smoothed topography with darker shading in 60 m increments of elevation. The landscape is dominated by the Cordillera Central Mountains in the interior and the Loquillo Mountains in the east. Elevations exceed 600 m in areas of the Cordillera. Average tropical-cyclone rainfall is shown on the middle map. Local maxima are found in the mountains associated with intense orographic lifting. Strong hurricane-force winds blow up the mountains enhancing convection and increasing the rate at which the rain falls. When the amount of rainfall from hurricanes and tropical storms is linearly correlated with elevation (at elevations above 10 m), orography explains about

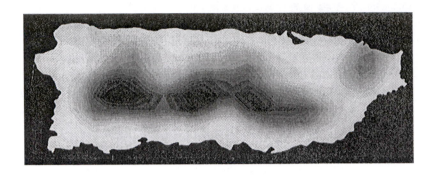

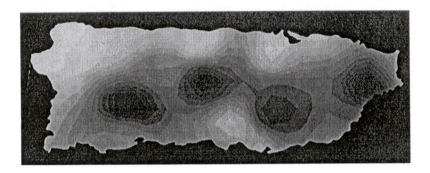

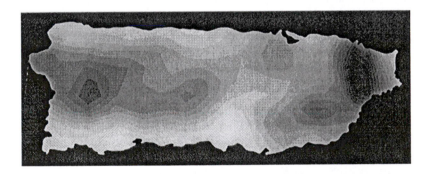

Fig. 9.10: Smoothed topography of Puerto Rico with shading every 60 m (top). Average rainfall from hurricanes and tropical storms within 500 km of the island (middle). Average June through November non-tropical-cyclone rainfall (bottom). Dark areas on the rainfall maps indicate regions of heavier precipitation. Adapted from Carter and Elsner (1997).

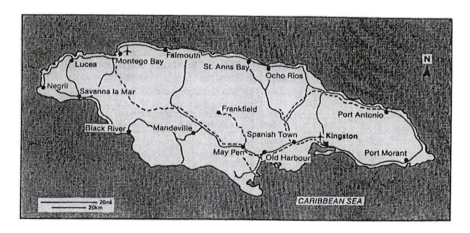

Fig. 9.11: Map of Jamaica showing the principal cities and roads. Kingston on the southeast coast is particularly vulnerable to hurricane strikes from the south. The Blue Mountains are located to the east of Kingston.

40% of the precipitation pattern. In contrast, orography explains less than 10% of the non-tropical-cyclone summertime rainfall (bottom map). Most of the correlation with elevation during the summertime is due to the Loquillo Mountains.

In both groups of days, rainfall is relatively light over the south-central coast of Puerto Rico near Poncé. This region is protected by the mountains immediately to the north and east. Improvements in predicting heavy rainfalls will help in warning the most vulnerable communities; yet according to Merrill (1993), the prediction of tropical cyclone rainfall is problematic for several reasons. In the first place, high winds make accurate measurement of precipitation difficult. Also, small errors in the forecast track of a hurricane substantially influence where the heaviest rain occurs. This is due to extreme spatial gradients of rainfall that often accompany hurricanes. Interactions between hurricanes and other weather systems are complex and not well understood. Indeed, detailed forecasts of the spatial patterns of heavy rainfall associated with tropical cyclones is an important research topic.

9.2   Hurricanes of Jamaica

Jamaica is an island nation of the Greater Antilles, located south of Cuba and west of Haiti (see Figure 9.1). With an area of 11,424 km^2—slightly smaller than Connecticut—Jamaica is the third-largest island in the Caribbean (Figure 9.11). Only Cuba and Hispaniola are larger. The rugged topography and karst geomorphology resembles that of Puerto Rico as does its latitude at 18°N. The Blue Mountains on the eastern half of the island rise to 2,256 m (7402 ft) at Blue Mountain Peak. Jamaica has 2.6 million people. Growth rates are less than in Puerto Rico. Population rose by 1.7% during the 1980s. The low growth rate (Puerto Rico had a 10% growth rate during the

Table 9.4: Hurricane landfalls in Jamaica. Approach refers to the direction from which the storm neared the island. S is an approach from the south. The category is an estimate of the Saffir/Simpson scale at landfall. TO and BE refer to tropical-only and baroclinically-enhanced hurricane types, respectively. Hurricanes occurring before 1886 do not have a Saffir/Simpson category and are not labeled as tropical-only or baroclinically-enhanced.

Year	Date	Approach	Category	Type	Name
1874	11/1	S	.	.	.
1879	10/12	SE	.	.	.
1880	8/18	S	.	.	.
1884	10/7	S	.	.	.
1886	8/19	SE	2	TO	.
1903	8/11	E	3	TO	.
1912	11/18	SW	4	BE	.
1916	8/15	SE	2	TO	.
1933	10/29	S	2	TO	.
1944	8/20	E	3	TO	.
1951	8/17	E	2	TO	*Charlie*
1988	9/12	E	3	TO	*Gilbert*

same period) is largely a result of migration. Migration was particularly high during the tough economic times of the 1960s and again during the late 1980s. Only about half the population lives in urban centers (680,000 in Kingston). Jamaica's tropical climate and attractive landscape draw more than 850,000 tourists annually. Besides tourism, the economy of Jamaica depends on agriculture, particularly the production of bananas and sugar.

9.2.1 Landfalls

Table 9.4 lists the twelve hurricanes that made direct landfall in Jamaica over the period 1851 through 1996. Jamaica was hit five times during the last half of the 19th century and five times again during the first half of the 20th century. Only two hurricanes made landfall in Jamaica over the period 1952 through 1996. Half of the twelve hurricane strikes occurred in August, but none before August 11th. Five hit after the first of October, including two in November. The tendency is for late-season hurricanes to approach from the south, originating over the southern and southwestern Caribbean Sea. Historically, Jamaican hurricanes have been rather formidable. All eight of the hurricanes since 1886 were category two or higher at landfall, while four were major hurricanes. The last 3 landfalling hurricanes approached Jamaica from the east.

Hurricane *Gilbert* is the most intense hurricane on record for the North Atlantic. *Gilbert* passed directly over Jamaica during September of 1988. The hurricane killed 45 persons and caused widespread destruction that crippled the Jamaican economy for months. An estimated 500,000 people were left homeless and agriculture production

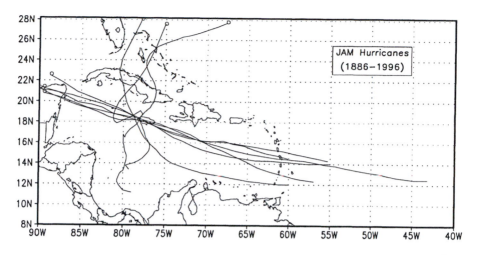

Fig. 9.12: Tracks of all Jamaican (JAM) hurricanes over the period 1886–1996. The tracks indicate the path at tropical storm strength or greater. The circle indicates the forward direction of tropical cyclone movement after striking Jamaica and not necessarily its point of dissipation.

was significantly curtailed. In fact, appraisals indicated that up to 30% of the sugar, 54% of the coffee, and more than 90% of the banana and cocoa harvests were destroyed (Pulwarty and Riebsame 1997). Total damage to Jamaica from *Gilbert* was approximately one billion U.S. dollars. Estimates put the insurance losses at $370 million (Eyre 1989). *Gilbert's* destructive path did not end with Jamaica. The storm demolished the Cayman islands and the Yucatan Peninsula before making final landfall on the northern Mexican coast. In its wake, *Gilbert* left a total of 318 dead in 7 countries.

Similar to hurricanes of Puerto Rico, Jamaica is hit most frequently by tropical-only hurricanes. The only baroclinically-enhanced hurricane was the November storm of 1912. Figure 9.12 shows the tracks of the eight hurricanes that made direct landfall in Jamaica over the period 1886 through 1996. Six of the eight hurricanes originated to the east of the Windward Islands as tropical-only hurricanes. The other two formed in the southern Caribbean Sea as late-season hurricanes. All eight of the hurricanes making landfall in Jamaica originated south of 15°N latitude. In general, the storms originating to the east of the Lesser Antilles continue westward toward Yucatan after striking Jamaica. The two storms that moved north through the Caribbean Sea affected the Bahamas. No tropical storm over the 111-year period has hit both Puerto Rico and Jamaica as a hurricane.

Jamaican hurricane strikes in 50-year intervals back through the 16th century are shown in Figure 9.13 (top). The frequency of direct hits ranges from a minimum of zero to a maximum of twelve. Similar to Puerto Rican hurricanes, Jamaican hurricanes were apparently more common during the 18th and first half of the 19th centuries compared to the 20th century. The abundance of hurricanes hitting Jamaica during the last half of the 17th century is comparable to the more recent epochs. Although fluctuations in hurricanes on the scale of centuries is likely, cautioned is advised in taking these numbers too literally.

Though similar to Puerto Rico in many respects, Jamaica shows a distinctly different seasonal pattern of hurricane activity. This is due to the variety of hurricane types that affect this part of the Caribbean Sea. The biweekly distribution of Jamaican hurricanes is presented in Figure 9.13 (bottom). July is split into two halves: no activity during the first 2 weeks and a relative abundance of activity during the second 2 weeks. Like Puerto Rico, hurricane activity in Jamaica is relatively rare until late July. There is a tendency for tropical storm activity around Jamaica during the last week of June into the first week of July (Naughton 1982), but this activity is weak or avoids Jamaica until later in the month. Activity peaks during the first half of August, but the peak is flat and runs through the middle of September. Early and middle-season hurricanes approach from the east developing from tropical easterly waves. A distinct break in Jamaican hurricane activity is noted during the second half of September before a secondary peak shows up during the last two weeks of October. The break during the second half of September creates a bi-modal distribution to the seasonal activity. The break comes at a time of significant overall activity in the North Atlantic basin.

The first peak of the Jamaican hurricane season occurs during August before the maximum in Puerto Rican hurricane activity. The second peak occurs during late October at a time when hurricanes in Puerto Rico are rare. The secondary peak in activity corresponds to the increase in October activity noted over the North Atlantic basin as a whole. The increase appears in both tropical-only and baroclinically-enhanced hurricane activity. The late-season tropical cyclones of Jamaica develop in the southwestern Caribbean Sea, often associated with an enhancement of convection in the monsoon trough. Widespread, deep convection may occur in association with the southward penetration of a middle-latitude baroclinic trough. Hurricanes and tropical storms developing between Nicaragua and Colombia tend to track north toward Jamaica, but typically stay well west of Puerto Rico. As a consequence, while it is uncommon to see a hurricane affect Puerto Rico in late October, late-October hurricanes in Jamaica are as common as late-August hurricanes.

9.2.2 Threats

Jamaican hurricanes are also examined by considering hurricane threats to the island. Similar to Puerto Rico, a hurricane that has at least one 6-hour observation within 500 km of the center of the island is defined as a hurricane threat to Jamaica. Based on this definition, Jamaica experienced 81 hurricane threats over the 111-yr period from 1886 through 1996 according to the best-track data set. The earliest hurricane threat to Jamaica occurred on May 20, 1970 from hurricane *Alma*, while the latest threat occurred on November 20, 1996 from hurricane *Marco*. Note that threats do not imply Jamaica felt any effects from the hurricane.

At least one threat occurred in 53 of the 111 years. This compares to 64 years with

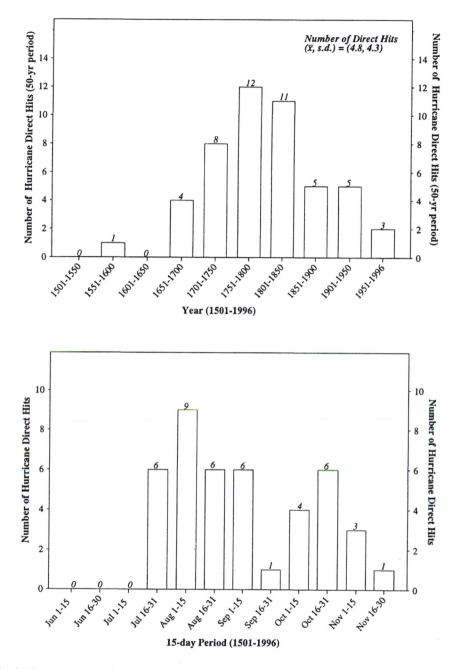

Fig. 9.13: Abundance of Jamaican hurricanes in 50-year intervals over the period 1501–1996 (top), and the distribution of direct hits in 15-day intervals over the same period (bottom). Note the frequency of Jamaican hurricanes during 18th and 19th centuries. Also note the two peaks in the distribution of seasonal activity. The average ($\bar{x}$) and standard deviation ($s.d.$) are given in the upper-right corner.

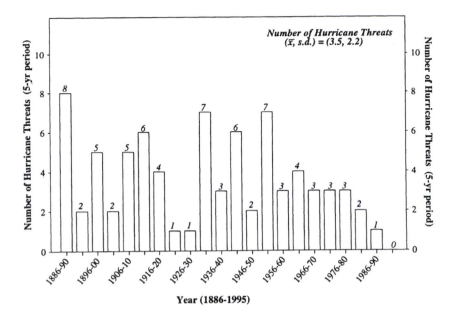

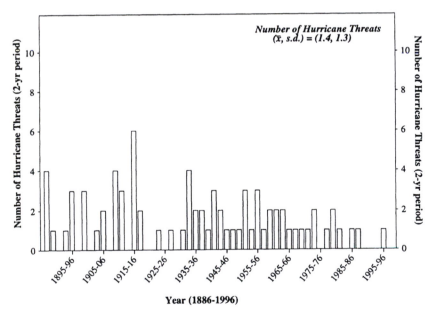

Fig. 9.14: Time series of the abundance of hurricane threats to Jamaica in 5-year intervals (top) and 2-year intervals (bottom). There are fewer threats over the more recent decades. The average ($\bar{x}$) and standard deviation ($s.d.$) are given in the upper-right corner.

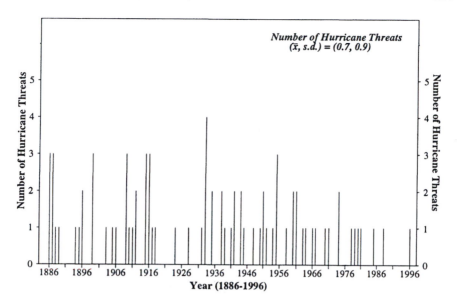

Fig. 9.15: Time series of the annual frequency of hurricane threats to Jamaica over the period 1886–1996. The average ($\bar{x}$) and standard deviation ($s.d.$) of the number of threats is given in the upper-right corner.

at least one threat to Puerto Rico over the same period. Figure 9.14 shows the time series of hurricane threats to Jamaica in 5-year and 2-year intervals. The number of threats in the 5-year intervals ranges from eight in the first pentad to zero in the last 5-year interval. The average number of threats per pentad is 3.5 with a standard deviation of 2.2. The middle 1950s featured 7 Jamaican-hurricane threats. Overall there is a decrease in the number of threats beginning in the 1960s. The most threats in any 2-year interval is six, which occurred during 1915 and 1916. These same two years produced a record eight hurricane threats to Puerto Rico. The correlation between the number of hurricanes threatening Jamaica to those threatening Puerto Rico in 5-year intervals is 0.402 (p-value = 0.06). The value of correlation suggests common factors influencing the frequency of hurricanes over two separate regions of the Caribbean. Despite the fact that no single storm has hit both Jamaica and Puerto Rico as a hurricane, the size of the areas used in defining hurricane threats yields to common threats from storms passing both Puerto Rico and Jamaica. This increases the correlation.

Single-year hurricane threats are shown in Figure 9.15. The most active year is 1933 with four threats. Seven years had 3 or more threats in a season before 1950 and only one year after 1950. Many years are without a threat. The scale of interannual threat variability is slightly smaller for Jamaica compared with Puerto Rico. This is indicated by the smaller standard deviation for Jamaican hurricanes. The longest streak of seasons with at least one threat is 4 years and the longest streak of no threats is 7 years from 1989–1995. The 1920s saw few Jamaican hurricanes. The early 1990s

Table 9.5: Hurricane threats to Jamaica. Values are frequencies and probabilities based on data over the period 1886–1996. Whole numbers are the count of years with that many hurricanes.

Number of Threats	Number of Years	Observed Probability
0	58	0.52
1	35	0.32
2	10	0.09
3	7	0.06
4 or more	1	0.01

Amount	Number of Years	Observed Probability
At least 1 threat	53	0.48
2 or more threats	18	0.16
3 or more threats	8	0.07
4 or more threats	1	0.01

were largely void of Caribbean hurricane activity (see Chapter 12). Jamaica has not been threatened by more than one hurricane in a season over the period 1974–1996.

The observed frequencies and empirical probabilities of annual hurricane threats to Jamaica are given in Table 9.5. The probability of exactly one threat is 32%. This probability is similar to the probability of one threat to Puerto Rico. However, the probability of two or more threats in a single season is only 16% compared to 27% for Puerto Rico. This indicates the assumption of hurricane threats as random, independent events might not be as realistic for Puerto Rico as it is for Jamaica. Ten years featured exactly 2 threats to Jamaica. Again, it should be emphasized that these are threat frequencies and do not represent strike probabilities, which are considerably lower.

9.3 Hurricanes of Bermuda

Bermuda is a group of 300 or so small islands in the western North Atlantic Ocean located approximately 1200 km southeast of New York City and 1700 km northeast of Miami. The largest islands include Bermuda, St. George's, St. David's, and Somerset (Figure 9.16). Twenty of the islands are inhabited. The islands have a total area of roughly 55 km^2. The ground consists of soft porous stone (karst) on top of an old volcano. There is little underground freshwater, so the residents depend on capturing rainfall on roofs for their water supply. The coral reefs surrounding the islands are a mixture of limestone, shell, and sand containing small amounts of true coral material. Small hills on Bermuda rise to 80 m (260 ft) above sea level.

Bermuda is a British dependency with a recorded history dating to the turn of the 17th century. In 1609, the ship *Sea Venture* captained by Sir George Somer wrecked on the coral reefs near Fort St. Catherine in a hurricane. Bermuda's coat-of-arms depicts the hurricane disaster that led to its first permanent settlement (Tucker 1996). Bermuda

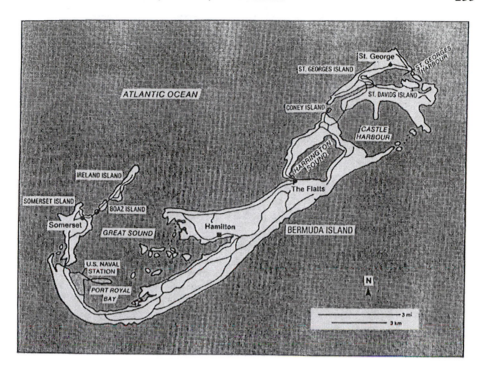

Fig. 9.16: Map of the islands of Bermuda. Although consisting of several hundred small islands, the four largest and most inhabited include Bermuda, St. George's, St. David's, and Somerset.

has a population of approximately 60,000, many of which are young. About 25% are under the age of 20, and 75% are native born. Bermuda's mild climate, beach resorts, and recreational facilities attract about 500,000 tourists a year.[6]

Bermuda is located outside the main development region for hurricanes. The islands are threatened quite frequently, however, by hurricanes that form at low latitudes and track northward. Terry Tucker provides an excellent chronology of Bermudan hurricanes in *Beware the Hurricane*. Her list contains tropical storms and hurricanes that have affected Bermuda since 1609. It includes dates of the technological advances leading toward effective early hurricane warnings for Bermudans. For instance, she reports that Bermuda began telegraphed warnings of tropical disturbances from the U.S. Weather Signal Service in 1906. In August of 1932 the Bermuda Meteorological Station was opened at Fort George. Aerial weather reconnaissance missions of the U.S. Air Force were flown out of Kindley Air Force Base from 1947 to 1963. Tucker's narrative is a mix of local history, architecture, and climate of Bermuda. Data on hurricane threats to Bermuda are obtained from the best-track data set. The chronology of Tucker (1996) is used for information on storms occurring before 1886.

---

[6]Tax laws accommodate insurance and investment companies from around the world.

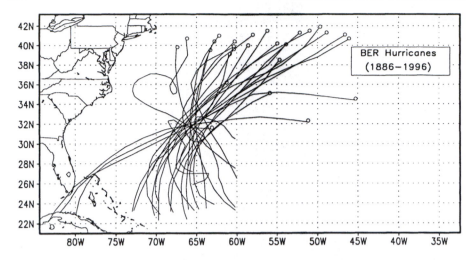

Fig. 9.17: Tracks of all Bermudan (BER) hurricanes over the period 1886–1996. The tracks indicate the path at tropical storm strength or greater. Note that the circle indicates the forward direction of tropical cyclone movement after striking Bermuda and not necessarily its point of dissipation.

9.3.1 Direct Hits

For the present climatology, we extract from Tucker (1996) those tropical cyclones that passed directly over Bermuda as a hurricane or near enough to give winds of hurricane force to the islands. The latter criterion is not used in defining landfalls previously for the United States, Puerto Rico, or Jamaica, which required a direct hit from all or part of the hurricane eye wall. The definition used here for Bermuda is consistent with the definition employed in Chapter 11 for examining return periods in coastal counties of the United States.

Figure 9.17 shows the tracks of hurricanes that directly affected Bermuda over the period 1886 through 1996. Bermuda is susceptible to hurricanes approaching from the south. There is a preference for storms to approach from a southwesterly direction. This is in contrast to Puerto Rico and Jamaica where hurricanes approach almost exclusively with an easterly component to direction. Most of tropical cyclones of Bermuda originate at low latitudes either in the central North Atlantic or the western Caribbean Sea. Tropical storms originating north of 28°N latitude are unlikely to affect Bermuda.[7] After passing Bermuda hurricanes continue recurving northeastward and do not affect the United States.

The frequency of hurricanes directly affecting Bermuda in 50-year intervals back through the 17th century are shown in Figure 9.18 (top). The frequency ranges from a minimum of 2 to a maximum of 11. There appears to be a slight increase in activity in the later decades of the record. Clearly there are significant century-scale fluctuations

---

[7]A few storms affecting south Florida have gone on to directly influence Bermuda.

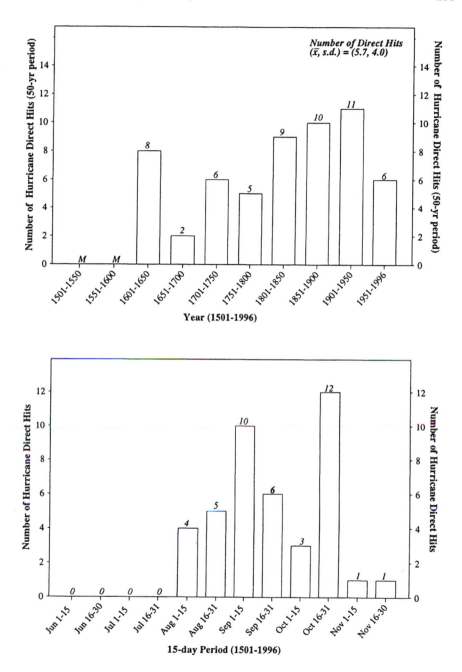

Fig. 9.18: Abundance of hurricanes directly affecting Bermuda in 50-year intervals over the period 1609–1996 (top) and the distribution of hurricanes in 15-day intervals over the same period (bottom). M denotes no data available. Note the frequency of Bermudan hurricanes during the last two weeks of October. The average ($\bar{x}$) and standard deviation ($s.d.$) are given in the upper-right corner of the top panel.

in hurricane activity over this part of the North Atlantic. The 18th century was less active then the 19th century, and the second half of the 20th century has been less active again. The reduction in hurricane activity during the last 50 years is noted in the frequency of both Puerto Rican and Jamaican hurricanes.

The seasonal distribution of hurricanes affecting Bermuda is presented in Figure 9.18 (bottom). Similar to Jamaica, the distribution is such that one can properly talk of two seasons. The first season begins in early August and peaks during the first two weeks of September. Bermudan activity is shifted a month later with respect to Jamaica's early season. Hurricane activity in Bermuda is virtually nonexistent before the start of August. A decline in activity during the first few weeks of October marks the end of the first season. The second season is shorter and occurs over the last two weeks of October. This corresponds to Jamaica's second season. By November hurricane activity is essentially over. Only two tropical cyclones of hurricane intensity have menaced Bermuda during November. The earliest hurricane in a season to affect Bermuda occurred on June 6, 1832. Like Jamaica, Bermuda is often visited by hurricanes in late October that develop in the western North Atlantic or western Caribbean Sea and track northeastward. Late-season hurricanes affecting Bermuda are often baroclincally-enhanced.

9.3.2   Threats

Hurricane threats to Bermuda are defined as before but for hurricanes that have passed within 300 km (compared with 500 km for Puerto Rico and Jamaica) of the island. Using this definition 71 hurricanes threatened Bermuda over the period 1886 through 1996. Hurricane threats occur in 55 of the 111 years for an average of at least one threat every 2 years. This is nearly identical to the number of threat years in Jamaica.

The time series of hurricane threats to Bermuda in 5- and 2-year intervals is given in Figure 9.19. The number of threats in the 5-year intervals ranges from seven in the second pentad (1896–1900) to zero in the period 1976 through 1980. No overall trend toward fewer hurricane threats is noted over the period of record. This is in contrast to Jamaican hurricane threats, which indicates a decreasing trend. The 5-year average number of Bermuda hurricane threats is 3.2 threats with a standard deviation of 1.6 threats. The most threats in any 2-year interval is five in the years 1891 and 1892. The years 1981 and 1982 featured at total of 4 hurricane threats to Bermuda. Although hurricanes of Bermuda and Jamaica have similar seasonal cycles, there is a weak negative correlation between their annual frequencies. In contrast the correlation between annual threats in Bermuda and Puerto Rico is 0.424 with a p-value of less than 0.05. This is indicative of the fact that frequently storms passing near Puerto Rico recurve and threaten Bermuda.

Single-year hurricane threats are shown in Figure 9.20. The most active year is 1891 with four threats, though only one produced hurricane-force winds in Bermuda. Half the years do not experience a single threat. The longest string of years with no threats is six from 1975 through 1981 and the longest consecutive string of years with

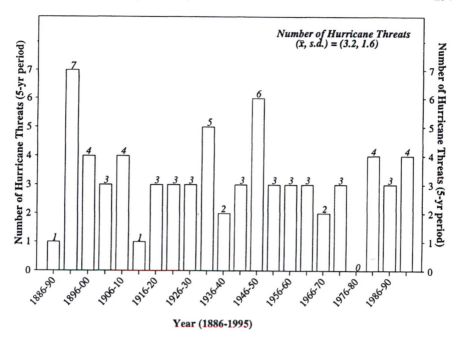

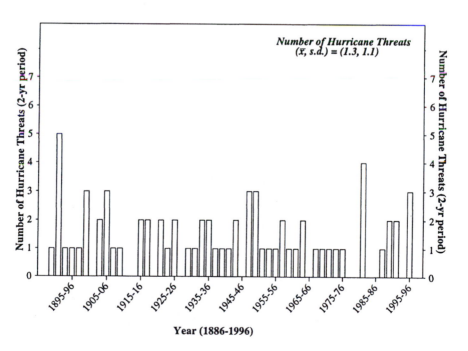

Fig. 9.19: Time series of the abundance of hurricane threats to Bermuda in 5-year intervals (top) and 2-year intervals (bottom). The frequency of hurricanes threatening Bermuda over the 111-year record shows no significant trend.

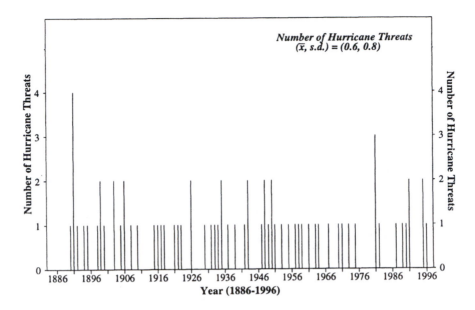

Fig. 9.20: Time series of the annual frequency of hurricane threats to Bermuda over the period 1886–1996. The average ($\bar{x}$) and standard deviation ($s.d.$) of the number of threats is given in the upper-right corner. The hurricane threat frequency for Bermuda is relatively constant over the period of record.

Table 9.6: Hurricane threats to Bermuda. Values are frequencies and probabilities based on data over the period 1886–1996. Whole numbers are the count of years with that many hurricanes. There is a 50% chance of at least one hurricane threat per year, but the probability of two or more threats per year is only 11%. Ten of the years had exactly two hurricane threats.

Number of Threats	Number of Years	Observed Probability
0	56	0.50
1	43	0.39
2	10	0.09
3	1	0.01
4 or more	1	0.01

Amount	Number of Years	Observed Probability
At least 1 threat	55	0.50
2 or more threats	12	0.11
3 or more threats	2	0.02
4 or more threats	1	0.01

at least one threat is five from 1947 through 1951. On average, Bermuda can expect to be threatened about once every 2 years. Table 9.6 gives the observed frequencies and empirical probabilities of annual hurricane threats to Bermuda. The probability of exactly one threat is 39%, which is somewhat higher than the probabilities for Jamaica and Puerto Rico. Recall that threats to Bermuda are defined using a smaller area. However, the probability of two or more threats in a single season is 11% compared to 27% for Puerto Rico. Thus multiple threats in a single season are not as common in Bermuda or Jamaica as they are in Puerto Rico. Ten years had exactly 2 hurricane threats.

10

Hurricane Cycles and Trends

The prospect of "global warming" leading to a future with more frequent and intense hurricanes has surfaced in the public imagination (see, e.g., Schneider 1989, Easterbrook 1995). Fluctuations in North Atlantic hurricanes have important implications for planning and development. Trends in abundance, duration, location, and intensity of hurricanes are of concern to both public and private economic sectors. Social benefits may be realized through accurate predictions of future activity. Practical considerations aside, there is a desire to understand nature in terms of simple explanations. Given a set of observations it is natural and practical to examine it for regular patterns and trends. This chapter looks at the case for oscillations and long-term trends in North Atlantic hurricane activity. The presentation follows the work of Elsner et al. (1999b).

In overview, biennial and semidecadal oscillations are identifiable in the record of North Atlantic annual hurricanes. These oscillations correspond to well-known physical processes implicated in modulating hurricane activity. The quasi-biennial shift in tropical stratospheric winds between an east and west phase (known as the quasi-biennial oscillation, or QBO) has a statistical relationship with North Atlantic hurricane frequencies. A semi-regular shift in equatorial Pacific Ocean sea-surface temperatures between a warm and cold phase (known as the El Niño-Southern Oscillation, or ENSO) appears to be linked to North Atlantic hurricane activity. Both these signals are found in the frequency of U.S. landfalling hurricanes. A near-decadal oscillation is noted in the frequency of baroclinically-enhanced hurricanes, but it is not apparent in the abundance of tropical-only hurricanes.

The annual number of North Atlantic hurricanes is relatively constant over the time period of record, though there is a slight trend. Decreases in tropical-only hurricane activity are somewhat compensated by increases in baroclinic hurricane activity. There is, however, a significant decrease in the duration of hurricanes over the past 111 years. This coincides with a trend toward more baroclinically-enhanced hurricanes, which tend to appear at more northerly latitudes. Despite the likely bias in the earlier part of the record toward fewer hurricanes, the average length of time a tropical cyclone spends as a hurricane and as a major hurricane is decreasing. This is consistent with

240

the claim that modern technology captures the weaker, shorter-lived hurricanes. More importantly, there is a shift toward increased numbers of baroclinically-enhanced hurricanes in the recent decades.

10.1   Time Series Analysis

Time-series analysis offers a rigorous approach to uncovering underlying cycles in a set of observations. Several methods are available for time-series analysis, including Fourier decomposition and autoregressive modeling. The method of singular spectrum analysis (SSA)—which is a special case of principal component analysis—is a tool in the geosciences for time-series analysis (Elsner and Tsonis 1996). SSA was first used in oceanography by Colebrook (1978) and in climatology by Fraedrich (1986) and Rasmusson et al. (1990). It was introduced as a useful method for time-series analysis by Broomhead and King (1986). Here various aspects of the North Atlantic hurricane record are examined using the SSA.

SSA begins with a lagged-covariance matrix (S) computed as

$$S_{ij} = \frac{1}{N_t - m + 1} \sum_{t=1}^{N_t - m + 1} x(i + t - 1)\, x(j + t - 1). \tag{10.1}$$

The principal components are computed from the eigenvectors of S. The original time series is reconstructed from the principal components as

$$x(i + j - 1) = \sum_{k=1}^{m} a_i^k e_j^k, \tag{10.2}$$

where x is the reconstructed component, a_i^k is the ith term of the kth principal component, and e_j^k represents the jth term of the kth eigenvector. The reconstructed components are limited in harmonic content. As such, they can be examined effectively using traditional time-series methods. Furthermore, since the time series is a sum of the individual reconstructed components, the removal of one or more components filters the record.

10.1.1   Annual Abundance

We consider first the record of annual frequency of North Atlantic hurricanes. To begin, it is important to remove the trends and ultra-low frequency components in the time series. For the hurricane record, trends and extremely low-frequency oscillations arise from changes in observing techniques over the years and from natural fluctuations, perhaps induced by changes in sea surface temperatures. The SSA is used as part of an algorithm for removing trends (Vautard and Ghil 1989). The algorithm includes a nonparametric test of significance for the trend (Kendall and Stuart 1977). Given a time series $x(t)$, $t = 1, 2, \ldots, N$, the number K_r of pairs of indices (t, u), with $t < u$ such that $x(t) < x(u)$ are counted. In general, if K_r is large there is an increasing

Table 10.1: Kendall statistics. Values are for the principal components of the hurricane record over the period 1886–1996. The 5% confidence interval is $(-0.135, +0.135)$. The null hypothesis of no significant trend is rejected only for the first principal component.

Mode	τ value	Trend ?	Acceptance
1	-0.457	Yes	Reject
2	-0.048	No	Accept
3	-0.016	No	Accept
4	$+0.018$	No	Accept
5	-0.012	No	Accept
6	-0.002	No	Accept
7	$+0.058$	No	Accept
8	$+0.003$	No	Accept

trend, and if K_r is small there is a decreasing trend. How large is determined by the test statistic

$$\tau = \frac{4K_r}{N(N-1)} - 1. \tag{10.3}$$

The distribution of τ for large N is normal with zero average and standard deviation

$$s = \sqrt{\frac{2(2N+5)}{9N(N-1)}}. \tag{10.4}$$

Allowing for a 5% chance of being wrong, the hypothesis of no trend is rejected outside the interval $(-1.96s, +1.96s)$. Results of the test for the first eight principal components (where $N = N_t - m + 1$) are shown in Table 10.1.

Here N_t is 111 years (1886–1996) and the window m is fixed at 15 years. Only the first principal component has a value of τ outside the range. The first principal component contains a significant ultra-low frequency oscillation. This oscillation is removed before additional analysis. Figure 10.1 shows the hurricane record without the low-frequency component (nonlinear trend). The annual abundance of hurricanes is expressed as a normalized departure from the average. The procedure is repeated on the detrended record to ensure against spurious trends resulting from the convolution. No principal components are outside the ± 5% limits.

The singular spectrum of the detrended time series using $N_t=111$ and $m=15$ is shown in Figure 10.2. Eigenvalues are shown for each of the time modes. The number of modes is equal to the window length. Every eigenvalue has an associated eigenvector representing a temporal pattern in the record. An important oscillation in the record— even if it is somewhat irregular—is represented as a pair of nearly equal eigenvalues in the singular spectrum. This is an approximation for records with noise.

The first three eigenvector pairs are shown in Figure 10.3. Each vector pair is a set of waves having a common frequency and phases that are in quadrature (one wave leads

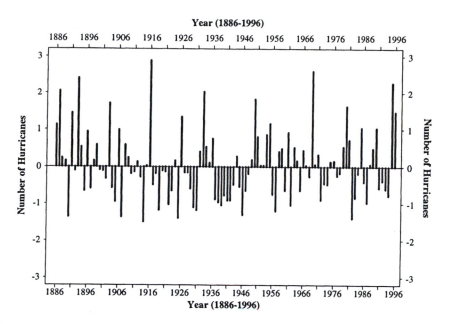

Fig. 10.1: Annual North Atlantic hurricanes over the period 1886–1996. The values are expressed as normalized departures from the mean. The nonlinear trend (i.e., first principal component) is removed.

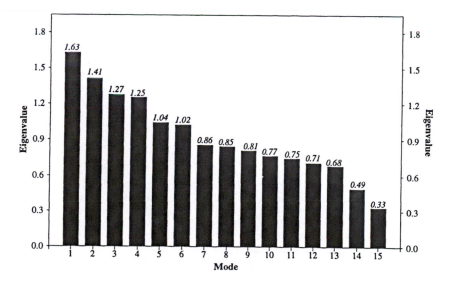

Fig. 10.2: Eigenvalues of the detrended hurricane record from the method of singular spectrum analysis (SSA). The first three pairs of eigenvalues (modes 1–6) are above the noise floor.

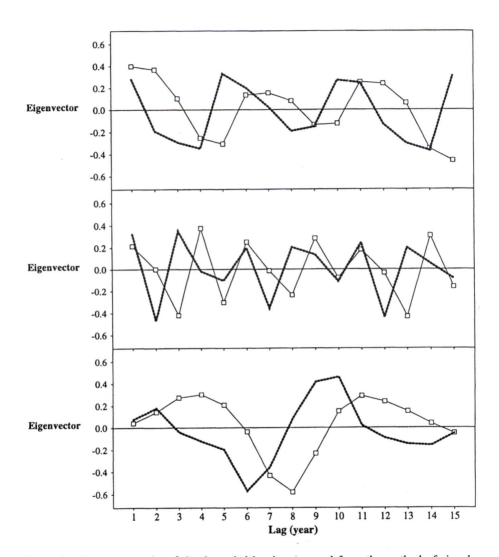

Fig. 10.3: Eigenvector pairs of the detrended hurricane record from the method of singular spectrum analysis (SSA). Record length is 111 years, and the lag is in years. The number of lags is determined by the window length. Amplitudes of the eignvectors are the same for each pair. Each eigenvector pair corresponds to a distinct oscillation in the annual abundance of North Atlantic hurricanes. The top eigenvector pair explains the largest percentage of variance in the detrended hurricane record. Each pair is in approximate quadrature, indicating one wave leads the other by $\frac{\pi}{2}$. In general, eigenvector pairs corresponding to higher modes do not share this characteristic.

the other by a quarter period). These characteristics indicate that the eigenvectors are associated with a meaningful oscillation in the hurricane record. Beyond mode 6 the eigenvalues are less than unity, and the eigenvector pairs are generally not in quadrature. This qualitative assessment does not speak to the statistical significance of the oscillations. Indeed, a similar analysis on a random, but correlated, time series will reveal pairs of nearly equal eigenvalues and associated eigenvectors in quadrature.

Reconstructed components are computed using Eq. 10.2.[1] Figure 10.4 shows the reconstructed components from the three leading eigenvector pairs. The components feature alternating positive and negative values. The frequency of oscillation is different for each. The reconstructed component from the second eigenvector pair indicates a high-frequency oscillation. The third pair suggests a low frequency intradecadal oscillation. The first component has the largest amplitudes. Amplitudes vary in time in each of the components so that different components are relatively more important to the total variability during different epochs. For example, during the 1950s and 1960s the higher frequency oscillations (first and second reconstructed components) have largest amplitudes. Beginning in the middle 1980s the low-frequency component is relatively more important indicated by higher amplitudes. Over the same period, the amplitudes of the second reconstructed component are lower.

Because of their restricted harmonic content, the reconstructed components are readily amenable to low-order autoregressive modeling. The frequency spectra can be examined effectively using the maximum entropy method (MEM) of spectral analysis. Furthermore, parsimonious models can be built for predictions (see Chapter 16). The MEM is capable of high spectral resolution making it possible to accurately pinpoint the underlying period of oscillation in a time record. If the time series is filtered, MEM can achieve this resolution without the problem of spurious peaks (Penland et al. 1991). Here MEM is applied to the three reconstructed components using the method of Press et al. (1989) with a maximum order of 15 to match the window length used in the SSA. Each reconstructed component has a distinct MEM spectrum with peaks corresponding to oscillations of 2.5, 5.6, and 7.4 years (Figure 10.5). As the oscillations are not pure waves, values of the periodicities represent a median over a range. The largest range occurs for the lowest frequency component. A check is made to ensure the nonlinear trend removed from the data contains no significant power at these frequencies. Moreover, a check against spurious peaks is made by repeating the analysis on a randomized record of hurricane occurrences. No significant peaks are found in the random data.

This two-step method of spectral analysis (SSA followed by MEM) is capable of resolving the low-frequency oscillations into distinct frequencies. This resolution is not possible if MEM is applied directly to the original record without the prefiltering afforded by SSA. The hurricane record is approximated by summing the contributions from each of the three reconstructed components. Figure 10.6 is a scatter plot of the filtered versus the detrended record. Most points are scattered along the diagonal. The reconstructed record is a good approximation to the actual record.

---

[1] The summation is limited to a single eigenvector pair.

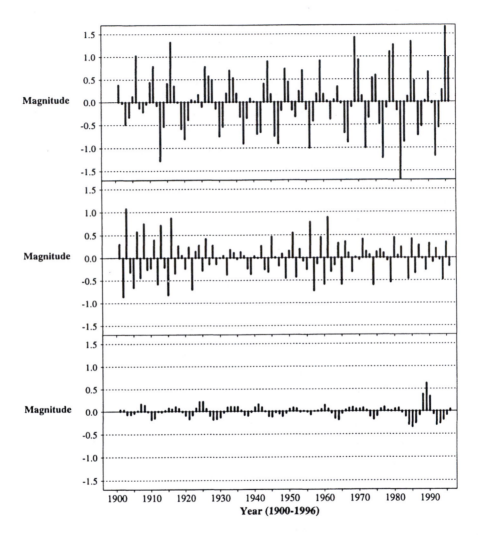

Fig. 10.4: Reconstructed components of the detrended hurricane record over the period 1901–1996 using the method of SSA. The components are computed by projecting the eigenvectors onto the hurricane record. Each reconstructed record corresponds to a distinct oscillation with limited harmonic content and amplitude modulation. The years from 1886–1900 are omitted because the oscillations are out of phase in the second and third components. In general the amplitudes decrease with successive components. Note the higher amplitudes in the low-frequency component during the 1980s and 1990s.

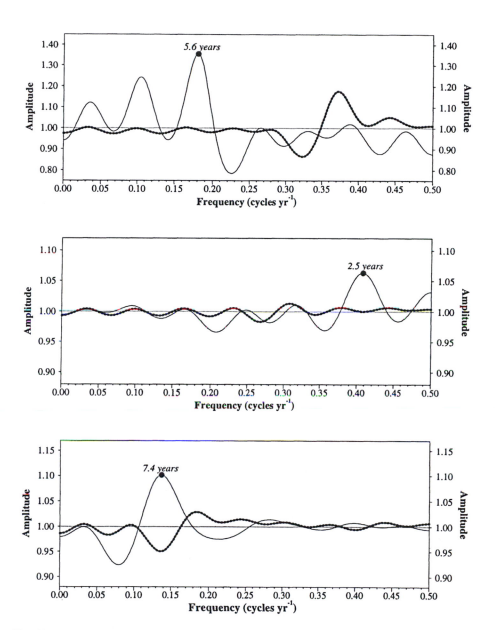

Fig. 10.5: Maximum entropy method (MEM) spectra of the three dominant reconstructed components of the detrended North Atlantic hurricane record over the period 1901–1996. The order of MEM is 15 years. The dot indicates the dominant frequency in the reconstructed component. The thick line corresponds to the MEM of a corresponding reconstructed component from a randomly permuted hurricane record. The ordinate scales are not the same in each graph.

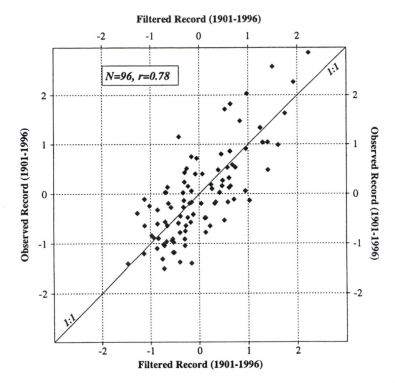

Fig. 10.6: Detrended North Atlantic hurricanes (normalized) versus a filtered record. The filtered record is obtained by adding the three dominant reconstructed components using the method of SSA. Number of cases (*N*) and the linear correlation (*r*) between the two records are provided. The linear correlation (*r*) between the two records is +0.78.

The 2.5-year periodicity reflects the well-established association of hurricane activity with the stratospheric quasi-biennial oscillation (QBO). The stratospheric QBO is a regular fluctuation in upper-level winds from strong easterlies to weak easterlies (or weak westerlies) that occurs over tropical latitudes at a periodicity of between 2 and 3 years. The semidecadal oscillation is likely tied to the El Niño-Southern Oscillation (ENSO) of the tropical Pacific Ocean, which has an irregular fluctuation in the range of 4 to 6 years and has been implicated in modulating major hurricane activity over the North Atlantic basin. Sir Gilbert Walker named the Southern Oscillation after his observations in the 1920s and 1930s that surface air pressures are high in the Pacific Ocean when they are low in the Indian Ocean from Africa to Australia. It is known that the Southern Oscillation (SO) is closely linked to changes in ocean temperatures over the tropical Pacific, referred to as El Niño events; thus the combined El Niño-Southern Oscillation (ENSO) event is widely known as ENSO. It is likely that both the QBO and ENSO modulate North Atlantic hurricane activity to some extent through changes in upper-troposphere winds. When the Pacific ENSO is in its warm (or El Niño) phase, enhanced convection over the central and eastern Pacific produces upper-tropospheric westerlies (winds blowing from west to east) across the tropical Atlantic. Anomalous westerlies above the equator generate upper-level convergence and sinking air, conditions which are detrimental to hurricane development. The increased subsidence is linked to enhanced low-level easterly winds across the tropical Atlantic. This leads to cooler than normal SSTs in the areas most favorable to tropical-only hurricane development.

The cause of the near-decadal oscillation in North Atlantic hurricanes is less certain. The oscillation might be related to fluctuations in sea-surface temperature. Tropical cyclone activity on the near-decadal scale has been linked to changes in SSTs over the western North Pacific (Walsh and Kleeman 1997). SSTs east of the Lesser Antilles are related to modulations in hurricane activity, with warm years (high SSTs) associated with more hurricanes (Kimberlain and Elsner 1998). Rainfall over portions of the Nordeste region of Brazil was heavy during the middle 1970s and again during the middle 1980s. In between, drought was common. These changes might be linked to near-decadal fluctuations in SSTs over the tropical Atlantic (Chang et al. 1997).

A plausible physical explanation centers on the atmosphere's response to observed variations in SST differences between the northern and southern hemispheres (SST gradient). Changes in SST gradients influence changes in SLPs and thus the direction of winds. Assuming hydrostatic equilibrium[2] and weak upper-air temperature contrasts, the SST gradient will result in lower SLPs in the southern hemisphere and higher SLPs in the northern hemisphere. The anticyclonic (clockwise) flow of air around the high pressure in the north will enhance the northeast trades, thereby increasing evaporation and keeping the ocean surface relatively cool. Similarly, the cyclonic (also clockwise) flow of air around low pressure in the southern hemisphere will restrict the southeast trades implying less evaporation and keeping the SSTs high. One way out of this positive feedback loop is through large-scale changes in ocean circulation aided by changes in atmospheric radiation and turbulent energy fluxes (Chang et al. 1997). Interestingly, Mehta (1998) finds a spectral peak at 8 to 9 years in tropical Atlantic SST anomalies and a statistically significant coherence with an index of North Atlantic tropical cyclone activity. The phase difference indicates tropical cyclone activity lags warmer SSTs by a few years. Moreover, Enfield and Mayer (1997) and Penland and Matrosova (1998) show a predictive relationship between El Niño and tropical North Atlantic SSTs.

Low-frequency changes in North Atlantic SSTs might have the largest impact on baroclinically-enhanced hurricanes by expanding the area conducive to hurricane development. Figure 10.7 shows the power spectra[3] of the annual number of tropical-only and baroclinically-enhanced hurricanes. As anticipated the high frequencies are dominated by rhythms corresponding to the QBO and the ENSO. This is especially true for tropical-only hurricanes, although there is a weak QBO signal in the occurrence of baroclinically-enhanced hurricanes. In contrast, significant periodicity in the range of 7 to 9 years is noted only for baroclinically-enhanced hurricanes. The above hypothesis

---

[2]Hydrostatic equilibrium is the balance between the vertical pressure gradient force and gravity.
[3]Here the Blackman-Tukey method (Blackman and Tukey 1958) is used.

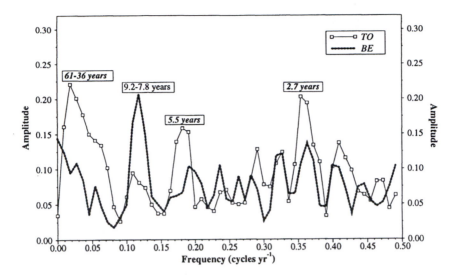

Fig. 10.7: Power spectra of the detrended annual tropical-only (TO) and baroclinically-enhanced (BE) hurricane records over the period 1886–1996. The Blackman-Tukey method is used. Largest peaks are marked with a corresponding period of oscillation.

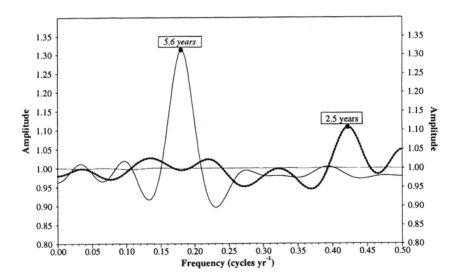

Fig. 10.8: Maximum entropy method (MEM) spectra of the three leading reconstructed components of the U.S. hurricane record over the period 1851–1996 using SSA. Note only the first two components show oscillations. The third component is flat with an amplitude near one.

concerning the role of North Atlantic SSTs in modulating hurricane activity hints that the variance would be concentrated at similar near-decadal frequencies. Much of this is speculation and awaits further research.

10.1.2 U.S. Hurricanes

Do similar oscillations exist in U.S. hurricane activity? To answer this question singular spectrum analysis is performed on the 146-year record of U.S. hurricanes (Chapter 8). Figure 10.8 shows the maximum entropy method spectra of the three leading reconstructed components of the U.S. hurricane record. Here N_t is 146 years and m is 15 years. Results show a quasi-biennial and a semidecadal rhythm. The 5- to 6-year oscillation corresponds to the ENSO as noted previously for all hurricanes. Not surprisingly, the strong influence of stratospheric QBO on North Atlantic hurricane activity appears to also modulate U.S. hurricane activity. The ENSO is a more dominant factor, however. The near-decadal component, associated with baroclinically-enhanced hurricanes, is not discernible in the frequency of U.S. landfalls. This is consistent with the fact that baroclinically-enhanced hurricanes tend to track away from the United States.

10.2 Superposed Epochs

The method of superposed epochs (composite analysis) is often used in climate studies. It consists of identifying occurrences of an event over time from an independent data set, then counting (or averaging) the variables from the dependent data at the event times. Here the dependent data are the various components of annual North Atlantic hurricane activity (e.g., the frequency of all hurricanes, major hurricanes, U.S. hurricanes, etc.). If the independent data are continuous values, like SSTs, the event is defined in terms of percentiles. The superposed epoch method is used here to examine annual North Atlantic hurricane activity vis-á-vis the QBO, ENSO, eastern North Pacific hurricanes, solar variations, and volcanic activity. Events are examined separately.

10.2.1 QBO

As mentioned, the stratospheric QBO and the Pacific ENSO are implicated in modulating hurricane activity over the North Atlantic through changes in upper-level winds. Stratospheric winds are available for upper-air stations in the Caribbean back to 1950. The 50 mb zonal wind averaged over values from stations reporting in the Caribbean from August through October provides an index of the QBO relevant to North Atlantic hurricane activity. The average 50 mb winds are from the east, but a particular year experiences a positive or negative departure from the average for the hurricane season. A departure from the mean is called an *anomaly*. A positive anomaly is defined as the westerly phase of the QBO while a negative anomaly is defined as the easterly phase. The oscillation is biennial so that easterly and westerly anomalies alternate approximately every year or so.

Table 10.2: North Atlantic hurricane activity with respect to extremes of the QBO over the period 1950–1996. Strength of the QBO is indicated by the upper-level (50 mb) zonal winds averaged from stations over the Caribbean from August through October. Q50 refers to the 50 mb zonal wind anomaly (m s^{-1}). The average ($\bar{x}$) for each category of activity is also given.

QBO East Phase					
Year	Q50	All	TO	BE	MH	U.S.
1952	-11.0	6	6	0	3	1
1954	-13.0	8	5	3	2	3
1956	-7.7	4	3	1	2	1
1968	-9.0	5	1	4	0	1
1970	-10.0	5	1	4	2	1
1972	-9.0	3	0	3	0	1
1977	-12.3	5	1	4	1	1
1984	-13.6	5	0	5	1	1
1992	-9.7	4	0	4	1	1
1994	-13.7	3	1	2	0	0
$\bar{x}$	*-11.1*	*4.8*	*1.8*	*3.0*	*1.2*	*1.1*

QBO West Phase					
Year	Q50	All	TO	BE	MH	U.S.
1955	7.7	9	9	0	6	3
1957	8.7	3	2	1	2	1
1959	8.7	7	1	6	2	3
1961	7.7	8	7	1	7	1
1964	10.7	6	5	1	6	4
1975	7.7	6	4	2	3	1
1978	9.0	5	2	3	2	0
1980	8.3	9	4	5	2	1
1985	10.0	7	4	3	3	6
1995	11.0	11	9	2	5	2
$\bar{x}$	*8.8*	*7.1*	*4.7*	*2.4*	*3.8*	*2.2*

Though the stratospheric QBO is close to being periodic, it is not a simple harmonic of the seasonal cycle. It is argued that the QBO is related to momentum transfer from the troposphere into the stratosphere. Vertically propagating waves transfer momentum upward. The momentum is absorbed by the horizontal air moving from east to west (zonal winds). It is suspected that the reduction of hurricane activity in east-phase years results from an increase of shear between winds in the lower stratosphere and winds in the upper troposphere. The shear disrupts the development of tropical cyclones by contorting the central column of warm air in the fledgling storm. Table 10.2 gives the average 50 mb zonal wind anomaly over the Caribbean region from a base period of 1950–1996.

Hurricane activity is given for the 10 extreme years of westerly anomalies and the 10 extreme years of easterly anomalies. The westerly phase is considerably more fa-

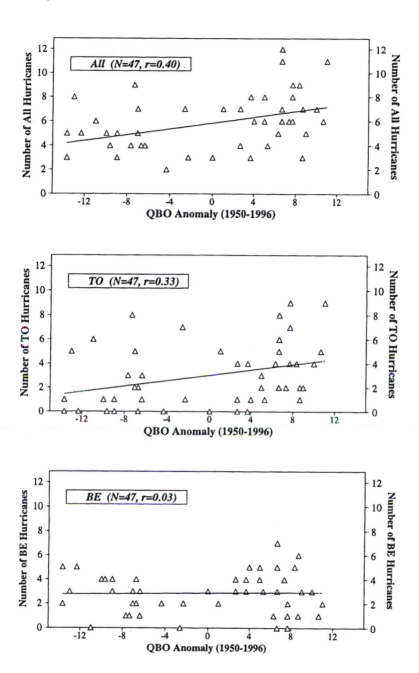

Fig. 10.9: North Atlantic hurricane activity versus the stratospheric QBO over the period 1950–1996. The QBO is the August through October average 50 mb zonal wind based on upper-air observations from stations over the Caribbean. N is the number of years and r is the linear correlation coefficient.

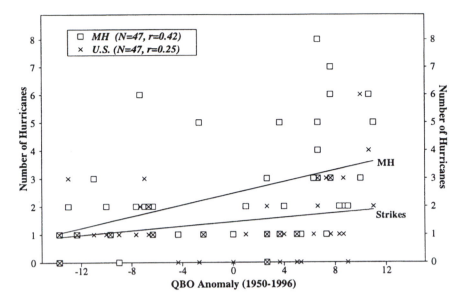

Fig. 10.10: Major and U.S. hurricanes versus the stratospheric QBO over the period 1950–1996. Major hurricane activity is depicted with squares and U.S. hurricane activity with crosses. The QBO is the August through October average 50 mb zonal wind based on upper air observations from stations over the Caribbean. N is the number of years and r is the linear correlation coefficient.

vorable for North Atlantic hurricane activity. The average number of hurricanes for the 10 extreme years of the westerly phase is 7.1, compared to 4.8 for the easterly phase. Moreover, extremes of the QBO discriminate very effectively between tropical-only and baroclinically-enhanced hurricanes as the QBO phase determines which type of storm is relatively most dominant. The QBO is strongly related to the frequency of tropical-only hurricanes. This is reflected in the differences in major hurricane frequency between the extreme QBO years, with a ratio of 3.5 to 1 in favor of major hurricanes during years of strong westerly anomalies. Recall that most major hurricanes are tropical-only. The number of U.S. hurricanes is also influenced by the extremes of the QBO. There are twice as many U.S. hurricanes during the west phase as there are during the east phase. The relationship between the stratospheric QBO over the Caribbean and hurricane activity can be seen in the scatter plots of QBO versus hurricane activity (Figures 10.9 and 10.10). Data are from the period 1950 through 1996. Increased hurricane activity with increasing values of QBO anomalies are noted for all categories of hurricane activity. As expected, very little correspondence is noted for baroclinically-enhanced hurricanes. Statistically significant positive relationships are found for all, tropical-only, major, and U.S. landfalling hurricanes. A QBO in sea-level pressures and other atmospheric variables may regulate the abundance and tracks of North Atlantic hurricanes (Shapiro 1982).

10.2.2   ENSO

It is well known that the Pacific ENSO has a relationship to North Atlantic hurricane activity (Gray et al. 1993, 1994). In fact much of the interannual variability in tropical climate can be traced to fluctuations in the ENSO. Hurricane activity over the Caribbean region, in particular, appears to be modulated by the ENSO (Caviedes 1991). Hess et al. (1995) show that the ENSO signal in the hurricane climate is more closely related to the set of hurricanes common over the deep tropics. During a typical warm ENSO event, which begins during a Northern Hemisphere spring, an area of unusually warm surface water appears along the eastern equatorial Pacific, expanding westward over several consecutive months (Philander 1983). The warming along the Peruvian coast is called El Niño or warm phase of ENSO.[4] Associated with the warm water in the eastern Pacific are above-average air pressures near the surface in the western Pacific, enhanced precipitation near the equator to the east of 160°E, and weaker easterly trade winds over the eastern Pacific.

ENSO represents a physical coupling between the tropical atmosphere and the waters of the Pacific Ocean (Figure 10.11). Under normal conditions warm surface waters and deep convection are located over the western tropical Pacific. In contrast, during an El Niño, the warmest waters are farther to the east as is the most vigorous convection. The compensating subsidence over the western North Atlantic inhibits hurricane development. The condition of colder than normal SSTs over the eastern and central Pacific is called La Niña; it represents the cold phase of ENSO. Warm and cold ENSO phases tend to alternate every several years or so. There is a tendency for warm ENSO conditions to develop or intensify when the QBO is in an easterly phase, triggered by concentrated and amplified convection over the equator. The enhanced equatorial convection promotes the occurrence of high frequency atmospheric pulses (Madden-Julian oscillation) that lead to oceanic Kelvin waves, which eventually change the SST patterns. The linkage between QBO and ENSO, though weak and subtle, confounds interpretations of their mutual relationship with hurricanes using simple statistical analyses. Changes in the tropical climate of the Pacific associated with ENSO are linked to climate fluctuations in other remote tropical locations and to portions of the extratropics as well.[5] Since the ENSO encompasses a large sector of the Pacific basin and influences much of the tropical climate, it can be characterized in several ways. Traditionally, the difference in sea-level pressures (SLPs) between Tahiti and Darwin ($SLP_{Tahiti} - SLP_{Darwin}$) is used as an indicator of the phase of ENSO. This Southern Oscillation Index (SOI) varies considerably from month to month. High pressures in the western Pacific (near Darwin, Australia) tend to be coupled with low pressures in the central Pacific (near the island of Tahiti) during a warm phase, creating a negative pressure difference. During a cold phase, surface-air pressures are reversed at the two

---

[4]The term *El Niño*, was originally applied to the annual warm ocean current that ran southward along the coast of Peru at Christmas time (hence Niño, Spanish for Christ child), and only later became associated with the intermittent warming of the central and eastern tropical Pacific (Trenberth 1997).

[5]Linkages of climate phenomenon over large distances are termed "teleconnections."

Normal Conditions

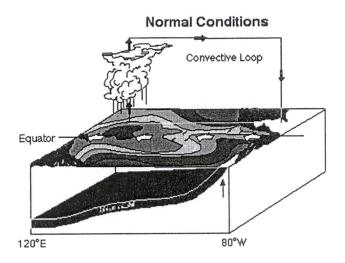

El Niño Conditions

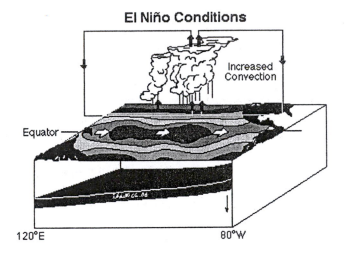

Fig. 10.11: Schematic of the warm phase of ENSO (El Niño). Normal conditions are shown for comparison. During normal conditions the atmospheric convection (showers and thunderstorms) are concentrated over the western Pacific with sinking air in the east. During warm ENSO events, the convection shifts with the warmest waters to the central Pacific. The eastern branch of sinking air is more pronounced over the western tropical Atlantic during warm ENSO events. This subsidence contributes to an unfavorable environment for hurricane origin and development over the North Atlantic.

Table 10.3: Phases of the Pacific ENSO. The phase is based on values of the JMA index. A reconstruction method is used to estimate SST patterns for earlier years. Y indicates the last digit of the year. Cold phases are marked, c, warm phases, w, and neutral phases, n.

Y	188	189	190	191	192	193	194	195	196	197	198	199
0	.	n	n	c	n	w	w	n	n	c	n	n
1	.	n	n	w	n	n	n	w	n	c	n	w
2	.	c	w	n	c	n	c	n	n	w	w	n
3	.	c	c	w	n	n	n	n	w	c	n	n
4	.	n	w	n	c	n	c	c	c	n	n	n
5	.	n	w	n	w	n	n	c	w	c	n	n
6	c	w	c	c	n	n	n	c	n	w	w	n
7	n	n	n	n	n	n	n	w	c	n	w	.
8	w	n	c	w	n	c	n	n	n	n	c	.
9	c	w	c	n	w	n	c	n	w	n	n	.

locations producing a positive SOI. Large month-to-month variations in the SOI make detecting the phase of ENSO difficult with this index alone.

Sea-surface temperatures (SSTs) in the eastern equatorial Pacific Ocean are a more reliable indicator of ENSO phase. SST values averaged over space and time for different tropical sectors are monitored for the appearance of significant warming heralding the onset of a warm ENSO phase. Figure 10.12 shows the correlation of SLP at Darwin with SLPs and SSTs across the globe. The strong negative relationship between SLPs across the tropical Pacific is clearly evident. When surface pressures are high over the western half of the tropical Pacific they are low over the eastern half. The strong positive relationship between Darwin pressure and equatorial SSTs over the eastern half of the Pacific is also evident. Higher than normal surface pressures over Australia are associated with warmer than normal SSTs over the eastern Pacific. SSTs over the eastern and central Pacific give a good indication of the ENSO phase.

The Japan Meteorological Agency (JMA) index based on a 5-month running mean of spatially-averaged SST anomalies over the tropical Pacific between 4°S and 4°N latitude and between 90° and 150°W longitude is an index that selects ENSO events with fidelity. If the JMA index is 0.5°C or greater for 6 consecutive months including October through December, the ENSO year of October through the following September is considered a warm phase. Index values less than $-0.5°C$ indicate a cold phase. Over the period 1949 through 1996, the JMA index is based on observed data. The SST fields are reconstructed back to 1868 using an orthogonal projection technique (Meyers et al. 1999). The reconstruction allows for an estimation of the JMA index to be made over the earlier years. Table 10.3 lists the ENSO years according to the JMA index over the period 1886–1996. In general, warm and cold events do not last for more than a single year. Neutral years are more persistent. Less than half of the years are considered to be either in a warm or cold ENSO phase.

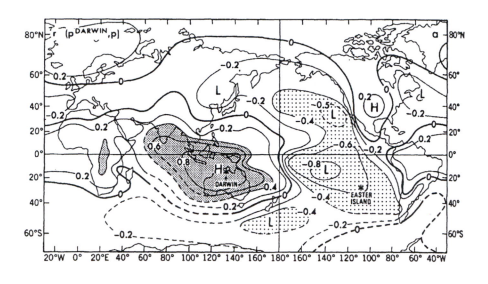

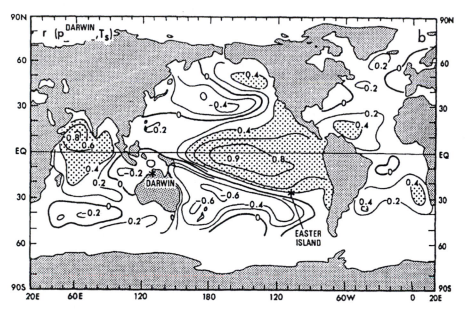

Fig. 10.12: Correlation between sea-level pressures at Darwin, Australia. Annual-average sea-level pressures (top) and the annual-average sea-surface temperatures over the globe (bottom). Values are the correlation coefficient with Darwin SLPs. Note the difference in SLPs across the Pacific and the SSTs over the central and eastern Pacific are both indicators of the phase of ENSO. Reproduced with permission from Peixoto and Oort (1992).

Table 10.4 shows the association between ENSO and North Atlantic hurricane activity based on the JMA index. Warm- and cold-phase years are identified as the year in which onset occurs; for instance, the warm phase of 1896 runs from October 1896 through September 1897. The influence of ENSO on hurricanes occurs during the hurricane season of onset. As an example, the 1896 North Atlantic hurricane season is considered in a warm-phase. In general, North Atlantic hurricane activity is substantially modified by the warm phase of the Pacific ENSO. In particular, fewer tropical-only hurricanes, fewer major hurricanes, and a reduced probability of multiple U.S. landfalls are features of El Niño years. On average there are twice as many tropical hurricanes during cold years as there are during warm years. The exception is baroclinically-enhanced hurricane activity which, as anticipated, has only a weak response to the phases of ENSO. Warm years feature a slight increase in baroclinically-enhanced activity. The ENSO cold phase increases hurricane activity over the North Atlantic, but not as substantially as the warm phase inhibits it. However, the cold phase does have a significant influence on U.S. landfalls.

The most noteworthy effect of the ENSO occurs for tropical-only hurricanes. Thus, it is not surprising that the probability of significant U.S. hurricane activity is reduced during warm ENSO years. The ratio of major U.S. hurricanes during non–El Niño years versus El Niño years is about three to one (Gray 1984a). Figure 10.13 shows the probability distributions of U.S. hurricanes with respect to the ENSO phases based on the assumption that the occurrence of U.S. hurricanes is a Poisson process. Clearly the cold phase is more conducive to U.S. hurricane activity. The annual probability of exactly 3 U.S. hurricanes is 20% during a cold year compared to only 6% during a warm year. Moreover, the average annual number of U.S. tropical-only hurricanes during the cold phase of ENSO is 1.5 compared with 0.2 during the warm phase.[6]

Differences in North Atlantic hurricane activity between the two extremes of the ENSO cycle are further illustrated in Figure 10.14. Points of origin for hurricanes suggest that early-season hurricanes tend to form over the Gulf of Mexico during a warm phase, whereas they tend to form off the southeast U.S. coast during a cold phase. Middle-season activity during a warm phase shifts toward the western North Atlantic, generally away from the United States. Higher latitude formations (north of 30°N) are more common during El Niño years. In contrast to the early season, middle-season hurricanes are more likely over the Gulf of Mexico during a cold ENSO than during a warm ENSO. Late-season hurricanes over the Caribbean Sea are much more common during a cold phase than during a warm phase (see also Caviedes 1991). In fact, during a cold ENSO phase, the average latitude of hurricane formation shifts farther to the south as the season progresses. This is to be expected as tropical-only hurricanes are more frequent after July. However, warm years are characterized by a northward shift of formation during October and November (see Table 10.5). This could be due to enhanced baroclinic factors along the east coast in early autumn during

---

[6]Note the probability of exactly one U.S. hurricane during warm years is larger than the probability during cold years.

Table 10.4: North Atlantic hurricane frequencies according to the phases of ENSO. Averages ($\bar{x}$) and standard deviations ($s.d.$) for all, tropical-only (TO), baroclinically-enhanced (BE), and major hurricanes (MH) are based on data over the period 1886–1996. Statistics for U.S. hurricane activity are based on data over the period of 1868–1996. N indicates the number of cases.

ENSO Event	Statistic	All	TO	BE	MH	U.S.
	N	26	26	26	26	32
Cold Phase	$\bar{x}$	5.8	4.2	1.6	2.3	2.3
	$s.d.$	2.2	2.3	1.4	1.8	1.7
	N	24	24	24	24	27
Warm Phase	$\bar{x}$	4.0	2.1	2.0	1.2	1.3
	$s.d.$	2.4	1.7	1.8	1.5	1.0
	N	61	61	61	61	70
Neutral	$\bar{x}$	5.0	3.3	1.8	2.0	1.7
	$s.d.$	2.5	2.4	1.7	1.8	1.2

El Niño years. Variations in latitude are generally larger during the warm years, but variations in longitude are larger during the cold years.

10.2.3   Eastern North Pacific Hurricanes

Hurricanes frequently develop over the eastern North Pacific Ocean near the coast of Mexico. The principal formation region is bounded by 90° and 120°W longitude and by 20° and 10°N latitude. Storms of this region generally track northwesterly, and stay out to sea. Occasionally they move inland over Mexico. On average, eastern North Pacific tropical cyclone activity begins several weeks before the start of the North Atlantic hurricane season, although there tends to be a break in seasonal activity around the middle of August. The average seasonal number of eastern North Pacific tropical cyclones is 15. The abundance during individual years varies considerably from this value. There were 8 North Pacific tropical cyclones during 1977 and 24 during 1992.

Interestingly, the seasonal abundance of tropical cyclones in the eastern North Pacific is inversely related to the abundance of hurricane activity in the Atlantic. Table 10.6 compares North Atlantic hurricane activity during active and inactive years in the eastern North Pacific. An active year is defined as 11 or more eastern North Pacific hurricanes. An inactive year has 6 or fewer hurricanes. The ratio of North Atlantic hurricanes for active and inactive years over the Pacific is 0.7 to 1. During years when activity is above normal over the eastern Pacific it tends to be below average over the Atlantic, and vice versa. Correlation does not imply causality, but the inverse relationship indicates that it is less likely than random chance to be extremely active over both basins in a given year. There is a tendency for North Atlantic tropical cyclone activity to be out of phase with tropical cyclone activity in other regions of the world (Willett 1955).

The inverse relationship is most pronounced for the category of tropical-only hur-

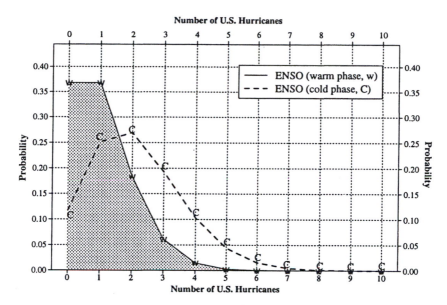

Fig. 10.13: Probability distributions for U.S. hurricanes with respect to ENSO. The distributions are based on a Poisson process for U.S. hurricane activity. The Poisson parameter is estimated from data over the period 1869–1996. Note that although overall U.S. hurricanes are more likely during a cold phase, the probability of exactly one U.S. hurricane is higher during a warm phase.

ricanes with a better than 2-to-1 ratio in favor of fewer tropical-only hurricanes during active eastern North Pacific seasons. This is because eastern North Pacific hurricanes often develop from tropical waves from the North Atlantic. Waves that develop under favorable conditions over the Atlantic rarely make it to the Pacific. Thus years featuring frequent tropical-only hurricanes over the deep tropics are years of fewer hurricanes along the western coast of Mexico. The frequency of eastern North Pacific hurricanes is unrelated to the frequency of baroclinic hurricanes over the Atlantic.

10.2.4   Solar Activity

Perhaps the most intriguing relationship is found with solar activity. It is fascinating in a couple of ways: (1) it represents an extraterrestrial influence and (2) it appears to explain (in a statistical sense) only the variability of baroclinically-enhanced hurricanes. Greater solar activity and higher solar irradiance can be expected to increase the tropical ocean temperatures by a few tenths of a degree Celsius. Greater evaporation leading to increased atmospheric moisture and instability might be a consequence of the warmer waters, particularly over the subtropical latitudes where the air is subsaturated.

The first modern account of the influence of the sun on tropical cyclones appeared in 1872 (Hoyt and Schatten 1997). Meldrum (1872) showed that sunspot numbers

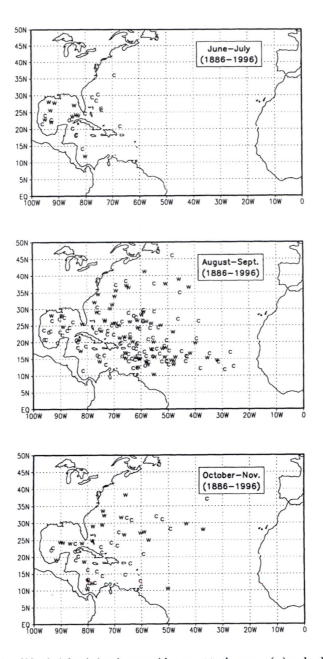

Fig. 10.14: Origin of North Atlantic hurricanes with respect to the warm (w) and cold (c) phases of ENSO over the period 1886–1996. Activity is broken down into early season (June–July), middle season (August–September), and late season (October–November).

Table 10.5: Points of origin for North Atlantic hurricanes. Values are averages ($\overline{x}$), standard deviations (*s.d.*), and number of cases (N) based on warm and cold ENSO events over the period 1886–1996.

		Warm ENSO Phase (1886–1996)			
Month	N	$\overline{x}$ (lat)	s.d. (Lat.)	$\overline{x}$ (Lon.)	s.d. (Lon.)
Jun-Jul	10	23.7	4.4	88.9	5.2
Aug-Sep	62	23.2	7.8	62.8	13.0
Oct-Nov	23	24.4	7.3	72.3	14.4
		Cold ENSO Phase (1886–1996)			
Month	N	$\overline{x}$ (Lat.)	s.d. (lat)	$\overline{x}$ (Lon.)	s.d. (Lon.)
Jun-Jul	17	23.8	5.1	82.0	8.1
Aug-Sep	98	20.9	6.5	64.9	16.6
Oct-Nov	31	20.8	7.0	72.3	12.9

are strongly correlated (positively) with the yearly number of Indian Ocean tropical cyclones, explaining slightly more than 50% of the year-to-year variation in cyclone activity. Later studies found that this relationship changed sign during the period 1910 to 1930. Poey (1873) examined the relationship of solar activity to tropical cyclones of the North Atlantic. He found a correspondence between the solar cycle and Caribbean hurricanes. Willett (1955) used an apparent relationship between North Atlantic hurricane activity and sunspot numbers to predict hurricane activity to the year 2020. His forecast called for an overall decline in activity until 1990, with a tendency for more northerly hurricanes resulting from an increased tendency for zonal flow in the middle latitudes. The increase in baroclinically-enhanced hurricanes over the period (Chapter 6) is consistent with his long-range forecast. His prognostications included the possibility of increased activity during the early decades of the 21st century, particularly along the east coast of the United States. A statistical sun-hurricane relationship is examined by the Weather Research Center of Houston. Their work is based on the idea that extraterrestrial influences causing the solar cycle on the sun might have similar effects on the large-scale circulation of the atmosphere reflected in the tracks and frequencies of tropical cyclones over the North Atlantic.

Cohen and Sweetser (1975) examine the sun-hurricane relationship for the entire North Atlantic basin. They find that spectra of both the abundance of tropical cyclones and the average length of the tropical cyclone season matches the spectrum of the solar cycle. Their analysis uses a 7-year running average of hurricane activity They offer no explanations as to how variations in the solar cycle influence tropical cyclone climate. The annually-averaged *Wolf* (or *Zurich*) sunspot numbers (R_z) are listed for the 10 years of maximum solar activity and the 10 years of minimum activity over the period 1886–1996 in Table 10.7. The table also gives the corresponding number of North Atlantic hurricanes in various categories. The annual-averaged Wolf sunspot numbers

Table 10.6: Years of extreme hurricane activity in the eastern North Pacific. North Atlantic hurricane activity (All, TO, and BE) in corresponding years is given. EP indicates the number of hurricanes observed in the eastern North Pacific. In order to make both extremes have the same number of years, 1995 (7 EP hurricanes) is included in the list of low activity years. There are 3 other years with 7 eastern North Pacific hurricanes that were not included.

	9 Maximum Years				9 Minimum Years			
Year	EP	All	TO	BE	Year	EP	All	TO	BE
1971	12	6	2	4	1967	6	6	2	4
1974	11	4	3	1	1968	6	5	1	4
1978	12	5	2	3	1969	4	12	5	7
1982	11	2	0	2	1970	4	5	1	4
1983	12	3	0	3	1977	4	5	0	5
1984	12	5	0	5	1979	6	5	2	3
1985	11	7	4	3	1988	6	5	4	1
1990	16	8	3	5	1995	7	11	9	2
1992	14	4	0	4	1996	5	9	8	1
Total	.	44	14	30	Total	.	63	32	31

are a commonly used index for examining relationships between the sun and climate. However, it should be kept in mind that sunspot number is not necessarily the best indicator of solar activity.

Total basin-wide hurricane activity appears only weakly related to the extremes in solar activity (the ratio is 1.3 to 1 in favor of more hurricanes with more sunspots), but baroclinic hurricane activity is strongly related, with a ratio of 2.1 to 1. The ratio is similar to the one found for tropical-only hurricanes with respect to eastern North Pacific storms. The sun-hurricane relationship is interesting because it accounts for oscillations in the hurricane record not explained by either the QBO or the ENSO. Note that 4 of the 10 minimum years occur in the reversal period of 1910 through 1930. If these years are replaced by the nonreversal minimum years of 1890, 1900, 1944, and 1964, then the relationship with baroclinically-enhanced hurricanes increases to 2.7 to 1.

No significant relationship between solar activity and U.S. hurricanes is noted as the solar extremes do not differentiate between years with many strikes and years with only a few.[7] Extreme caution must be used in interpreting this analysis since the 10 years of minimum solar activity are all before 1955 while the 10 years of maximum solar activity are all after 1946. Given the fact that a significant proportion of the baroclinically-enhanced hurricanes occur over the western North Atlantic and track away from land, years before aircraft reconnaissance or satellites might be biased toward fewer such hurricanes. It is safe to conclude, however, that if there is a relationship between solar output and North Atlantic hurricane climate, as this and other analyses have suggested,

---

[7]This is consistent with the fact that only 25% of U.S. hurricanes are baroclinically-enhanced.

Table 10.7: Years of extremes in the solar cycle. The cycle is indexed by annually-averaged values of Wolf sunspot numbers (R_z). North Atlantic hurricane activity (All, TO, and BE) in corresponding years is given.

	10 Maximum Years				10 Minimum Years			
Year	R_z	All	TO	BE	Year	R_z	All	TO	BE
1947	151.6	5	5	0	1888	6.8	5	3	2
1956	141.7	4	3	1	1889	6.3	5	4	1
1957	190.2	3	2	1	1901	2.7	3	2	1
1958	184.8	7	7	0	1902	5.0	3	3	0
1959	159.0	7	1	6	1911	5.7	3	3	0
1979	155.4	5	2	3	1912	3.6	4	1	3
1980	154.6	9	4	5	1913	1.4	3	1	2
1989	157.7	7	5	2	1923	5.8	3	2	1
1990	141.8	8	3	5	1933	5.7	10	10	0
1991	145.2	4	0	4	1954	4.4	8	5	3
Total	.	59	32	27	Total	.	47	34	13

it is likely linked through changes in baroclinically-enhanced hurricane activity (see Table 10.7).

Cross-correlation between sunspots and North Atlantic hurricane activity is shown in Figure 10.15. Values of cross-correlation are low. The lag is given in years. Baroclinically-enhanced hurricane activity is apparently driving the correlation between solar cycle fluctuations and hurricanes as the cross-correlation functions for all hurricanes and baroclinically-enhanced hurricanes are nearly identical. Peaks in the sun-hurricane relationship occur near 7 and 14 years. The interaction between two or more physical cycles may generate harmonics. When two interacting cycles occur at distinct frequencies, harmonic generation produces a signal at the sum and difference of these frequencies. The forcing frequencies of the 11-year and 22-year solar cycles are 0.091 cycles per year and 0.045 cycles per year, respectively. The sum produces a frequency at 0.136 cycles per year corresponding to a periodicity of 7.4 years (Burroughs 1992). This periodicity is commensurate with the periodicity of the oscillations in the abundance of baroclinically-enhanced hurricanes. As expected, the cross-correlation function between tropical-only and baroclinically-enhanced hurricane activity is predominately negative. Years of above normal tropical-only activity tend to be years with fewer baroclinically-enhanced hurricanes; although the relationship is weak.

Total variation in solar radiant output associated with the 11-year solar cycle might be as much as 0.15%. Though small, this amount can have significant impacts on climate (Hoyt and Schatten 1997). It is speculated that baroclinically-enhanced hurricane activity might be sensitive to small changes in solar output through increased evaporation at the ocean surface. Increased evaporation leads to an atmosphere more

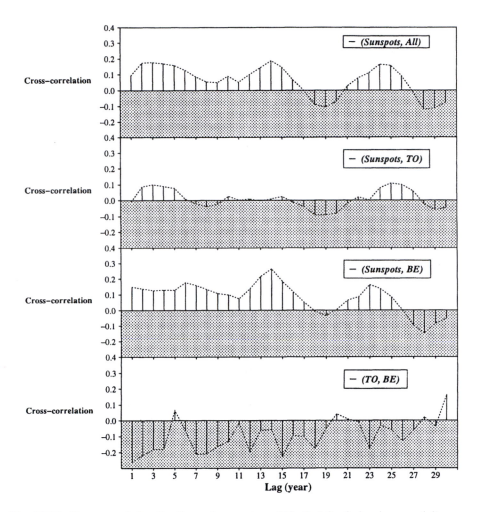

Fig. 10.15: Cross-correlation functions of sunspots and North Atlantic hurricane activity over the period 1886–1996, except for the bottom panel which shows the cross-correlation between tropical-only (TO) and baroclinically-enhanced (BE) hurricanes. (Sunspots, All) denotes cross-correlation between sunspot numbers and the number of all North Atlantic hurricanes. Same for (Sunspots, TO), etc. The lag is in years. Correlation values are generally less than ±0.2. Largest correlation is noted for all and baroclinically-enhanced hurricane activity. Tropical-only and baroclinically-enhanced hurricane frequencies are negatively correlated.

Table 10.8: U.S. hurricane frequencies with respect to the solar cycle. Values are based on data over the period 1851–1996. Minimum and maximum years are denoted as zero years and minus and plus indicate the number of years before and after the extrema, respectively. N is number of the U.S. hurricanes.

	Years Relative to the Extrema				
Relative Year	−2	−1	0	+1	+2
N (U.S. Hurricanes)	41	44	55	36	38

conducive to hurricanes through an increase in moisture and a decrease in stability (see Chapter 1). Warmer temperatures at lower altitudes are associated with a larger lapse rate of temperature. This is particularly true at higher latitudes where baroclinically-enhanced hurricanes tend to form. At lower latitudes, where most of the tropical-only hurricanes originate, the atmosphere is typically close to water-vapor saturation, so an increase in radiant energy does not significantly increase the evaporation rate or change the atmospheric stability.

The 7-to-9-year cycle of baroclinically-enhanced hurricane activity can be explained as a modulation frequency between the 11-year sunspot cycle and the 18–19 year Saros lunar cycle.[8] Drought occurrences in the western United States have a significant rhythm at a periodicity of 7 to 8 years (Cook et al. 1997). Lunar tidal forcing might influence atmospheric circulation and pressure fields through the non-equilibrium state of the atmosphere and the associated pressure gradients (O'Brien and Currie 1993). The detection of this relationship is not proof of any significant physical mechanism. However, the coincidence of the lunar and solar-cycle beat frequency with baroclinically-enhanced hurricanes and western United States drought occurrences is intriguing and merits further study.

It has been suggested that the solar influence on climate is through the ultraviolet (UV) radiation, which significantly changes the rate in which stratospheric ozone is produced. A modeling study done at the Imperial College in London by Joanna Haigh finds that changes in ozone at solar maximum explain the northward shift in the path of winter storms across Europe. The results indicate that increases in stratospheric temperature in response to increases in solar radiation push stronger summer easterly winds into the upper troposphere of the tropics. This forces middle-latitude circulation patterns in the troposphere to retreat toward the poles (Haigh 1996). The northward migration might be linked to increased baroclinic-type hurricanes. Another explanation centers around the fact that during solar maximum, the solar wind effectively blocks some of the galactic cosmic rays streaming toward the earth. The cosmic rays might be linked to the production of clouds and precipitation in the troposphere.

Another interesting relationship emerges when considering U.S. hurricanes and solar activity. Table 10.8 shows the number of U.S. hurricanes with respect to either

---

[8]The Saros lunar cycle modulates tidal forcing.

Table 10.9: North Atlantic hurricane activity with respect to major volcanic eruptions. Values are averages over all eruptions for the period 1878–1996. Years of eruption are denoted as zero years and minus and plus indicate the number of years before and after the explosion, respectively. Years relative to the eruptions are marked as integers in the top row. The **bold** numbers indicate the highest value in the row. TO and BE refer to tropical-only and baroclinically-enhanced hurricanes, respectively. U.S. indicates U.S. landfalling hurricanes. Data for All, TO, and BE hurricanes extend back only to 1886.

	\multicolumn{11}{c}{Years Relative to the Major Volcanic Eruptions}										
Category	−5	−4	−3	−2	−1	0	+1	+2	+3	+4	+5
All	5.0	5.2	5.0	5.1	4.2	4.3	5.7	3.8	**6.3**	6.0	4.2
TO	3.1	2.1	3.2	3.1	2.0	2.4	2.5	1.4	**3.9**	3.5	2.1
BE	1.9	**3.1**	1.8	2.0	2.2	1.9	2.9	2.3	2.4	2.5	2.1
U.S.	1.1	1.6	2.4	1.2	1.5	1.5	1.6	1.5	**2.6**	2.1	1.6

extreme of the solar cycle. The zero year represents the year of either a maximum or minimum in sunspot numbers. The minus and plus years indicate years before and after the extreme, respectively. Hurricanes affecting the United States are somewhat more frequent in the years leading to a minimum or maximum in solar activity and considerably more abundant in the extreme year compared to the years immediately following the extreme. We offer no physical explanation for this apparent relationship. However, for the past 300 years or so the solar cycle has featured periods of high and low sunspot numbers with regularity. The next solar maximum is expected in the year 2000 or 2001.

10.2.5   Major Volcanic Eruptions

Volcanic eruptions can significantly alter the weather and climate of the planet. World-wide global surface-air temperatures are noticeably cooler a few years after the eruption of a major volcano (Angell and Korshover 1985, Mass and Portman 1989). The effect is attributed to the absorption of incoming solar radiation by ash high in the atmosphere. The cooling is sometimes widespread and dramatic. Following the explosion of Tambora in Sumbawa, Indonesia on April 13, 1815 the atmosphere cooled significantly. In fact, 1816 is commonly referred to as the "year without a summer" in New England and in western Europe. It is estimated that the supereruption of the Toba volcano in Sumatra 73,000 years ago led to annual hemispheric cooling at the surface of 3 to 5°C, which may have triggered world-wide glacial conditions. It is even speculated that such a dramatic climate disruption can feed back to influence volcanoes themselves (Rampino and Self 1992). In any event, the absorption of solar radiation creates a warmer stratosphere several years after the explosion. Table 10.9 shows North Atlantic hurricane activity with respect to 10 significant volcanic eruptions of the last

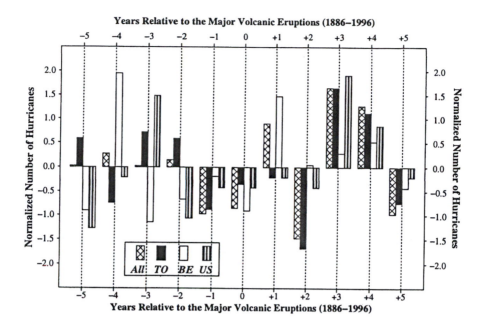

Years Relative to the Major Volcanic Eruptions (1886–1996)

Fig. 10.16: North Atlantic hurricanes with respect to major volcanic activity over the period 1886–1996. Values are normalized departures from the mean. Years of eruption are denoted as zero with minus (−) and plus (+), indicating the number of years before and after the eruption year, respectively. Values are plotted for all, tropical-only (TO), baroclinically-enhanced (BE), and U.S. hurricanes (US).

130 years.[9] Categories of activity are listed in the first column. Average frequencies are given with respect to the volcano years.

Data suggest major volcanic eruptions may lead to a small increase in the number of North Atlantic hurricanes 3 to 4 years after the explosion. The relationship is strongest for total number of hurricanes, however there is a small signal in both the tropical-only and U.S. hurricanes. On average there is a 26% increase in the abundance of hurricanes during the third year following an eruption. The increase is 20% during the fourth year. The signals are clear in Figure 10.16, which shows the normalized number of all, tropical-only, baroclinically-enhanced, and U.S. hurricanes with respect to major volcanic eruptions. If $\bar{x}$ is the average number of hurricanes over the 111-year period (−5 to +5 years), then the normalized number of hurricanes for the year is given as $x' = \frac{(x - \bar{x})}{s.d.}$. The signal is quite noisy for years prior to the eruption, and even until +1 year after. However, 2 years after the explosion there is a drop in hurricane activity followed in years three and four by an increase in activity.

It is noteworthy that there is no discernible change in baroclinically-enhanced hur-

---

[9]The 10 major volcanic eruptions are Krakatau in 1883, Tarawera in 1886, Mount Pelee/Soufriere/Santa Maria in 1902, Ksudach in 1907, Katmai in 1912, Agung in 1963, Awu in 1966, Fuego in 1974, El Chichón in 1982, and Pinatubo in 1992. The first nine were selected by Mass and Portman (1989) as producing a significant stratospheric dust veil.

ricane activity after a major stratospheric dust loading. This is consistent with our assumption that baroclinic hurricanes are influenced by solar brightness through changes in evaporation rates occurring over the subtropical oceans, but are not appreciably influenced by changes occurring in the stratosphere. Changes in the stratosphere occur because of the increased absorption of infrared (IR) radiation in the dust layer. Changes in the absorption of ultraviolet radiation (UV) can occur through changes in the amount of ozone. Tropical-only hurricanes are related to fluctuations in stratospheric UV and IR absorption because these effects change the temperature profile of the stratosphere. Temperature changes affect the winds, which influence the QBO. It is argued that tropical-only hurricane activity is influenced by process that change the dynamics or thermodynamics of the lower stratosphere. Stratospheric processes can induce significant changes in the dynamics of the troposphere. On the other hand, baroclinic hurricane activity is influenced by processes that change the thermodynamics of the lower troposphere. It should be kept in mind that the composite analysis did not remove the portion of interannual hurricane variability associated with ENSO or QBO, and that the volcanic signal in the hurricane record is very small. However, as noted in Mass and Portman (1989), removal of the ENSO signal in the composite analyses of surface temperatures actually enhances the apparent volcanic effect of the largest eruptions.

10.3 Hurricane Trends

Trends in hurricane activity may indicate significant changes in the vulnerability of people and property to a hurricane disaster. A reduction in the hurricane threat will be felt by some, while an increase in the threat will occur for others. Here various components of the historical North Atlantic hurricane record are examined for long term trends. Mention is made of the potential for using general circulation climate models to diagnose the potential for significant changes in hurricane climate.

10.3.1 Annual Frequency

Are hurricanes becoming more frequent? The answer is a qualified yes. Figure 10.17 shows the annual abundance of hurricanes as a scatter-plot along with the least-squares linear regression line. A small upward trend is noted with a correlation of 0.22. Assuming each year is independent, the p-value on the regression coefficient is 0.022 indicating the trend is marginally significant. The annual frequency of hurricanes increases from 4 at the beginning of the record to 6 at the end. The trend is consistent with the assumption that not all hurricanes were counted in the earliest years.

The trend is also within the range of natural variability in North Atlantic hurricane climate. This can be seen by considering the average difference in the number of hurricanes occurring in years ending with a six (1886, 1896, 1906, ...) and the number of hurricanes occurring in years ending with a seven (1887, 1897, 1907, ...). The average number of hurricanes in years ending in a six is 6.6 compared to 3.9 in years ending

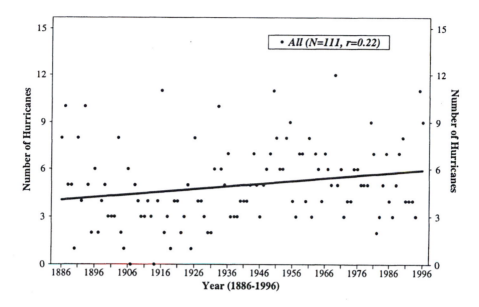

Fig. 10.17: Scatter plot of the annual number of North Atlantic hurricanes over the period 1886–1996. The ordinary least-squares linear regression line is also shown. The number of years (N) and the regression correlation (r) are given in the upper-right corner.

in a seven (see Chapter 4) This large natural variation is of similar magnitude to the variation in hurricane numbers before and after satellites and aircraft reconnaissance. It is unrelated to observational biases. Moreover, there are large swings in hurricane activity within the modern era. For instance, the average number of hurricanes from 1950 through 1969 is 6.6 compared to 5.0 over the same length period from 1970 through 1989. The trend in annual frequency is partly due to the lack of activity during the 1910s and 1920s.[10]

The increase in North Atlantic hurricane activity is not due to an increase in the number of tropical-only hurricanes as tropical-only activity is decreasing over the period. The North Atlantic basin is apparently experiencing a greater number of hurricanes due an increase in the number of baroclinically-enhanced hurricanes. Here we find a significant positive trend (p-value < 0.001) that explains a bit more than 25% of the annual variance for this category. A significant change in the ratio of baroclinically-enhanced to tropical-only hurricanes occurred during the middle and late 1960s as seen in Figure 10.18. Baroclinically-enhanced hurricanes are more frequent during the 1960s, 1970s, and 1980s. The dearth of tropical-only hurricanes during the early decades of the 20th century are not accompanied by an increase in baroclinic hurricanes. The recent increase in abundance of baroclinically-enhanced hurricanes is co-

---

[10]Interestingly, the number of tropical cyclones around Australia has decreased dramatically since the middle 1980s (Nicholls 1992).

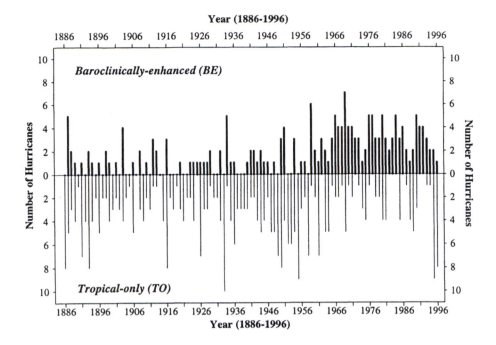

Fig. 10.18: Baroclinically-enhanced (BE) (dark impulses) and tropical-only (TO) hurricanes over the period 1886–1996. The values are annual abundance of storms. The ordinate is reversed for the graph of TO hurricanes. Note the increase in the number of BE hurricanes during the middle and late 1960s is coincident with a decrease in the number of TO hurricanes.

incident with a salinity anomaly in the North Atlantic Ocean. The salinity anomaly of the late 1960s, which is examined in more detail in Chapter 16, represents an important component of low-frequency ocean variability.

10.3.2   Annual Average Duration

Despite the modest increase in the number of North Atlantic hurricanes over the 111-year period, the average duration of hurricanes appears to be decreasing rather significantly as seen in Figure 10.19. The least-squares linear trend explains more than 25% of the annual variation and is significant with a p-value less than 0.0001 on the regression coefficient. The decrease in duration amounts to approximately three days per century and is most pronounced for the category of tropical-only hurricanes. These negative trends support the contention that the number of missing hurricanes over the earlier part of the record is not likely very large. If there is a serious sampling artifact in the earlier portion of the record it would likely be such as to under sample hurricane duration. That is, during the earlier years, a particular hurricane may not have been detected until later on—only after tracking closer to land or over the main shipping lanes. It is clear, however, that average hurricane duration is longer for the earliest

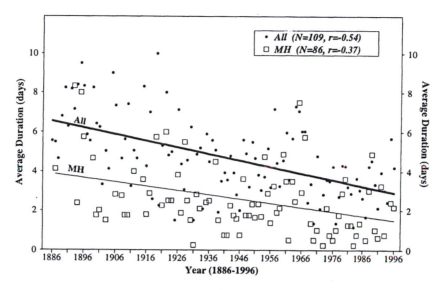

Fig. 10.19: Duration of hurricanes and major hurricanes over the period 1886–1996. Values are annual averages. The ordinary least-squares linear regression lines are also shown. The number of years (N) and the regression correlation (r) are given in the upper-right corner. Only years in which there is at least one hurricane are used.

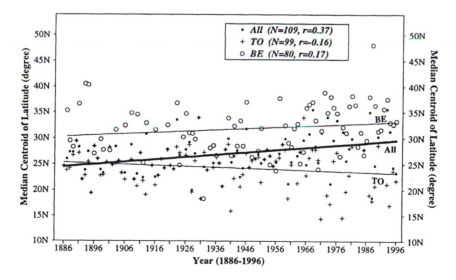

Fig. 10.20: Median hurricane positions over the period 1886–1996. Values are the annual median centroid of latitude for all, tropical-only (TO), and baroclinically-enhanced (BE) hurricanes. Number of years (N) and the regression correlation (r) are provided. Only years in which there is at least one hurricane in the category are used. A significant positive trend for all hurricanes is noted.

years. The decrease in annual average duration is mirrored for major hurricanes. The trend is statistically significant and amounts to about 2 days per century. The trend is also consistent with the likelihood that shorter-duration storms far from land were not detected in early times. A missing short-duration hurricane in a given year will bias the average toward longer duration. Note that the trend exists even when only the more reliable period of years from 1950 through 1996 are examined.

10.3.3 Median Latitude

No strong theoretical evidence exists to support the contention that a warmer climate will greatly expand the hurricane-prone ocean areas of the North Atlantic basin. Even with an increased area of critically warm SSTs ($> 26°$C), the atmosphere will not likely maintain a correspondingly larger area of critically low vertical shear of the horizontal winds necessary for hurricane development. This idea is substantiated by output from general circulation model (GCM) simulations that show warmer SSTs under a scenario of double of carbon dioxide concentrations (CO_2), but not a measurable increase in the area supporting tropical cyclones (Haarsma et al. 1992). As indicated previously, the historical hurricane record indicates only a modest positive trend in the frequency of North Atlantic hurricanes.

It is, however, possible that a warmer atmosphere will support hurricane development farther north or increase the frequency at which hurricanes move out of the tropics. Figure 10.20 shows a plot of the median centroid of hurricane latitude as a function of year computed from the 6-hourly best-track locations. The least-squares linear regression line ($r = 0.415$) indicates a significant trend (p-value < 0.001) toward hurricane activity occurring at higher latitudes. This trend is significant under remedial transformations intended to correct for unequal error variances (*heteroscedasticity*). Nonparametric regression using a rank-based residual dispersion measure yields nearly identical results as the least-squares test.[11] There is no significant trend in the annual median latitude for tropical-only or baroclinically-enhanced hurricanes. By saying North Atlantic hurricanes are occurring farther north with time does not mean that it is due to other trending factors, like global warming. There is no significant trend in the median (or average) longitude, though there is a tendency for hurricane activity to shift slightly farther east over time. This is due to the geography of the North Atlantic basin, which limits higher latitude hurricanes from forming too far to the west.

The positive trend in median latitude is attributed to the increase in baroclinically-enhanced hurricane activity. The average median latitude of baroclinically-enhanced hurricanes is 31.9°N compared with 24.4°N for tropical-only hurricanes over the period 1886–1996. No significant trends are found for tropical-only or baroclinically-enhanced hurricanes separately. Thus, the trend is related to the increase in baroclinically-enhanced hurricane activity. A future of more baroclinically-enhanced hurricanes raises the question of whether the potential for hurricane damage will shift north-

---

[11]The trend is significant even if only the most recent 50 years are used.

ward over the populated northeast. However, tropical-only hurricanes are responsible for most of the strikes along the northeast U.S. coastline.[12]

10.3.4   Intensity

The scenario of global warming resulting in more intense hurricanes is a debated topic in global change research. Factors that control the occurrence and frequency of storms appear to be different from those that determine their maximum intensity. The historical hurricane record from the North Atlantic contains some circumstantial evidence against the claim that more intense hurricanes will be a consequence of a warmer planet. Hurricane *Andrew* in 1992 came during the relatively cool summer following the eruption of Mount Pinatubo. The category five hurricane that devastated the Florida Keys in 1935 occurred before significant amounts of anthropogenetic greenhouse gases where in the atmosphere. The catastrophic hurricane *Camille* in 1969, that demolished parts of coastal Mississippi and Louisiana, occurred during the relatively cool decades of the 1960s and 1970s.

On the other hand, if the oceans were to warm significantly the potential for more intense hurricanes would increase. The upper bound on a hurricane's intensity is determined by the underlying ocean temperature and the thermodynamic conditions of it's atmospheric surroundings. Rarely do storms reach this limit because of inhibiting environmental factors such as wind shear. A warmer ocean would increase the frequency at which the thermodynamic conditions for the most intense storms were met. If no changes occur to increase the occurrence or extent of inhibiting factors, then average hurricane intensities would increase. Besides radiative changes, warming of the oceans could result from a bolide impact[13] or from a giant explosion of an undersea volcano. Spectacular examples of such events might warm the tropical oceans to as much as 50°C. Theoretical calculations predict that excessively warm oceans will support relatively small but extremely intense hurricanes (*hypercanes*) with winds in excess of 300 m s^{-1} (Emanuel et al. 1995).

A closer look at the historical hurricane data suggest that when global surface temperatures are used as an indicator of warm and cold years across the globe, warm years tend to support slightly more hurricanes, but of weaker intensity. Conversely, cool years produce fewer hurricanes of higher intensity. The evidence is not strong. The increase in the number of North Atlantic hurricanes during warmer years is attributed to heightened baroclinically-enhanced hurricane activity. Baroclinically-enhanced hurricanes form at higher latitudes and are weaker. During cool years the somewhat fewer hurricanes are confined mainly to the deep tropics. These storms are stronger. In fact, over the period from 1950 through 1996, the annual-averaged maximum sustained winds have declined (Figure 10.21). Over the same period global temperatures have risen. The decreasing trend for all hurricanes is significant at a *p*-value less than 0.05. The

---

[12]See Chapter 8.

[13]It is speculated that a giant meteorite struck the Caribbean Sea some 6 million years ago. The impact changed the earth's atmosphere (perhaps including more intense storms), which likely contributed to mass extinction of species.

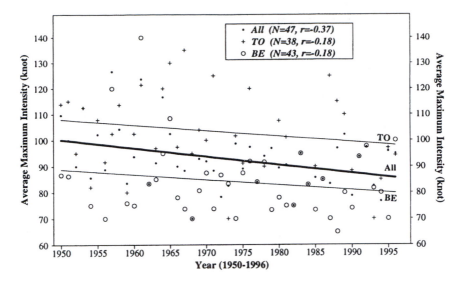

Fig. 10.21: Maximum intensity for all, tropical-only (TO), and baroclinically-enhanced (BE) hurricanes over the period 1950–1996. The values are annual averages. The ordinary least-squares linear regression lines are also shown. The number of years (N) and the regression correlation (r) are given in the upper-right corner. Only years in which there is at least one hurricane are used.

trends are not significant for either tropical-only or baroclinically-enhanced hurricanes separately. There is no statistically significant correlation between global surface air temperatures and North Atlantic hurricane intensity or frequencies (All, TO, and BE). The early years of the 20th century are considered unreliable with respect to precise intensity estimates for many of the hurricanes. Therefore caution is advised in interpreting the trends.

10.3.5 General Circulation Models

A sophisticated tool for examining future climate scenarios is the *general circulation model* or GCM. A GCM is a computer algorithm consisting of many lines of code used in simulating the general circulation of the atmosphere and ocean. The model includes the fundamental equations of motion along with equations for a myriad of physical processes on various time and space scales. Utility of a GCM arises from the ability to use it for simulating the climate under conditions that are different than they are at present. A popular simulation involves a doubling of the carbon dioxide levels. Carbon dioxide is a principal greenhouse gas that helps keep the atmosphere in radiation balance at moderate temperatures. Increasing carbon dioxide through industrialization creates a new balance for the earth's atmosphere at a higher temperature. The question is, how will climate respond to this global warming? In particular, will the warming affect either the frequency or intensity of hurricanes?

Simulations using the Goddard Institute for Space Studies (GISS) GCM suggest that the global frequency of tropical cyclones increases with a doubling of the carbon dioxide environment (Druyan and Lonergan 1997). In particular, the eastern and western North Pacific basins are two to four times more active over the Gulf of Mexico, increasing by about 50% under global warming. As with other studies of this type increases in tropical cyclone activity are linked to higher ocean surface temperatures (Carson 1992 and Haarsma et al. 1992). However, because of model complexity, it is difficult to analyze the physical relationships directly. Other realistic GCM simulations of global tropical cyclone activity are performed by L. Bengtsson and colleagues at the Max-Planck-Institute für Meteorologie, using a T106 model resolution.[14] The model generates hurricanes over the globe with a frequency and spatial distribution commensurate with observations. The model is consistent in showing a suppression of hurricanes in the western North Atlantic during warm ENSO events due to enhanced subsidence caused by the proximity of the tropical Pacific convection. They find that a double carbon dioxide concentration leads to a reduction in the number of tropical cyclones (Bengtsson et al. 1995 and 1996). The main reasons for the reduction in apparent hurricane activity are changes in the large scale circulation patterns including a weaker *Hadley circulation*,[15] stronger upper troposphere westerlies and reduced low-level vorticity. Projections showing a reduction in tropical cyclone frequency under global warming are opposite to the conclusions drawn based on results from the GISS model.

Caution must be urged against taking the GCM simulations to literally vis-á-vis hurricane activity. Though GCMs are an effective tool for projections of global climate change, horizontal resolution is inadequate for resolving the full dynamics of developing or intensifying tropical cyclones (Lighthill et al. 1994). A strong argument against the integrity of some of the results implied by model simulations is the apparent lack of connection between warmer SSTs and hurricane activity. The fact that warm SSTs ($> 26°C$) are a necessary condition for tropical cyclone formation does not imply that warmer SSTs lead to more or stronger hurricanes. There are other important conditions for tropical cyclone formation (see Chapter 1), some of which are more critical than SSTs (Lighthill et al. 1994). For instance, though SSTs are warm enough over the South Atlantic Ocean to support hurricanes, none occur there.[16] This is due in part because of the lack of low- and middle-troposphere disturbances, and in part because of significant amounts of convection over equatorial Africa and Brazil. The convection over land limits convection over the adjacent waters of the South Atlantic Ocean due to compensating sinking motions.

Model simulations of global climate under a scenario of more solar radiation in-

---

[14]T106 is shorthand for triangular truncation at wavenumber 106, which corresponds to a horizontal spatial distance between values of roughly 125 km.

[15]The Hadley circulation is a time-averaged vertical circulation of the tropics featuring rising motion near the equator and subsidence (sinking air) near 30°N latitude. The air near the ground in the Hadley circulation of the Northern Hemisphere is characterized by a steady northeasterly flow toward the equator.

[16]Occasionally during the Austral autumn, a weak tropical cyclone will form off the southwest coast of Africa. Winds are unlikely to exceed 30 kt with these systems.

dicate an increase in the land-sea thermal contrast, with the strongest signal occurring during summer over the subtropics (Cubasch et al. 1997). Large land-sea contrasts are associated with increased baroclinity. This is consistent with an environment more favorable for baroclinically-enhanced hurricane activity over the North Atlantic, and it may help explain the apparent relationship between baroclinic hurricanes and the solar cycle noted previously. The climate signal in response to increased solar radiation is similar to the effect noted in models having more greenhouse gases.

Thus a broadening area of warm ocean encompassing the subtropical waters of the North Atlantic will not necessarily lead to an increase in hurricane frequency due to constraints from other conditions. As described early in the chapter, some of the natural interannual variability of hurricane abundance over the North Atlantic is tied to the QBO and ENSO rather than directly to Atlantic Ocean sea-surface temperature changes. Yet it is intriguing to note that the apparent increase in the frequency of baroclinically-enhanced hurricanes is consistent with a northward expanding area of warm SSTs. Relatively small changes in the area covered by warm SSTs might result in significant changes to hurricane activity on decadal or longer time scales (Wendland 1977). GCMs do not have sufficient horizontal resolution to realistically model hurricanes. It is likely that future developments will lead to finer resolution and more complete physics in the models. The new models will be capable of answering fundamental questions regarding the effect of climate change on hurricane activity. Higher-resolution GCMs that include a dynamic ocean coupled with post-model-run statistical equations could be the future for extended-range hurricane predictions.

11

Hurricane Return Periods

The occurrence of a hurricane is a rare event. The likelihood of rare events that are not periodic in time are expressed in terms of *return periods*. Return periods are used to describe the occurrence of earthquakes and extreme flood events. The qualification that events are not periodic is important. A phenomenon that occurs regularly—even if it is rare—is best described in terms of periodicities (see Chapter 10). Return periods are appropriately used when no important oscillations occur in the record. Various definitions for return periods lead to a vagueness concerning concepts associated with their use. For present purposes, both return periods and recurrence intervals are utilized. A *recurrence interval* is the time between successive events. The recurrence interval is also called the *interarrival time*. Recurrence intervals may vary widely and nonuniformly over the period of record. A *return period* is defined as the inverse of the annual probability. A hurricane which has a probability of once in 10 years is referred to as a 10-year event. A return period of 10 years should not be understood as a forecast of an event happening about every 10 years.[1] Instead, over a long period of record, say 150 years, fifteen hurricanes can be expected to have occurred. Or, in any year, there is a 10% chance that a hurricane will occur. Unfortunately, there is no universally agreed upon definition of return period. The average of the recurrence intervals in the record serves as a rough estimate of the return period.

The accuracy of a return-period estimate depends on the length of the record. The longest available record should be used whenever possible. Low-probability extreme events, like the occurrence of a major hurricane strike along a small stretch of coastline, are particularly sensitive to record length. The relative infrequency of hurricanes is such that no climate summary should be considered a stable assessment of the risk of hurricane recurrence at any single point (Simpson and Lawrence 1971). However, return period estimates for areas along the coastline are necessary for planning and other purposes (e.g., designing and implementing hurricane-resistant building codes, planning evacuation routes, setting insurance rates, and so on). Return periods of hurricanes in U.S. coastal counties indicate large variability from region to region. Coastal

---

[1]Return periods are commonly misinterpreted in this way.

counties of south Florida are most vulnerable to the destructive winds of North Atlantic hurricanes. Counties in eastern North Carolina are also quite vulnerable to the ravages of hurricanes. Return periods of major hurricanes show similar variations. In this chapter the concepts of recurrence interval and return period are used to describe hurricane activity along the coasts of the United States. Interarrival times are based on data over the period 1851–1996. Estimates of return periods in individual counties are based on data over the period 1900–1996. Successive recurrence intervals are examined for segments of coastline extending from southern Texas to Maine. A discussion on the limitations associated with return-period estimates is given. A look at a method for quantifying trends in return periods is described. The possibility of shifts in landfall probabilities along the U.S. east coast is examined.

11.1   Recurrence Intervals

Recurrence intervals provide a way to examine the frequency of hurricanes in a particular area. Successive interarrival times are checked for patterns or trends. Here interarrival times for hurricanes along the U.S. coastline are examined. Recurrence intervals for coastal regions from Texas to Maine are examined first. Figure 11.1 shows consecutive recurrence intervals for hurricanes in Texas by region over the period 1851 through 1996. Data sources are discussed in Chapter 8. The abscissa is the ordinal number of recurrence intervals and the ordinate is the length in years between consecutive hurricane landfalls. For northern Texas, the first recurrence interval (the time between the first and second strike) is 13 years. The average recurrence interval is 5.9 years with a standard deviation of 5.6 years. The longest time between successive hurricane strikes is 21 years. Return periods are shortest for northern and southern Texas, although southern Texas went more than 30 years without a hurricane strike. Variations in recurrence intervals decreases moving north along the coast. Neither of the regions indicate a significant trend in the time between hurricane strikes.

Interarrival times for coastal regions of Florida are shown in Figure 11.2. Recurrence intervals vary widely between northwestern and southeastern Florida. With the exception of northeastern Florida, the interarrival times are generally shorter than for coastal regions of Texas. Strict comparisons cannot be made as recurrence intervals depend on the length of coastline considered. Northwest Florida, covering the largest stretch of coastline, has the shortest average recurrence interval of 3.2 years, followed by southeast Florida with an average recurrence interval of 4.2 years. The longest interarrival time is 21 years for southwest Florida. Northwest Florida shows a small trend toward longer wait times between hurricanes. In southwest Florida, recurrence intervals do not appear to be completely independent as successively shorter intervals tend to follow longer intervals. Northeast Florida, covering the smallest stretch of coastline, has the fewest hurricane strikes and the longest average wait between successive strikes. As can be seen from Figure 11.2, the shortest interarrival time was 3 years, and the longest was 17 years.

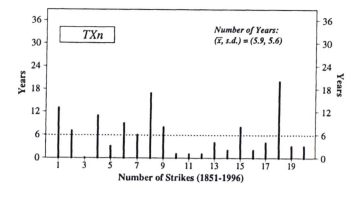

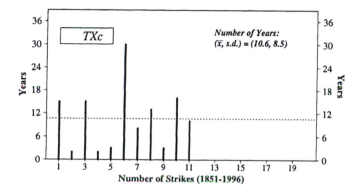

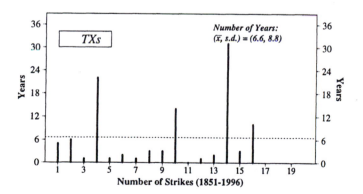

Fig. 11.1: Recurrence intervals for coastal regions in Texas over the period 1851–1996. The intervals are defined as the time in years between successive hurricane strikes (interarrival time). The average recurrence interval (dotted line) is an approximation of the return period. A zero indicates more than one hurricane in a year. The average ($\bar{x}$) and standard deviation ($s.d.$) are also provided. Shortest recurrence intervals are noted in northwest Florida.

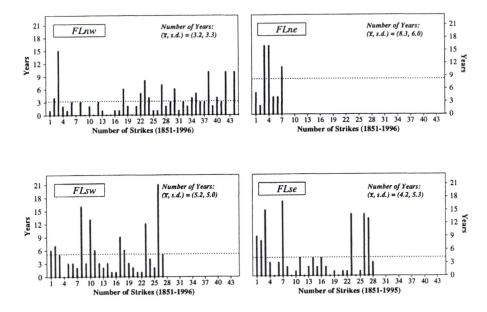

Fig. 11.2: Recurrence intervals for coastal regions in Florida over the period 1851–1996. The intervals are defined as the time in years between successive hurricane strikes (interarrival time). The average recurrence interval (dotted line) is an approximation of the return period. A zero indicates more than one hurricane in a year. The average ($\bar{x}$) and standard deviation ($s.d.$) are also provided.

Recurrence intervals for other coastal states are shown in Figures 11.3 and 11.4. As expected, interarrival times for Louisiana and North Carolina are short with averages less than 4 years. Only rarely do these states wait more than a dozen years between hurricanes. No significant change in interarrival times is noted for Louisiana or North Carolina. North Carolina shows a similar tendency as southwestern Florida. Successively longer intervals tend to follow shorter intervals, suggesting the time between hits is not strictly independent. In particular, North Carolina often gets hit more than once in a year. Alabama and Georgia, with shorter coastlines, have considerably longer recurrence intervals. Variations in the interarrival times are large. Multiple-hit years have occurred along both coastlines as have multidecadal interarrival times. The average interarrival time for hurricanes in South Carolina is 5.5 years with a standard deviation of 4.6 years. Recurrence intervals appear to be increasing along this section of the U.S. coast. The three recurrence intervals for Virginia preclude meaningful statistics. A hurricane visit to the northeast United States is typically felt in more than one state. Visits are more rare. New York has the shortest interarrival average of 10 years. Generally, New York waits at least 6 years between strikes. A slight tendency for shorter return intervals over time is noted. The average recurrence interval is 12 years in Connecticut and almost 27 years in Rhode Island.

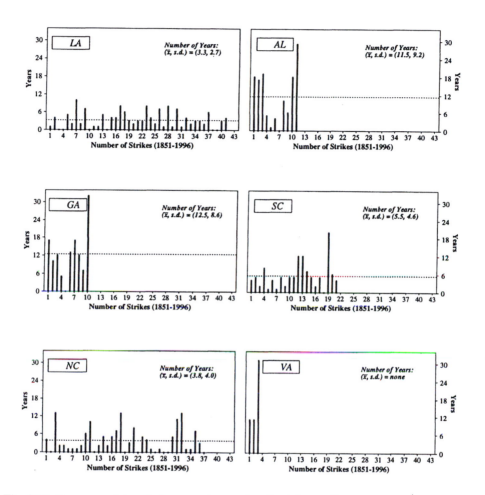

Fig. 11.3: Recurrence intervals for coastal states from Louisiana to Virginia, excluding Florida over the period 1851–1996. The intervals are defined as the time in years between successive hurricane strikes. The average recurrence interval (dotted line) is an approximation of the return period. A zero indicates more than one strike in a year. The average ($\bar{x}$) and standard deviation (s.d.) are also provided. State abbreviations are LA for Louisiana, AL for Alabama, GA for Georgia, SC for South Carolina, NC for North Carolina, and VA for Virginia. Note there have been only 4 hurricane landfalls in Virginia in the 146-year period. Interarrival times for hurricanes in Louisiana are relatively constant at approximately once in 3 to 4 years.

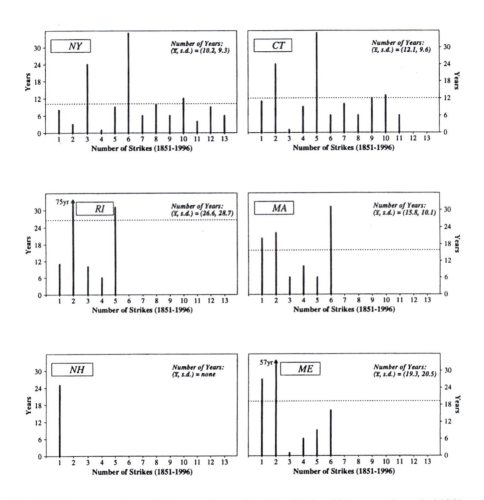

Fig. 11.4: Recurrence intervals for coastal states from New York to Maine over the period 1851 – 1996. The intervals are defined as the time between successive hurricane strikes. The average recurrence interval (dotted line) is an approximation of the return period. A zero indicates more than one strike in a year. The average ($\bar{x}$) and standard deviation ($s.d.$) are also provided. State abbreviations are NY for New York, CT for Connecticut, RI for Rhode Island, MA for Massachusetts, NH for New Hampshire, and ME for Maine. An arrow with the year next to it (see RI and ME) indicates that the recurrence interval exceeds the ordinate scale. Note that New Hampshire has been hit only twice in the 146-year period. Interarrival times for hurricane affecting New York and Connecticut are relatively constant at about once in 10 to 12 years.

11.2   Return Periods

11.2.1   Definitions

The probability of hurricane-force winds occurring within a political jurisdiction is important for preparation and emergency management. Here the return periods of hurricane-force winds in coastal counties from Texas to Maine are examined. The occurrence of hurricane winds in the county is called landfall for this analysis. Given the total number of landfalls in a county (L) over a record of length N in years, the average annual probability of a strike (p) is estimated by the ratio:

$$p = L/N. \tag{11.1}$$

A county that is hit by a hurricane 10 times in 100 years has an annual probability of 0.10 (10%). More frequent landfalls mean a larger annual probability. The return period is the inverse of the annual probability given as:

$$T = 1/p. \tag{11.2}$$

Thus, the return period for an annual probability of 0.10 is 10 years. Note that an unlikely event over a relatively short time interval is more probable over a longer interval. The annual probability does not change over the interval, but the probability of observing a hurricane increases if more years are included.

The probability that a county will not experience a hurricane at the end of year t is

$$Y_t = (1 - p)^t. \tag{11.3}$$

It follows that the probability the county will not have been hit at the beginning of year t, but gets hit during that particular season is

$$Z_t = pY_{t-1}. \tag{11.4}$$

In a related way, it is interesting to compute the wait time (W) in years within which there exists a 50% chance of at least one landfall. This is done using the equation

$$(1 - p)^W \geq 0.50, \tag{11.5}$$

where p is the annual probability of landfall, as before. Wait time is the number of consecutive years over which the probability of observing at least one hurricane increases to exceed 50% over the interval. With an annual probability of 10%, a wait time (W) of 7 years gives a probability of better than 50% for observing at least one landfall in the consecutive-year interval. That is, one can expect to wait an average of 7 years before the probability of getting hit sometime in the interval first exceeds the probability associated with flipping a fair coin. The probability of getting hit in any particular year remains constant. The equality in the equation gives the wait time expressed as a real number. The above formalism does not preclude other plausible definitions of return periods and wait times.

Here we present return-period statistics for individual coastal counties in the United States based on data from Jarrell et al. (1992). These data stratify hurricanes according to direct and indirect strikes for each coastal county. The data of Jarrell et al. (1992) is an update of the data presented in Hebert and Taylor (1975). The radius to maximum winds (R) is defined as the distance from the hurricane center to the circle of maximum winds in the eye wall (see Chapter 1). A county is regarded as receiving a direct hit when all or part of the county falls within $2R$ to the right and R to the left of a hurricane's landfall with respect to the direction of motion (from the perspective of someone at sea looking toward the shoreline). On average the direct hit zone extends roughly 80 km (50 mi) along the coastline ($R \approx 26$ km). Not all hurricanes are the same size so storms are judged individually for their spatial extent.

An indirect hit occurs on either side of the direct hit zone where winds still exceed hurricane force or where tides are at least one meter above normal. Again, a degree of subjectivity is necessary in making these decisions. Hits to coastal counties are counted for storms that retain their hurricane intensity as they move from land to sea. The chronological list extends from 1900 through 1990. Here we update this list to include hurricane landfalls through 1996. Seven U.S. hurricanes occurred over the period 1991 through 1996, including *Bob* in 1991, *Andrew* in 1992, *Emily* in 1993, *Erin* and *Opal* in 1995, as well as *Bertha* and *Fran* in 1996. The inclusion of these storms is based on analyses made by forecasters at the National Hurricane Center (NHC).

Statistics for hurricanes (direct plus indirect) and major (direct only) are given for 175 coastal counties from Texas to Maine. The distribution of counties examined is shown in Figure 11.5. The symbols are located at the geographic center of each county. Symbol type indicates the size of the county in square kilometer intervals. Coastal counties are defined by the National Oceanic and Atmospheric Administration (NOAA) as counties (or parishes) with at least 15% of their land area in a coastal watershed or coastal cataloging unit based on information from 1992. Not all counties that fit this definition are used here. In general, coastal counties are largest along the Gulf coast and in Florida. The eastern counties of Maine are quite large. Everything else being equal, a larger county will have a shorter return period. A direct or indirect hurricane strike indicates that hurricane-force winds (> 33 m s^{-1}) were felt somewhere in the county. For major hurricanes, a direct hit suggests that part of the county had winds blowing in excess of 50 m s^{-1}. The relative infrequency of hurricanes and especially major hurricanes in areas the size of counties makes the estimates of return periods sensitive to a few storms.[2]

Hurricane and major hurricane return periods and wait times in coastal counties of the United States are presented in Figures 11.6 through 11.19. Numbers in the counties are given in tenths of a year for return periods and wait times of less than 50 years. Whole numbers are used for return periods and wait times longer than this. In the text rounded numbers are used. For example, a return period of 32.3 years is referred to as 32 years and a return period of 48.5 years is referred to as nearly (or approximately)

---

[2]It is advisable to consider only the broad-scale patterns.

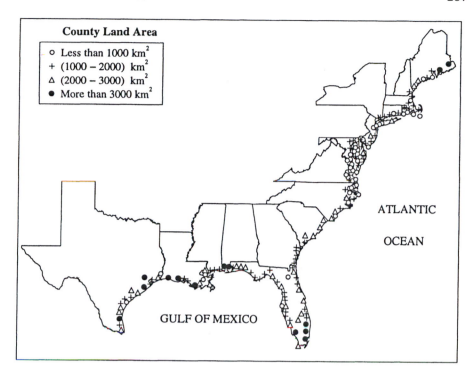

Fig. 11.5: Location of the 175 coastal counties from Texas to Maine used in determining hurricane and major hurricane return periods. The symbol locates the geographical center of each county and symbol type indicates the county land area in square kilometers (km^2). In general, county size decreases northward along the U.S. east coast.

50 years. A return period exceeding 97 years indicates the county was not hit by a hurricane during the 20th century. Overall the ratio of the number of accounts of major hurricanes to hurricanes is 21%. As mentioned, the larger the county the shorter the return period, other factors being equal.

It is observed that the estimates of hurricane return periods presented here are similar to those given in Simpson and Lawrence (1971). They divided the Gulf and Atlantic coast into 58 nonoverlapping 50-mile segments and estimated tropical storm and hurricane frequencies in each sector based on data over the period 1886 through 1970. Estimated values of hurricane return periods presented here match previous estimates (Simpson and Lawrence 1971, Ho et al. 1987). The spatial variability of hurricane frequencies along the Gulf and Atlantic coastlines are based on data over similar record lengths. Storms with uniform track orientation at landfall are grouped along coastal segments.[3] Information on hurricane incidence is used to develop stochastic wind-loss projection models for planning and decision making. Commercial models have been

---

[3]The assumption is that if the coastal segments are chosen properly, the data will be homogeneous with respect to the meteorology of the hurricanes that pass through.

developed by several different companies for use in the financial and insurance industries (see Chapter 19). The probabilities generated from more sophisticated models, such as these, can be checked against the coastal county return periods presented here.

11.2.2   Texas to Alabama

Return periods and wait times for the occurrence of hurricanes and major hurricanes in coastal counties of Texas are shown in Figure 11.6. Note that not all the counties are on the immediate coastline. Return periods for hurricane-force winds range from a high of 14 years in Cameron County to a low of 5 years in Galveston County (Galveston). The longer the return period the lower the annual probability. In general, return periods and wait times are slightly lower in central and northern Texas, although the difference with south Texas is only marginal. Eight of the 17 counties have wait times of less than 5 years. The five southern counties all have wait times exceeding 5 years. The counties most frequently hit by major hurricanes are Brazoria and Galveston, both in the north. In fact, Brazoria has a wait time of less than 10 years for a major hurricane strike. Kenedy is the largest county along the south Texas coastline. It has the shortest wait time for major hurricanes of any county in the south. The annual probability of a major hurricane in Nueces County (Corpus Christi) is 4.1%. Jefferson (Beaumont) and Orange counties have hurricane return periods close to 10 years, but the wait times for major hurricane-force winds exceed 65 years. Twelve of the counties have wait times less than 25 years. Harris County (Houston) has a return period of 7 years for hurricanes and 32 years for major hurricanes.

Return periods and wait times of hurricanes and major hurricanes in the parishes of Louisiana are given in Figure 11.7. Hurricane return periods range between 8 and 20 years. Terrebonne and Plaquemines are the most frequently affected parishes. Both parishes extend into the Gulf of Mexico. Orleans Parish (New Orleans) has a wait time of approximately 10 years. The return periods of hurricanes in Louisiana parishes are typically a few years longer than return periods in the counties of Texas, but the return periods for major hurricanes are a bit shorter in Louisiana. This is because late-season major hurricanes are rare over Texas. With the exception of Vermillion and Iberia, wait times for intense hurricane winds in the parishes of Louisiana are generally between 13 and 17 years. New Orleans can expect to get hit by a major hurricane with a probability exceeding 50% in any 16-year time interval. Return periods and wait times in the five counties of Mississippi and Alabama are shown in Figure 11.8. Return periods range between 8 and 12 years for hurricane-force winds and between 20 and 50 years for major hurricane-force winds. Harrison (Gulfport) and Jackson (Pascagoula) counties in Mississippi have the shortest hurricane wait times of approximately 5 years. Mobile County (Mobile) in Alabama has the shortest wait time (approximately 13 years) for intense hurricanes along this part of the coast.[4] The larger Baldwin County in the east has a wait time for major hurricanes of 33 years.

---

[4]The counties of Mississippi and Alabama were affected by hurricane *Danny* in 1997 and by hurricane *Georges* in 1998.

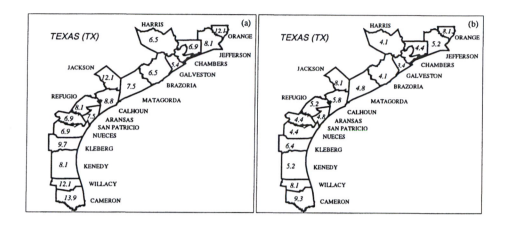

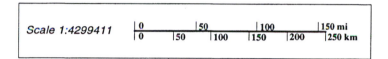

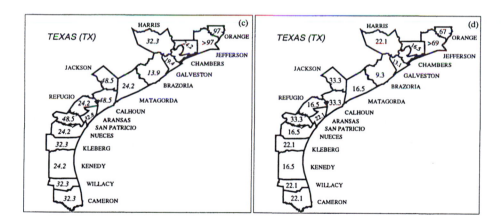

Fig. 11.6: (a) Hurricane return periods and (b) wait times in coastal counties of Texas based on data over the period 1900–1996. (c) Major hurricane return periods and (d) wait times in coastal counties are also given. The number in each county is the return period or wait time in years. Names of the counties and a map scale are provided.

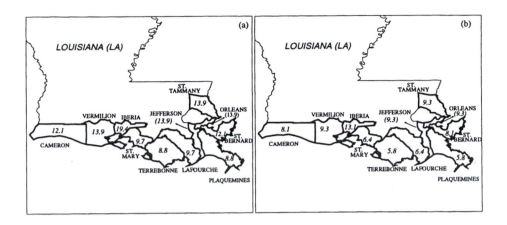

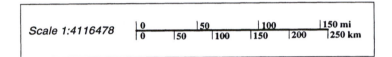

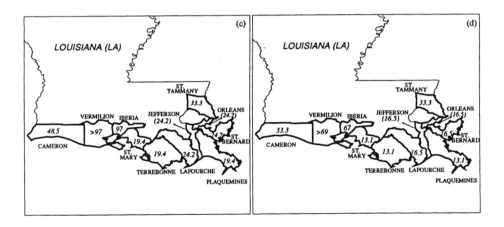

Fig. 11.7: (a) Hurricane return periods and (b) wait times in coastal parishes of Louisiana based on data over the period 1900–1996. (c) Major hurricane return periods and (d) wait times in coastal parishes are also given. The number in each parish is the return period or wait time in years. Names of the parishes and a map scale are provided.

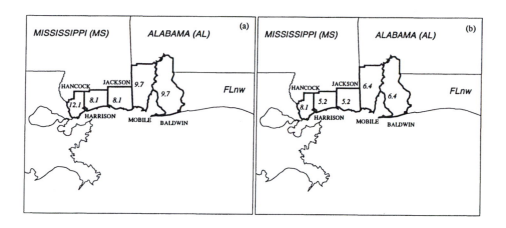

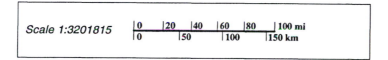

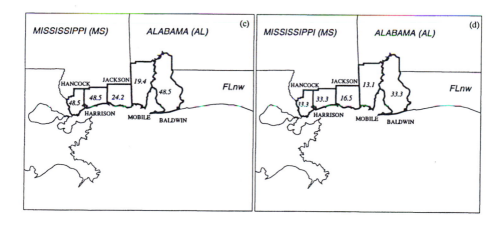

Fig. 11.8: (a) Hurricane return periods and (b) wait times in coastal counties of Mississippi and Alabama based on data over the period 1900–1996. (c) Major hurricane return periods and (d) wait times in coastal counties are also given. The number in each county is the return period or wait time in years. Names of the counties and a map scale are provided.

11.2.3   Florida

The counties of northwest Florida show a large range of return periods depending on location (Figure 11.9). Counties along the Florida Panhandle have return periods close to 10 years. In contrast, counties farther to the east, along the Big Bend region, have return periods between 20 and 50 years. Wait times are shortest for Escambia (Pensacola) and Bay (Panama City) counties and longest for Jefferson and Taylor counties. Wait times for major hurricane winds in counties along the northwestern Florida coastline are longer than wait times in counties over much of the Gulf coast. The exceptions are Okaloosa (Crestview) and Walton counties in the Panhandle, where wait times are 22 and 16 years, respectively. The return period for a major hurricane in Pasco County (New Port Richey) is about 50 years. Bay, Gulf, and Franklin counties are unusual in that hurricane frequencies are quite high yet the frequencies of major hurricanes are low. The counties of Wakulla, Jefferson, Taylor, Dixie, and Levy have not had a direct strike from a major hurricane this century. This part of Florida is without barrier islands or significant beaches indicative of the infrequent occurrence of strong hurricanes. On the other hand it is certain that northwestern Florida was struck by powerful hurricanes during the 19th century (see Ludlum 1963). If the record is extended to include these earlier storms, then return periods would be shorter.

Hurricane return periods in counties of southwestern Florida are depicted in Figure 11.10. Estimates of average recurrence intervals range between 8 and 16 years. Hillsborough (Tampa) and Pinellas (Clearwater) have wait times of approximately 8 years for hurricane-force winds. Lee County (Fort Myers) has a return period of 11 years and Collier County (Naples) has a wait time of 5 years. The shortest return period is noted for Monroe County (Key West), which includes the Florida Keys. Monroe County experiences the highest frequency of direct hurricane landfalls of any U.S. county. The return period for Monroe County is 4 years and the wait time is 2 years. The frequency of major hurricanes strikes in southwestern Florida increases from a low in Pinellas (St. Petersburg) and Hillsborough counties in the north to a high in Collier and Monroe counties in the south. The probability of major hurricanes in Monroe County exceeds 50% over any 7-year interval. Frequencies of hurricanes in counties along the southeast coast of Florida are considerably higher (Figure 11.11). Overall counties in this part of Florida have the shortest return periods of anywhere between Texas and Maine. Palm Beach (West Palm Beach), Broward (Fort Lauderdale), and Dade (Miami) counties each have land areas exceeding 3000 km^2. Dade County rivals its neighbor to the west with a return period of only 5 years. Wait times generally increase going north. Indian River County (Vero Beach) has a wait time of 5.8 years. Dade County has the shortest wait time between intense hurricane winds at 13 years, which is similar to Galveston and Mobile counties along the Gulf of Mexico. Return periods range from approximately 10 years in Brevard County (Melbourne) to 32 years in Nassau County on the Georgia border. No major hurricane has made landfall in northeastern Florida between Indian River County and Nassau County during the 20th century.

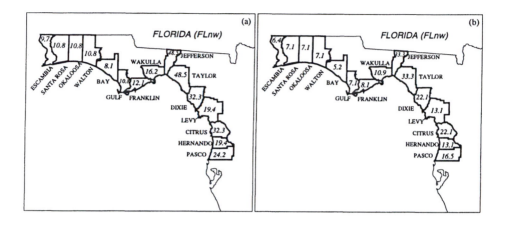

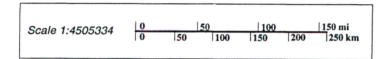

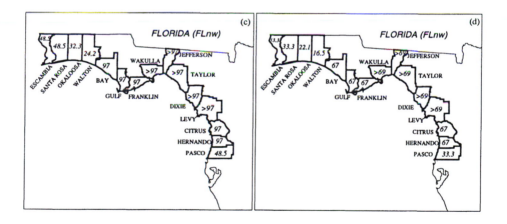

Fig. 11.9: (a) Hurricane return periods and (b) wait times in coastal counties of northwestern Florida based on data over the period 1900–1996. (c) Major hurricane return periods and (d) wait times in coastal counties are also given. The number in each county is the return period or wait time in years. Names of the counties and a map scale are provided.

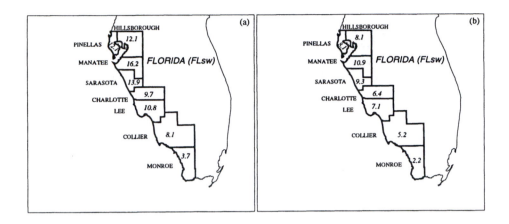

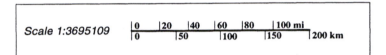

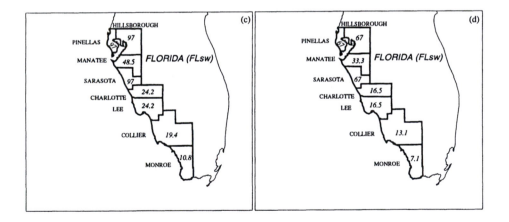

Fig. 11.10: (a) Hurricane return periods and (b) wait times in coastal counties of southwestern Florida based on data over the period 1900–1996. (c) Major hurricane return periods and (d) wait times in coastal counties are also given. The number in each county is the return period or wait time in years. Names of the counties and a map scale are provided.

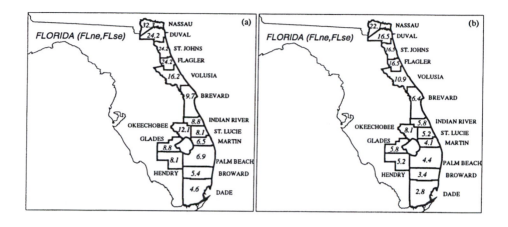

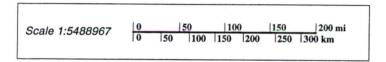

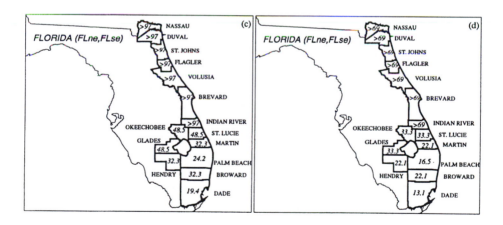

Fig. 11.11: (a) Hurricane return periods and (b) wait times in coastal counties of eastern Florida based on data over the period 1900–1996. (c) Major hurricane return periods and (d) wait times in coastal counties are also given. The number in each county is the return period or wait time in years. Names of the counties and a map scale are provided.

11.2.4 Georgia to North Carolina

Return periods and wait times for hurricanes and major hurricanes in counties along
the Georgia and South Carolina coasts are shown in Figure 11.12. Return periods for
hurricanes range between 11 and 97 years. A return period of 97 years indicates the
county was hit only one time over the 97-year period. Hurricane wait times are generally longer than 30 years for the counties in Georgia. The exception is Chatham
County (Savannah) which has a wait time of 13 years. Glynn County (Brunswick) has
a wait time of 67 years for hurricane-force winds. Camden County along the Florida
border has a wait time of 33 years. Wait times are considerably shorter in the counties
of South Carolina. Values range between 7 years in Charleston County (Charleston)
to 16 years in Horry County in the north. The annual probability of a hurricane in
Beaufort County is 7%. Georgetown County has a return period of 14 years. Similarly, Colleton County has a return period of 14 years and a corresponding wait time
of 9 years. Major hurricane strikes along this part of the U.S. Atlantic coastline are
relatively infrequent. This is due, in part, to the concave shape of the coastline. Recurving storms tend to remain offshore. Hurricane *Hugo* in 1989 was the last major
hurricane to affect the counties of South Carolina. *Hugo* remained on a northwesterly
track without significant recurvature. In general, return periods exceed 95 years except
in Charleston County (Charleston). Charleston County encompasses a long stretch of
the South Carolina coast. Here the return period is about 50 years. Sampling errors
increase for larger values of return period.

Farther to the north, along the North Carolina coastline, hurricane and major hurricane frequencies are noticeably greater (Figure 11.13). The coastline of North Carolina
extends eastward into the western North Atlantic to intercept the recurving hurricanes
that miss Georgia and South Carolina. In particular, Dare County, which includes
the Outer Banks, has the shortest hurricane return period at approximately 6 years.
Hurricane-force winds can be expected on the Outer Banks with a probability exceeding 50% in any 3-year consecutive time interval. In general, return periods are 10 years
or less from Dare County southward. Carteret County (Morehead City) has a wait time
for hurricane-force winds of 4 years. Onslow County has a wait time of 7 years. Values
of return period and wait time are not independent between counties as a single hurricane typically affects more than one county. New Hanover County (Wilmington) has
a wait time for hurricanes of 6 years.[5] North of the Outer Banks, hurricane strike frequencies are typically lower with return periods greater than 10 years. Wait times for
major hurricanes range from a low of 9 years in Dare and Hyde counties to more than
70 years in Bertie and Chowan counties. Brunswick County in the south has a return
period for major hurricane-force winds of 24 years. New Hanover has a return period
of 32 years. Washington County has a return period of 97 years and a corresponding
wait time of 67 years. Camden County in the north has a wait time for major hurricanes
of 33 years.

---

[5]The counties of southern North Carolina were affected by hurricane *Bonnie* in 1998.

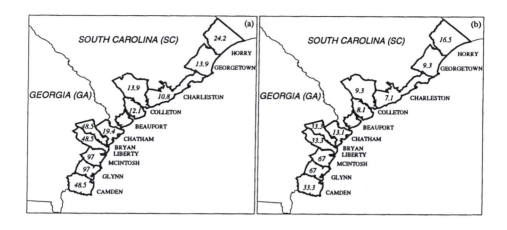

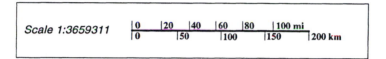

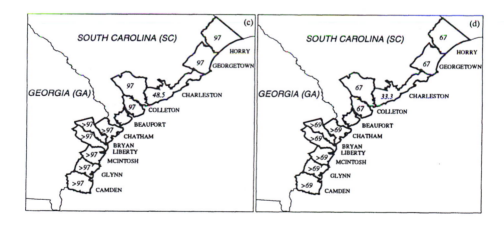

Fig. 11.12: (a) Hurricane return periods and (b) wait times in coastal counties of Georgia and South Carolina based on data over the period 1900–1996. (c) Major hurricane return periods and (d) wait times in coastal counties are also given. The number in each county is the return period or wait time in years. Names of the counties and a map scale are provided.

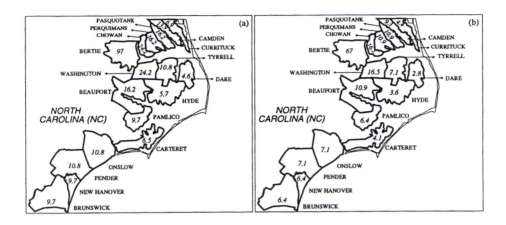

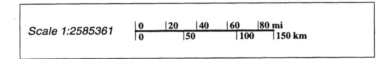

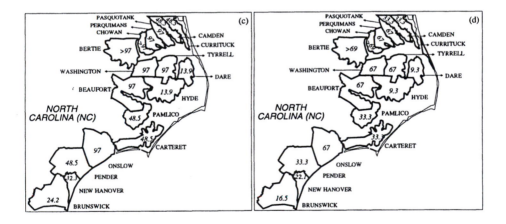

Fig. 11.13: (a) Hurricane return periods and (b) wait times in coastal counties of North Carolina based on data over the period 1900–1996. (c) Major hurricane return periods and (d) wait times in coastal counties are also given. The number in each county is the return period or wait time in years. Names of the counties and a map scale are provided.

11.2.5 Virginia to Maine

Hurricane and major hurricane frequencies in counties and independent cities of Virginia and Maryland are significantly lower than the frequencies in the counties of North Carolina (Figures 11.14 and 11.15). Chesapeake and Virginia Beach have the shortest return periods for hurricane-force winds of any area in Virginia. Virginia Beach has a wait time of 6 years and Chesapeake has a wait time of 8 years. Return periods generally increase with increasing latitude. Accomack County has a return period of 24 years. In Maryland, hurricane return periods exceed 45 years for all areas with the exception of Worcester County in the southeast, which has a return period of 32 years. Coastal areas along the western bank of the Chesapeake Bay, including Baltimore experience hurricane-force winds more frequently than the areas along the eastern bank. No major hurricane has occurred in this region since the turn of the 20th century. Hurricane frequencies in the counties of Delaware and New Jersey are provided in Figure 11.16. Wait times range between 13 and 22 years for counties along the immediate coastline to about 70 years for counties along the Delaware Bay. Sussex County in Delaware has a return period for hurricanes of 32 years. Ocean County in New Jersey has a wait time of 13 years. Middesex County (New Brunswick) has a wait time of 67 years for hurricane-force winds. As with Maryland, major hurricane activity has been nonexistent along the Delaware and New Jersey coasts since 1900.

Return periods and wait times of hurricanes and major hurricanes in coastal counties of New York and Connecticut are shown in Figure 11.17. Hurricane and major hurricane strike frequencies increase substantially across the counties of western Long Island. Suffolk County in the east is the most vulnerable county of the region with a wait time of 8 years. New Haven County (New Haven) has a wait time of 11 years. Major hurricanes can be expected in this region about once in 20 years on average. The preferred track is from the south. The differences in wait times between Suffolk County and the counties of southern Connecticut indicate that major hurricanes diminish in intensity as they cross Long Island. Hurricane and major hurricane frequencies along the Rhode Island and Massachusetts coastlines are given in Figure 11.18. Wait times of 8 years are noted for the southeastern counties of Massachusetts. Barnstable (which includes Cape Cod) and Nantucket counties have 12-year return periods for hurricane strikes. Return periods between 14 and 20 years are noted for counties over the rest of the region. Providence County (Providence) has a wait time of 13 years. Major hurricane wait times show a slightly different pattern. Shortest wait times are noted in the counties of Rhode Island. Hurricane and major hurricane return periods and wait times in the counties of New Hampshire and Maine are shown in Figure 11.19. Cumberland County (Portland) has a wait time of 33 years. Wait times for hurricanes exceed 30 years in these areas. The exception is Washington County, which has a return period of 16 years. Washington County covers a large area and extends eastward into the North Atlantic. The county is susceptible to fast moving hurricanes recurving northward out of the tropics. No major hurricanes have struck northern New England during the 20th century.

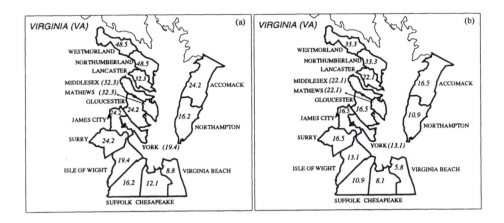

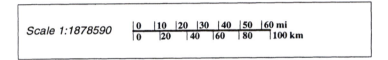

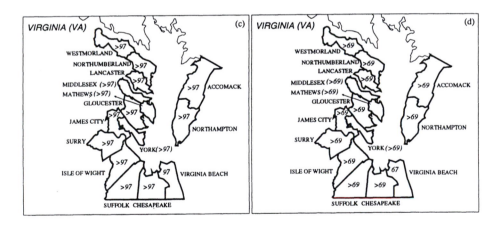

Fig. 11.14: (a) Hurricane return periods and (b) wait times in coastal counties and independent cities of Virginia based on data over the period 1900–1996. (c) Major hurricane return periods and (d) wait times in coastal counties and cities are also given. The number in each area is the return period or wait time in years. Names of the counties and independent cities along with a map scale are provided.

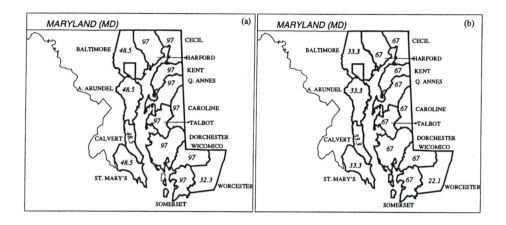

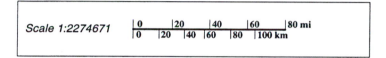

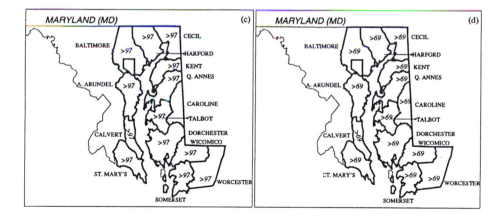

Fig. 11.15: (a) Hurricane return periods and (b) wait times in coastal counties and independent cities of Maryland based on data over the period 1900–1996. (c) Major hurricane return periods and (d) wait times in coastal counties and cities are also given. The number in each area is the return period or wait time in years. Names of the counties and independent cities along with a map scale are provided.

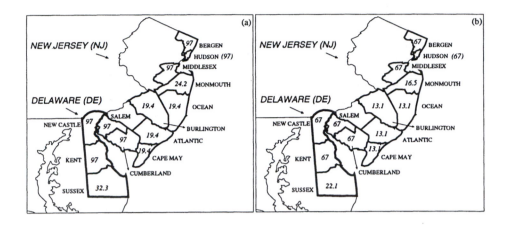

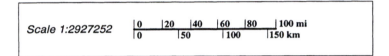

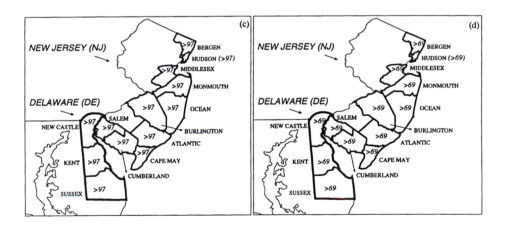

Fig. 11.16: (a) Hurricane return periods and (b) wait times in coastal counties of Delaware and New Jersey based on data over the period 1900–1996. (c) Major hurricane return periods and (d) wait times in coastal counties are also given. The number in each county is the return period or wait time in years. Names of the counties and a map scale are provided.

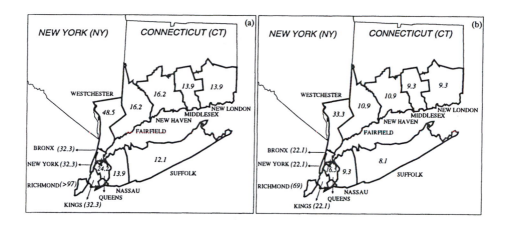

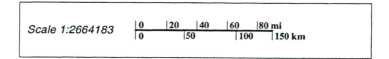

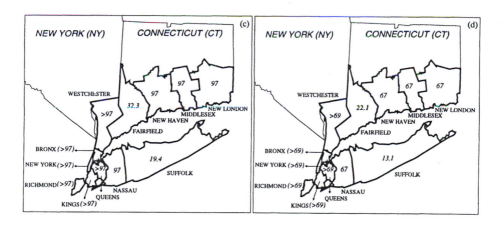

Fig. 11.17: (a) Hurricane return periods and (b) wait times in coastal counties of New York and Connecticut based on data over the period 1900–1996. (c) Major hurricane return periods and (d) wait times in coastal counties are also given. The number in each county is the return period or wait time in years. Names of the counties and a map scale are provided.

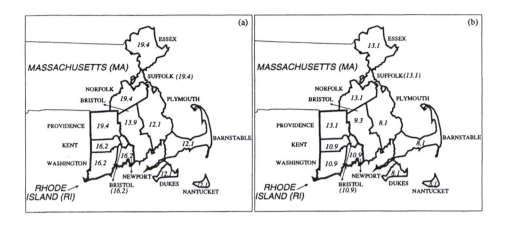

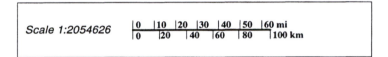

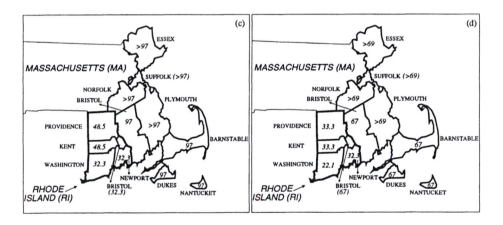

Fig. 11.18: (a) Hurricane return periods and (b) wait times in coastal counties of Rhode Island and Massachusetts based on data over the period 1900–1996. (c) Major hurricane return periods and (d) wait times in coastal counties are also given. The number in each county is the return period or wait time in years. Names of the counties and a map scale are provided.

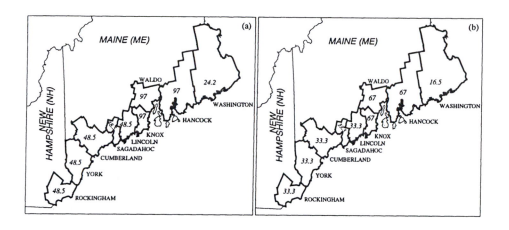

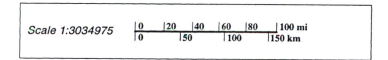

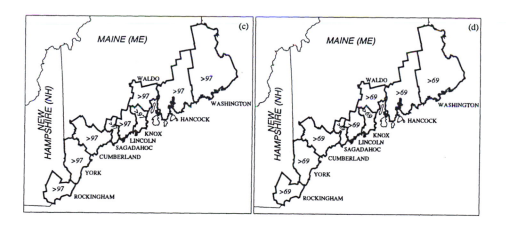

Fig. 11.19: (a) Hurricane return periods and (b) wait times in coastal counties of New Hampshire and Maine based on data over the period 1900–1996. (c) Major hurricane return periods and (d) wait times in coastal counties are also given. The number in each county is the return period or wait time in years. Names of the counties and a map scale are provided.

11.3   Limitations

Uncertainty in the estimates of return periods increases significantly for values close to the length of record. In general, greatest incertitude occurs in counties along the northeast coast. The rare occurrence of major hurricane-force winds along the eastern seaboard makes the estimates of return periods quite uncertain. Furthermore, most coastal locations have never experienced a hurricane of magnitude four or five on the Saffir/Simpson hurricane destruction potential scale over the period of record. For major hurricanes, uncertainty is greatest in counties between northeastern Florida and Maine. In regions where there are only a few events a parametric model is often employed for estimating return periods. The estimation of return periods is accurate under two assumptions: Each year has the same probability of a landfall and each landfall is an independent event. The assumption of constant probability can be examined with the aid of cumulative distributions like those provided in Chapter 8. Compare for instance the cumulative frequency distributions of hurricanes occurring in Louisiana and North Carolina. For Louisiana, the cumulative frequency suggests a constant annual probability; whereas for North Carolina the frequency indicates the annual probability changes from decade to decade. As such, the annual probability of hurricane occurrences in North Carolina depends on whether the decade is active or inactive. An annual probability from the whole record will be an over or under estimate depending on the epoch. The fact that hurricane probabilities in North Carolina are a function of time is not surprising. It was noted in Chapter 10 that hurricane strike probabilities in the United States are related to fluctuations in the climate, particular fluctuations in the phases of ENSO. Fluctuations and changes in ENSO are likely related to changes in hurricane frequencies along limited sections of the coast.

The assumption that hurricane strikes in an area are independent events can also be checked. Consider the occurrence of hurricanes over Florida. The total number of Florida hurricanes over the period 1851–1996 is 91. This puts the annual probability at 62%. In any given year Florida is more likely than not to be hit. However, over the 146-year period there are only 63 different years with hurricane strikes, making the annual probability 43% (63/146) of at least one Florida hurricane. This suggests the annual probability of not experiencing a hurricane is 57%.[6] To continue, given that Florida gets hit with a hurricane in a particular year, there is 41% (26/63) chance of getting hit again in the same year. The return period depends on what annual hit rate is used. If the average annual probability is used then the return period is 1.6 years, but if the annual probability is estimated from the percentage of years with hurricanes, then the return period is 2.3 years. Multiple-hit years are also relatively common for North Carolina. In general, as the area of consideration increases, the likelihood that hurricane events are independent in time decreases. This is because climate patterns responsible for hurricane development and steering can be persistent. This is especially

---

[6]Clustering of strikes indicates that multiple-hit years in Florida might be more common than expected under the assumption of independent events.

true when the monthly circulation is dominated by stable, global scale waves. Features of a persistent atmosphere are conspicuous on climate maps of height departures from normal. Certain patterns can last for several weeks to several months, or longer. The climate conditions responsible for initiation and movement of tropical cyclones over a particular corridor may remain in place long enough to allow several hurricanes through the same region.

11.4   Trends

Areas with the shortest return periods (highest landfall probabilities) are coastal locations that jut into the sea. Dare County, North Carolina and the counties of southern Florida have the highest landfall probabilities. The greatest probability is noted in Monroe County, Florida with annual odds against a hurricane strike at 3 to 1 (annual probability of a hit at about 25%). As the occurrence of a hurricane at any one location is a rare event, it is possible to consider strikes as the outcome of a Poisson random process (see Chapter 14). As mentioned, parametric modeling of return periods is often useful. A characteristic of the Poisson distribution is the equality of mean and variance. The mean is the annual hurricane frequency. The ratio of the sample variance to the sample mean ($R = \sigma^2/$mean) is used to test for a Poisson process. The test is made by noting the ratio (R) can be converted to a chi-squared random variable (e.g., Keim and Cruise 1998). A test is made on the annual probability of landfalls in the southeastern and northeastern United States. The southeast includes the states of Georgia and South Carolina, and the northeast includes the coastal states from New York to Maine. A failure to reject the null hypothesis that the distribution is Poisson ($\alpha = 0.10$) occurs in both regions.

With the occurrence of hurricanes described as Poisson, the interarrival times are distributed exponentially. As such, a regression of the natural logarithm of recurrence intervals on the cumulative number of days from the first landfall allows for an examination of trends. Figure 11.20 shows the values of recurrence intervals for hurricane strikes over the southeastern and northeastern United States (Elsner et al. 1999a). The least-squares trend line for the southeast is positive indicating that the interarrival times are increasing. Said another way, the frequency of landfalls along the southeast coastline is decreasing. In contrast, a negative trend is noted for the northeast indicating an increasing frequency of hurricane landfalls. Fewer hurricane landfalls in the southeast are matched by more landfalls in the northeast. Though neither trend is statistically significant, the results suggest possible longterm shifts in landfall probabilities.

Shifts in landfall probabilities are consistent with the hypothesis that the frequency of warm ENSO events is on the rise. ENSO data show a change in the El Niño/La Niña cycle starting in the middle 1970s. Over the period 1950–1975 La Niña events were strong, and they tended to last longer than El Niño events. Subsequently, El Niño events have become more frequent and longer. In fact, the warming in the tropical Pacific from 1990 to 1995 is unprecedented in its duration over the 20th century (Tren-

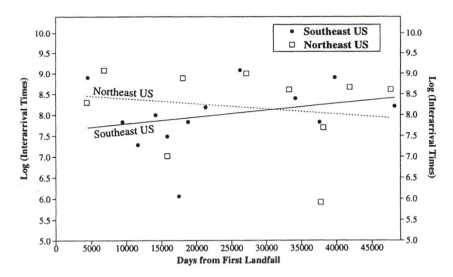

Fig. 11.20: Recurrence intervals for hurricane landfalls along the southeast (SC and GA) and northeast (NY to ME) coasts. The ordinate is the logarithm of interarrival times and the abscissa is the cumulative number of days from the first strike. Values based on data over the period 1851–1996.

berth and Hoar 1996). While warm ENSO years tend to inhibit the formation of deep tropical hurricanes there is a corresponding increase in the probability of more northerly baroclinically-enhanced hurricane activity (see Chapter 10). Thus it might be the case that changes in ENSO can explain the apparent decrease in landfall probabilities along the southeast coast. Based on nonlinear time-series models, it is argued that ENSO is a self-adjusting regulator of the climate system, whereby El Niño events are activated to reverse positive global trends and La Niña events are initiated to reverse negative trends (Tsonis et al. 1998). In fact a physical hypothesis is invoked to suggest a role for the ENSO in balancing global temperature (Sun and Trenberth 1998). As the planet warms, the frequency and intensity of warm ENSO events increases to help redistribute the excess heat. Thus global warming may actually reduce the occurrence of deep tropical hurricanes indirectly through an increase in El Niño events, thereby shifting hurricane landfall probabilities along the coastline. A geographic redistribution of hurricane probabilities is cause for considerable concern regardless of whether there are changes in overall activity.

12

Hurricanes of the Early 1990s

Hurricane activity in the North Atlantic basin during 1995 and 1996 was well above the long-term average. Cumulatively the two seasons featured a record 20 hurricanes. Eleven hurricanes formed during 1995 and nine more followed in 1996. Many of the storms formed deep in the tropics from easterly waves. The extensive activity marked an abrupt change from the early 1990s, when consecutive seasons had below normal levels of activity; and much of the activity was baroclinically-enhanced. Tropical-only hurricanes were more common during the 1940s and the 1950s. The return of tropical-only hurricanes to the North Atlantic basin is likely the result of several global and local factors including cool SST conditions in the equatorial central and eastern Pacific (cold phase of ENSO) and warm SSTs in the tropical Atlantic. This chapter reviews the record-setting nature of the 1995 and 1996 North Atlantic hurricane seasons. The ample activity during the brief 2-year period is compared to the dearth of activity during the previous 4 years. Speculations on climatological factors related to the heightened activity are given. The presentation in this chapter follows the work of Kimberlain and Elsner (1998).

12.1 Two Consecutively Active Hurricane Seasons

Both the 1995 and 1996 North Atlantic hurricane seasons had significantly more hurricanes than average. Several long-standing records were broken. The two seasons combined for a total of twenty hurricanes—the most active 2-year period on record. The average annual number of North Atlantic hurricanes over the period 1886 through 1996 is five. Over the shorter, more reliable period 1950 through 1996, the average is approximately six (see Chapter 4). In contrast to the activity during 1995 and 1996, the previous 4 years between 1991 and 1994 averaged 3.8 hurricanes per season. Table 12.1 shows the monthly frequency of North Atlantic hurricanes over these two consecutive periods. The total number of hurricanes is about the same, but the character of the two periods is remarkably different. Most impressive is the difference in frequencies between tropical-only and baroclinically-enhanced hurricanes. The earlier period featured 13 baroclinically-enhanced and 2 tropical-only hurricanes. This

Table 12.1: Frequencies of North Atlantic hurricanes. The top set of numbers are for the combined seasons over the period 1991 through 1994. The bottom set is for the combined seasons of 1995 and 1996. Values are given for each month and by hurricane type. Totals are also given.

Category	\multicolumn{6}{c}{1991–1994}	

Category	Jun	Jul	Aug	Sep	Oct	Nov	Total
All	.	.	4	6	2	3	15
TO	.	.	1	1	.	.	2
BE	.	.	3	5	2	3	13
Category	Jun	Jul	Aug	Sep	Oct	Nov	Total
			1995–1996				
All	1	2	8	4	4	1	20
TO	.	2	8	4	2	1	17
BE	1	.	.	.	2	.	3

compares with 3 baroclinically-enhanced and 17 tropical-only systems during the later period. Even the November hurricane of 1996 (*Marco*) was tropical-only. Seasonally, the 1995–1996 period featured 3 hurricanes before August. This compares with no early-season hurricanes from 1991 through 1994. September was most active during the earlier period, while August was most active during the later period.

The frequency of North Atlantic hurricanes over the 6-year period from 1991 through 1996 is shown as a time series in Figure 12.1. The top panel shows the annual abundance of all hurricanes and the bottom two panels show the annual numbers of tropical-only and baroclinically-enhanced hurricanes. Each year from 1991 through 1994 was below normal. There were exactly 4 hurricanes in 1991, 1992, and 1993. In contrast both the 1995 and 1996 seasons were above the average. Most of the activity during 1995 and 1996 was tropical-only. This compares strikingly to the lack of tropical-only hurricanes during the earlier period. However, the inactive period was characterized by an above normal frequency of baroclinically-enhanced hurricanes. Collectively, 1995 and 1996 saw less than the average amount of baroclinic activity. Clearly the 6-year period featured an unusual variation in North Atlantic hurricane activity. The variation also appears in social impacts. Estimated direct economic losses from North Atlantic tropical cyclones during 1994 totaled $1 billion. This contrast with $10 billion in losses during 1995 (Pielke and Pielke 1997b). Surprisingly, death tolls were reversed. The 1994 season resulted in 1175 deaths (many due to flooding in Haiti from hurricane *Gordon*) compared to 118 deaths during the active 1995 season. A closer scrutiny of the heightened activity during 1995 and 1996 is warranted.

12.1.1   North Atlantic Hurricanes of 1995

Total hurricane activity over the North Atlantic during 1995 was well above the long-term average. The season began early, with hurricane *Allison* appearing in June. *Allison*

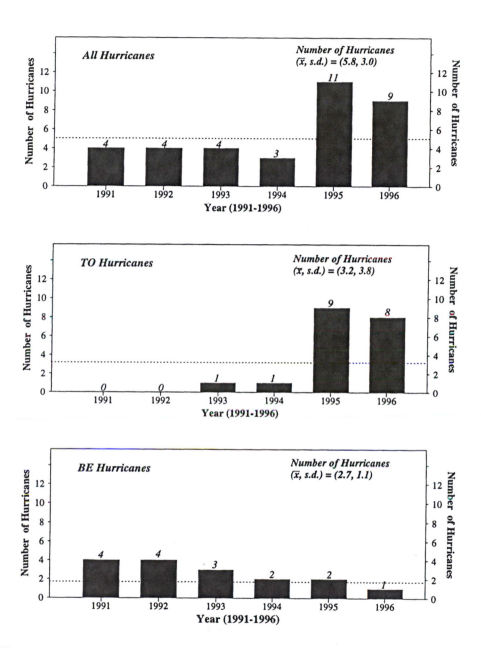

Fig. 12.1: Annual abundance of North Atlantic hurricanes over the period 1991–1996. The frequency of all, TO, and BE hurricanes are shown in the top, middle, and bottom panels, respectively. The averages ($\bar{x}$) and standard deviations ($s.d.$) are also provided. The dotted lines show the average frequency over the period 1886–1996.

was the first June hurricane since *Bonnie* in 1986. The average frequency of June hurricanes amounts to about one hurricane every 5 years. June hurricanes account for 4% of all North Atlantic storms (Chapter 4). During August, an outbreak of hurricanes occurred to the east of the Lesser Antilles including five tropical storms originating within 7 days. Three of the five (*Humberto*, *Iris*, and *Luis*) were hurricanes at the same time. Since the turn of the century, the only other season to featured 3 hurricanes simultaneously occurred in 1961.[1] This outbreak made August an extremely active month; in fact, it was the single most active month since September of 1949.

The 1995 season also featured five major (or intense) hurricanes. This activity is well above the long-term average of two intense hurricanes per year (Chapter 7). The 1995 hurricane season represents the most active outbreak of major hurricanes over the North Atlantic since 1964. All the major hurricanes during 1995 originated from tropical easterly waves and developed over tropical latitudes south of 20°N. This is consistent with the observation that greater than 80% of all intense North Atlantic hurricanes originate from African waves (Landsea 1993). Most of the intense hurricanes of 1995 remained east of the United States. Only late-season hurricane *Opal* made landfall in the United States. Hurricane *Opal* was the most intense hurricane to visit the Gulf of Mexico since hurricane *Allen* of 1980. It was one of the strongest hurricanes on record during the month of October. Hurricane *Hattie* in 1961 was a powerful hurricane that originated over the western Caribbean during late October.[2] Hurricane *Opal* intensified rapidly over the central Gulf of Mexico, dropping 49 mb in 16 hours. However, the winds around *Opal* decreased considerably before it made landfall near Pensacola. *Opal* gets the distinction of being the most intense October Gulf hurricane, the most intense hurricane to make landfall in northwestern Florida during the 20th century, and one of the costliest hurricanes to date (Hebert et al. 1996).

The 1995 season saw the first hurricane in the Caribbean region since *Hugo* in 1989. As one of the five intense hurricanes, *Luis* lashed the island communities of Antigua and Barbuda in early September. The absence of Caribbean hurricanes over the period 1990 through 1994 is part of a longer interval of reduced activity going back to 1982. In fact, over the 13-year period from 1982 through 1994, 10 years had no Caribbean hurricanes. Given an annual probability of 40% that no Caribbean hurricanes will occur, (1950–1995 base period) and assuming that years are independent, the probability of 10 no-hurricane years out of 13 is given by the binomial expansion as $p = \frac{13!}{10!3!}(0.4)^{10}(0.6)^3$. The value of p is less than 1%. The recent downturn in the number of hurricanes affecting the Caribbean basin has been mentioned previously (Landsea et al. 1996). A summary of each individual tropical cyclone during 1995 is given in Lawrence et al. (1998). Speculations as to the possible factors responsible for the heightened activity are given in Saunders and Harris (1997) and in Landsea et al. (1998).

---

[1]There were 4 hurricanes at one time during September of 1998. This tied the record set back in August of 1893.

[2]Hurricane *Mitch* in October of 1998 reached an intensity of 905 mb over the western Caribbean making it the most intense October hurricane on record for the North Atlantic.

12.1.2   North Atlantic Hurricanes of 1996

The above normal activity of 1995 continued through the 1996 season.[3] Thirteen tropical storms occurred during 1996 and nine become hurricanes. Six of the hurricanes
went on to become major hurricanes. Only 4 other years during the 20th century (1916,
1926, 1950, and 1961) featured at least 6 major hurricanes. The combined two-year
total of 11 major hurricanes ranks as the most abundant over the 111-year record.
The 1996 season also continued the trend from the previous season as most of the
storms originated from African easterly waves. Six of the nine hurricanes visited the
Caribbean Sea, the most since 1916. Hurricane *Bertha* originated in early July and
reached major hurricane intensity a day later as it passed north of Puerto Rico. This
was the earliest intense hurricane in the North Atlantic since hurricane *Alma* in 1966.
Bertha tracked northwestward and made landfall in North Carolina. Of the six intense
hurricanes, only *Fran* made landfall in the United States as a major hurricane. This
was similar to the previous year during which most intense hurricane activity remained
east of the United States. Like *Bertha* earlier in the year, hurricane *Fran* made landfall
along the North Carolina coast. In fact, both hurricanes hit near Wilmington. Multiple-
hit years for North Carolina are not uncommon. Data back to 1851 indicate an annual
probability about 20% for the occurrence of at least one North Carolina hurricane landfall. The probability of two or more hurricane landfalls in a given year is less than 5%.
However, if North Carolina gets hit by a hurricane, there is a 24% chance that it will
get hit again that same year. The clustering of landfalls in time makes physical sense
considering that a season often displays preferred paths of tropical cyclone movement.
The preference for certain tracks in a given year is based on the position and persistence
of the large-scale weather features across the North Atlantic basin.

   Similar to the 1995 season, a significant outbreak of hurricanes occurred near the
peak of the season. During a week-long span in late August, three tropical cyclones
developed and two became hurricanes. This outbreak was part of a 4-week long stretch
of nearly continuous tropical cyclone activity. Hurricane *Hortense* hit the southwest
corner of Puerto Rico during September causing considerable damage and several fatalities. *Hortense* was the first Puerto Rican hurricane since *Hugo* in 1989.[4] Late
October was active in 1996 as it had been during 1995. Hurricanes *Lili* and *Marco*
developed over the western Caribbean Sea after the middle of October.

12.2   Comparisons to the Early 1990s

The heightened activity during the middle 1990s came on the heals of an extended
period of below average activity. In terms of abundance, the average annual number of
hurricanes over the previous 4 years was less than four per year compared to ten per
year during 1995 and 1996. The contrast between the two periods is not a consequence
of extreme conditions in one or two months. The below average activity was sustained

---

[3]The 1997 North Atlantic hurricane season featured three hurricanes and one major hurricane.
[4]Hurricane *Georges* in 1998 crossed the length of Puerto Rico as a formidable category two hurricane.

over most months during the period. The same can be said about the above average activity during 1995 and 1996. Hurricanes during the early 1990s were generally short lived and less intense. Most dramatic is the observation that the earlier hurricanes developed much farther to the north compared with the hurricanes of the later 2 years. This suggests different mechanisms responsible for the activity during the two periods. Interestingly, the earlier 4-year period was harder on the catastrophe insurance industry (see Chapter 19). Hurricane *Andrew* in 1992 was responsible for a significant portion of the losses during the early 1990s.[5] It developed during an otherwise quiet hurricane season. *Andrew* made landfall as a category four hurricane in southern Dade County during August and devastated the community of Homestead. Hurricane *Bob* the year before made landfall along the northeast coast.

Hurricane origins during the periods 1991–1994 and 1995–1996 are shown in Figure 12.2. Significant differences are noted in both the latitudes and the longitudes of formation. During the earlier period fourteen of the fifteen hurricanes formed north of 20°N latitude compared with three of the twenty during the latter period. Moreover, 47% of the earlier hurricanes formed north of 30°N compared to 5% of the later hurricanes. One of the early storms originated near 42°N latitude. Three of the earlier hurricanes formed west of 70°W longitude compared with eight of the later storms. Only one of the earlier hurricanes formed over the Gulf of Mexico, and none originated over the Caribbean Sea. This tendency toward fewer hurricanes originating in the Caribbean region began in the 1980s (Reading 1989).

The seasonal variation of hurricanes over the two periods is compared in Figure 12.3. The years from 1991 through 1994 were characterized by the absence of early season activity. No hurricanes formed during the months of June or July. Cumulatively, the early period activity peaks in September. The slow start to the seasons of the early 1990s translated into below average number of hurricanes. The early period featured 3 November hurricanes. In comparison, the years 1995 and 1996 were characterized by early-season hurricane activity and a pronounced peak during the month of August. This translated into above average activity. The 1995 season featured three October hurricanes. Hurricane activity in October is a signature of active North Atlantic hurricane seasons. It should be noted that the observed differences in hurricane activity for each month suggest sustained disparities in climate factors between the two consecutive time periods.

Along with their higher latitudes of development, hurricanes during the early 1990s were generally less intense than the long-term average. For instance, only four major hurricanes occurred over the period 1991 through 1994, compared to an average of about two per year and in stark contrast to the eleven over the period 1995–1996. Figure 12.4 shows the annual-average maximum intensity for each year in the period with a comparison to the long-term average. As before, the maximum intensity for each hurricane is an average over all storms during the year. The time average over the period 1991–1994 is 87.6 kt, which compares to 95.4 kt over the period 1995–1996.

---

[5]*Andrew* was the strongest and most devastating hurricane of the early 1990s.

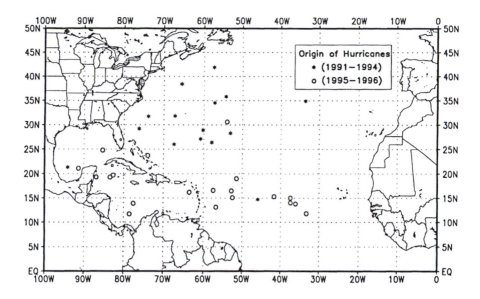

Fig. 12.2: Origins of North Atlantic hurricanes over the period 1991–1996. Asterisks indicate the hurricanes during the period 1991–1994 and circles indicate the hurricanes during the period 1995–1996. The 25°N latitude separates the points of origin between the two periods.

Average maximum intensities were particularly low during 1993 and 1994, but were near the long-term average during 1995 and 1996. Maximum intensities by hurricane type indicate that, in general, tropical-only hurricanes were weaker and baroclinically-enhanced hurricanes were stronger over the 6 years. In particular, the two tropical-only hurricanes during the earlier period remained below category three intensity. In contrast the 8 baroclinically-enhanced hurricanes during 1991 and 1992 had an average maximum intensity of 96 kt. Hurricane *Andrew* in 1992 was a strong category four baroclinically-enhanced hurricane. Of interest is the fact that tropical-only hurricanes of 1995 and 1996 are near the long-term average intensity for this category of storms, but the baroclinically-enhanced hurricanes of 1991 and 1992 are well above the average intensity for this category.

Baroclinically-enhanced hurricanes are typically of shorter duration. Hurricanes during the early 1990s followed this pattern. A *hurricane day* is defined as four 6-hourly observations of a hurricane. A single hurricane lasting 24 hr results in one hurricane day. Two hurricanes each lasting 12 hr also results in one hurricane day. Total hurricane days over the period 1995–1996 was 107 compared to 41 over the period 1991 through 1994. The difference was more pronounced for major hurricane days. In particular, hurricane *Luis* of 1995 ranked as the third longest-lived intense hurricane since 1950 (Landsea 1998). Only hurricanes *Donna* in 1960 and *Esther* in 1961 lived longer as intense hurricanes. Hurricane *Edouard* in 1996 was also an exceptionally robust major hurricane. Figure 12.5 shows the average-annual duration

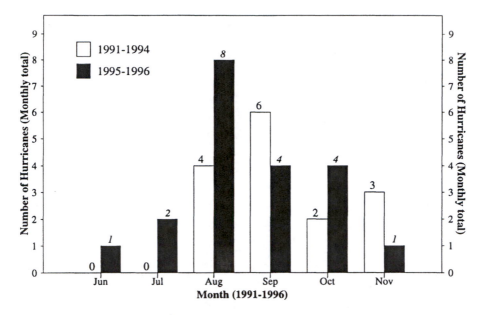

Fig. 12.3: North Atlantic hurricane activity by month. Total frequencies are given for the two consecutive periods of 1991–1994 and 1995–1996. The later period featured early-season hurricanes with a peak during August.

for hurricanes over the 6-year period. Hurricane duration is shown for all, tropical-only, and baroclinically-enhanced hurricanes. The long-term mean over the period 1886–1996, shown as a dotted line, indicates the 6-year period was below average by about a day and a half. This period is part of a longer trend toward shorter-lived hurricanes (Chapter 10). In particular, the period 1991–1994 featured hurricanes that lasted an average of only 2 to 3 days. The average duration of hurricanes during 1995 was above the long-term mean. In fact, both 1995 and 1996 featured hurricanes more typical of the longer-lived hurricanes from the late 1940s through the 1950s. As expected, years with frequent higher-latitude baroclinically-enhanced activity tend to be years with shorter average duration.

The short duration of hurricanes during the early 1990s is reflected in their paths. Tracks of the individual hurricanes over the period 1991–1994 are shown in Figure 12.6. In general the higher latitude storms track a shorter distance before dissipation. The direction of motion is either north or northeast away from the United States. Bermuda was threatened by two hurricanes during the period. Hurricane *Andrew* took a somewhat unusual due westward course north of 25°N latitude. The low-latitude hurricane *Gert* formed in the Bay of Campeche and moved inland over Mexico. The only hurricane that formed south of 15°N during this 4-year period was hurricane *Chris* in 1994. *Chris* made only a brief appearance before dissipation. No hurricanes formed or crossed through the Caribbean Sea. Paths of the individual hurricanes during 1995

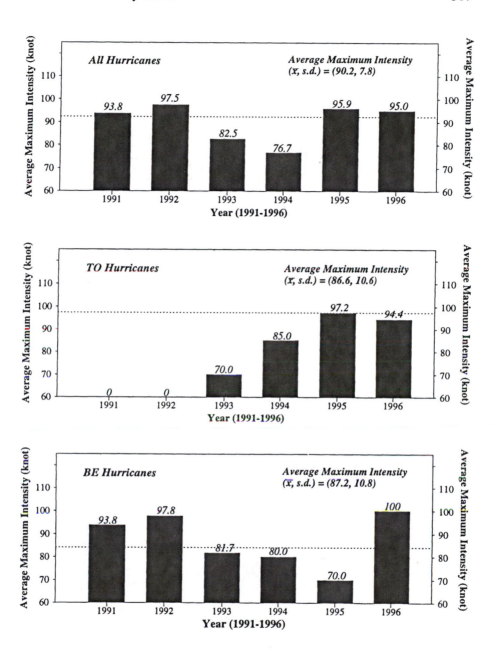

Fig. 12.4: Maximum intensity for North Atlantic hurricanes over the period 1991–1996. Values are annual averages for all (top), tropical-only (middle), and baroclinically-enhanced (bottom) hurricanes. A zero indicates no tropical-only hurricanes. The dotted line indicates the average maximum intensity over the period 1886–1996. Hurricanes during 1995 and 1996 were generally more intense than hurricanes during 1991–1994.

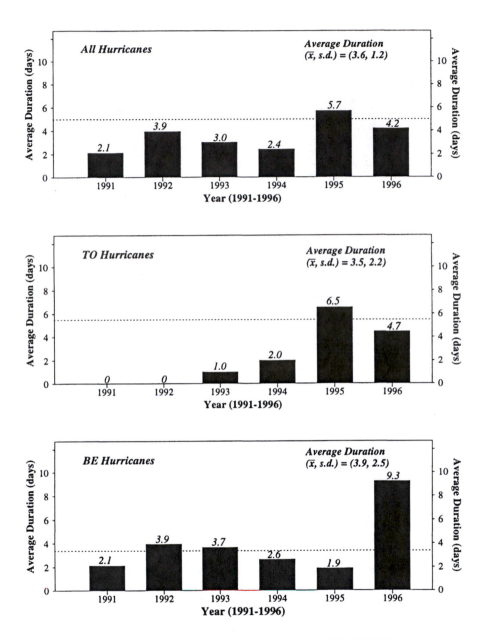

Fig. 12.5: Duration of North Atlantic hurricanes over the period 1991–1996. Values are annual averages for all (top), tropical-only (middle), and baroclinically-enhanced (BE) hurricanes. A zero indicates no tropical-only hurricanes. The dotted lines indicates the average duration over the period 1886–1996. Hurricanes during 1995 and 1996 were generally of longer duration than hurricanes during the period 1991–1994.

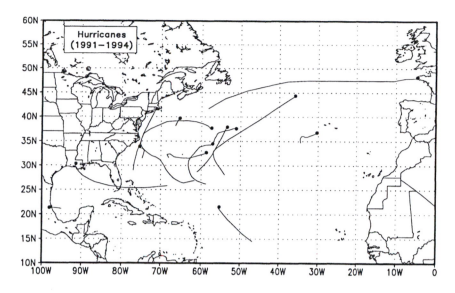

Fig. 12.6: Tracks of North Atlantic hurricanes during the period 1991–1994. The dot indicates the point of dissipation. Hurricanes over this period had a tendency to move northeastward away from the United States. A notable exception was hurricane *Andrew* in 1992, which crossed extreme southern Florida before making final landfall in Louisiana.

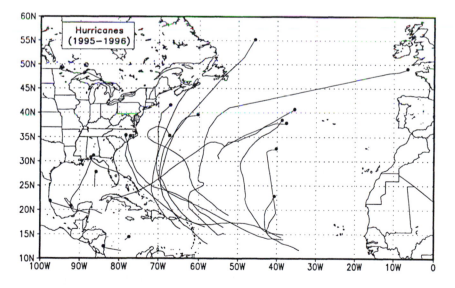

Fig. 12.7: Tracks of North Atlantic hurricanes during 1995 and 1996. The dot indicates the point of dissipation. Despite the abundance of low-latitude formations, no hurricanes passed directly through the Caribbean Sea.

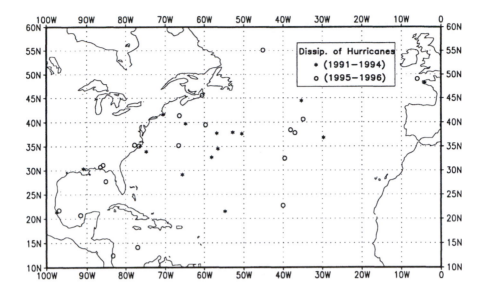

Fig. 12.8: Dissipation points for North Atlantic hurricanes over the period 1991–1996. Asterisks indicate hurricanes over the period 1991–1994 and circles indicate hurricanes over the period 1995–1996. Only two hurricanes (*Andrew* in 1992 and *Gert* in 1993) visited the Gulf of Mexico between 1991 and 1994. This compares with five hurricanes during 1995 and 1996.

and 1996 are shown in Figure 12.7. In general, hurricanes during this period took the classic parabolic sweep out of the tropics. There was a strong tendency for recurvature during this later period. Most of the hurricanes remained east of the United States. However, Bermuda was again threatened by several storms. The Caribbean was active, particularly the Leewards and Puerto Rico. The western Caribbean was also active, but no hurricane passed directly through the region.[6]

Dissipation points for hurricanes during the 6-year period are shown in Figure 12.8. In contrast to latitudes of origin, latitudes of dissipation cannot be separated by year of activity. There are, however, some differences between the periods in terms of dissipation longitude. Three of the 15 early-period hurricanes dissipated east of 50°W longitude. This compares with 7 of the 20 later-period storms. The Gulf of Mexico experiences fewer hurricanes during periods of greater high latitude formations. Only 2 hurricanes dissipated west of 80°W during the early period compared to 6 during 1995 and 1996. Interestingly, only hurricane *Gert* of 1993 formed over the Gulf during the early 1990s. There were no hurricanes or tropical storms in the Gulf of Mexico during 1991. This happened only two other times during the 20th century (1927 and 1962). Major hurricane formation over the Gulf of Mexico was below average during the period from the early 1970s through 1995 (Lehmiller et al. 1997). Yet overall Gulf activity remained nearly constant over these years as hurricanes tracked into the region,

---

[6]Hurricane *Georges* in 1998 passed directly over the Greater Antilles from Puerto Rico to eastern Cuba.

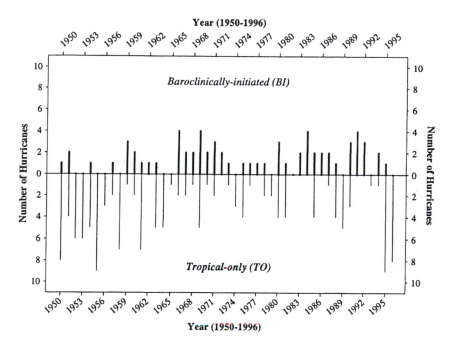

Fig. 12.9: Annual abundance of tropical-only and baroclinically-initiated hurricanes over the period 1950–1996. Values for tropical-only hurricanes are plotted on the reversed axis. Baroclinically-initiated hurricanes are more frequent beginning in 1966.

despite the decline in formations. Two hurricanes made landfall in the United States during the earlier period compared to four during 1995 and 1996.[7] The factors that produce favorable conditions for tropical-only hurricanes are likely related to factors that make U.S. threats more probable.

12.3   Baroclinically Initiated Hurricanes

Variations in the frequency of tropical-only hurricanes were noted in Chapter 5. Figure 12.9 shows the interannual variability of tropical-only and baroclinically-initiated hurricanes over the period 1950 through 1996. No tropical-only hurricanes occurred in 9 of the 47 years, including the years 1962, 1972, 1977, 1982, 1983, 1984, 1986, 1991, and 1992. Tropical-only hurricanes were more abundant during the 1950s and early 1960s. A reduction in tropical-only activity is noted since the middle 1960s. However, tropical-only hurricanes made a significant return during the combined 1995 and 1996 seasons. In particular, only two tropical-only hurricanes occurred during the period 1991–1994. Of these, hurricane *Chris* in 1994 struggled against upper-level wind shear, and developed only into a minimal hurricane. The other, hurricane *Gert* in 1993,

---

[7]The 1997 and 1998 hurricane seasons featured one and three U.S. hurricanes, respectively.

formed over the Bay of Campeche, but was a hurricane for less than a day before it moved inland over Mexico. Both the 1991 and 1992 seasons were void of tropical-only hurricane activity. The hurricanes that formed during the early 1990s tended to originate from baroclinic disturbances. Hurricanes that develop from baroclinic systems are termed baroclinically-initiated (see Chapter 6). The heightened baroclinic activity of the early 1990s started in 1990 when 3 of the 8 hurricanes were baroclinically-initiated. This activity foreshadowed the 7 baroclinically-initiated hurricanes during 1991 and 1992. All 4 of the North Atlantic hurricanes during the 1991 season originated from baroclinic disturbances (Avila and Pasch 1992). Both *Bob* and *Claudette* became major hurricanes. Three of the 4 hurricanes during 1992 were baroclinically initiated as where 2 of 3 during 1994.

Baroclinically-initiated hurricanes do not occur every season. The variations in baroclinically-initiated hurricanes from one year to the next are less extreme than the variations in tropical-only hurricanes. The average number of baroclinically-initiated hurricanes per season is 1.5. The 1991–1994 period, with its total of nine storms, was above the long-term average. In comparison, hurricane *Tanya* in 1995 was the only baroclinically-initiated storm during 1995 and 1996. Notwithstanding the dearth in baroclinically-initiated hurricane activity over the later 2-year period, a trend toward more baroclinically-initated hurricanes over the 47-year period (1950–1996) is noted. The trend is not statistically significant, however. The t-value on the regression coefficient is 1.403 (p-value = 0.167). The combination of relatively few baroclinically-initiated and many tropical-only hurricanes during 1995 and 1996 is reminiscent of seasons during the 1950s.

12.4   Climate Considerations

A necessary condition for hurricane formation is warm sea-surface temperatures (SSTs). The tropical Atlantic waters to the east of the Lesser Antilles are typically warm enough to support tropical cyclone formation. This area is part of the so-called "main development region," and is a source region for many tropical-only hurricanes. SSTs in this region were warmer than normal beginning in 1995. Considering the month of September only, the SST in a 5° latitude-longitude area centered at 12.5°N and 47.5°W averaged over 1995 and 1996 was 29°C. This compares to the 4-year average September temperature of 28.4°C. None of the individual years during the early 1990s featured September temperatures warmer than 29°C. Though the average temperature difference is less than 1°C, the relationship between SSTs in this region and tropical-only hurricane activity is robust over the 47-year period from 1950 through 1996. Observations and theory support the idea that hurricane formation is sensitive to small changes in SST in the range between 27°C and 29°C (Carlson 1971, Emanuel 1991, Evans 1993). This is because the trade-wind temperature inversion keeps the development of deep cumulus in check. The cap is broken rather abruptly when enough heat and moisture have accumulated in the boundary layer to erode the inversion. It therefore seems plausible that a modest 1°C difference in SST can be the difference between an active and an inactive year.

Hurricane activity with respect to the 7 warmest and 7 coolest years east of the Lesser Antilles is summarized in Table 12.2. Values in bold denote years over the 6-year period 1991–1996. Note that 1995 is ranked seventh warmest in the record. The average SST for 1996 was only slightly cooler than the value for 1995. In fact, 1995 featured the warmest SSTs on record over the main development region as a whole (Saunders and Harris 1997). The records extend back to 1865. Both 1991 and 1992 rank in the top five for the coolest on record. The 7 warmest years featured 40 tropical-only hurricanes compared with only four during the 7 coolest years. Conversely, the cool years support more baroclinically-initiated hurricanes relative to the warm years. The differences in hurricane type are reflected in the differences in median hurricane latitude. During warm years the average median latitude is 25.1°N compared with 32.6°N during the cool years. The hurricanes during 1996 had a median latitude of 23.3°N. It is speculated that the warm SSTs are responsible for lower sea-level pressures resulting in increased convection over the deep tropics. The enhancement of convection occurs in association with a strengthening of the intertropical convergence zone and a shift in the trade winds over the central North Atlantic (Carton et al. 1996).

Variations in hurricane activity across latitudes from year to year have been noted by others (Goldenberg and Shapiro 1996). Years of high-latitude activity similar to the 1991–1994 period are commonly associated with below-normal rainfall over portions of western Africa and warmer than normal SSTs over the equatorial central and eastern Pacific (El Niño or warm ENSO conditions). These two factors are linked to hostile environmental conditions for the initiation and development of North Atlantic hurricanes over the deep tropics (south of 20°N). Changes in vertical shear of the horizontal winds over the subtropical and middle-latitude western North Atlantic are unrelated to, or are in an opposite phase to, vertical shear south of 20°N. Thus during years when tropical-only activity is suppressed, hurricane development over the subtropical latitudes is more likely. Hurricanes at higher latitudes frequently originate from baroclinic disturbances or are favorably enhanced by their presence. The few tropical-only hurricanes are strong enough to have survived the hostile environment.

Large-scale changes in surface pressure patterns might also be responsible for modulations in hurricane activity. For instance, the paucity of hurricanes during the early 1990s is coincident with below-average sea-level pressures over parts of the Northern Hemisphere. Sea-level pressures over the central Arctic were substantially below the long-term average during the early 1990s (Walsh et al. 1996). The movement of sea ice at polar latitudes is related to fluctuations in the *geostrophic wind* through changes in sea-level pressure patterns over the arctic.[8] Sea ice which moves toward the equator mixes with the subpolar waters of the North Atlantic Ocean leading to deep-water overturning that drives the global thermohaline circulation (see Chapter 16). The at-

---

[8]The geostrophic wind relationship describes the balance between the Coriolis force and the force exerted by the pressure-gradient.

Table 12.2: Tropical-only hurricane activity. Values are give for the seven warmest (top) and seven coolest (bottom) Septembers over the period 1950–1996 based on SSTs in a region to the east of the Lesser Antilles (5° latitude × 5° longitude box centered at 12.5°N and 47.5°W). TO is the number of tropical-only hurricanes during the season. The 7-year average ($\bar{x}$) of SST, TO, and median latitude are also provided. Median latitude is computed from the 6-hourly best-track positions for all hurricanes for a given year. Bold numbers represent years in the period 1991–1996 (after Kimberlain and Elsner 1998).

Year	SST (°C)	TO	Median Lat. (°N)
1952	29.32	4	25.9
1955	29.26	8	24.6
1969	29.12	6	28.5
1958	29.11	7	28.8
1966	29.09	2	24.9
1988	29.09	4	18.5
1995	**29.02**	**9**	**24.3**
$\bar{x}$	*29.14*	*5.7*	*25.07*
1984	28.36	0	31.1
1974	28.33	3	21.2
1992	**28.32**	**0**	**35.6**
1972	28.31	0	38.8
1976	28.25	1	33.4
1991	**28.22**	**0**	**31.6**
1982	28.15	0	36.4
$\bar{x}$	*28.23*	*0.6*	*32.59*

mospheric mass loss in the Arctic is accompanied by a mass gain in the eastern North Atlantic and western Europe. The mass gain is realized through higher sea-level pressures over the region. The above-normal pressures are most pronounced during the North Atlantic hurricane season. Higher pressures over western Europe ensure a flow of dry air into the tropics along the northwestern African coast. The dry air inhibits the development of tropical easterly waves. by limiting the amount of associated deep convection. The abundance of tropical-only hurricanes under these conditions is kept in check. Future research into the factors that contribute to the variations in incidence of baroclinically-initiated hurricanes will lead to a better understanding of the relationships between hurricanes and climate.

13

History of Seasonal Hurricane Forecasting

Regional and global climate patterns determine the character of a hurricane season. Locally, warm sea-surface temperatures coupled with frequent vigorous tropical waves signal an active season. The development of a warm ENSO (see Chapter 10) over the central equatorial Pacific with its accompanied global teleconnection patterns portend fewer hurricanes and a reduced threat of strikes to the United States. A systematic understanding of the linkages between climate and hurricanes provides a basis for forecasting. This chapter is a chronicle of some of the research contributions that have been made over the years toward our understanding of climate and hurricanes. It serves as a brief historical account of the contributions that led to the current seasonal forecast models of North Atlantic hurricane activity. The presentation follows the work of Hess and Elsner (1994b).

Scientific contributions over the past century are grouped into four distinct but overlapping chronological periods. Early work focused on uncovering physical relationships between hurricane activity and persistent weather patterns across the Atlantic. These efforts stressed the importance of sea-level pressure patterns in determining the character of a given hurricane season. The interaction between tropical and middle-latitude weather regimes in determining hurricane activity was also considered. An effective tool for understanding the relationship between climate and hurricanes is the composite weather map. The widespread use of the composite map in highlighting important upper-level features associated with hurricane activity defines the second period of research. Significant differences in the height patterns were observed for seasons with contrasting activity. Modern satellites provide a comprehensive view of weather events in the tropics. In particular, the movement of tropical easterly waves can be monitored. The link between easterly waves and the character of the hurricane season was investigated with the aid of satellite pictures. The combined knowledge of various climate factors led to rules for predicting hurricane activity several months in advance. These rules are formalized into a set of statistical models. Grouping the historical periods is subjective, but it provides a useful way to understand the improvements made in our understanding of hurricanes and climate.

13.1   Physical Relationships

Hurricanes make a large depression in the atmospheric pressure field at low altitudes. Therefore it might be reasonable to expect a relationship between sea-level pressure patterns and hurricane activity. A relationship along these lines is recognized in Garriott (1895 and 1906). He finds that sea-level (SLP) variations tend to occur over extended geographical areas. Changes to the large-scale SLP patterns over the North Atlantic are linked to variations in the abundance of hurricanes. When SLP is unusually low over the Caribbean Sea, it tends to be above normal over the British Isles. The development of hurricanes is aided by a strengthening or displacement of the Azores high. This change produces a freshening of the easterly trade winds at low latitudes across the central North Atlantic. A deep layer of easterly winds creates an environment conducive to the maintenance of pre-hurricane disturbances.[1] An early study by Bowie (1922) noted that a freshening of the trades could provide a proper environment for hurricanes, but their intensification is ultimately tied to the amount of latent heat released from the embedded thunderstorms. Favorable conditions for a fledgling storm are thus linked to the co-location of strong convection with an incipient disturbance embedded in a flow of deep easterly winds.

These early studies were limited by the amount of available data. Later investigations of Brennan (1935) and Ray (1935) reveal that antecedent weather conditions in the Caribbean region are sometimes related to disturbed weather during August and September. Based on data over the period 1903–1934, Brennan (1935) notes in particular that when rainfall is above normal, and surface winds and pressures are below normal in June and July, there is an increased potential for hurricane development during the peak of the hurricane season. Seasonally lagged relationships between SLP over the Caribbean and hurricane frequency are also described in Ray (1935) using data over the period 1899–1933. Surface-pressures at San Juan during early summer show a strong relationship with hurricane frequency over the North Atlantic. May through July pressures provide the most significant relationship with hurricanes during August through October. When early summer pressures are low hurricane activity in the following months is high. This is consistent with the contemporaneous relationship between SLP and hurricanes noted previously. It was also discovered that warm sea-surface temperatures (SSTs) over the Caribbean during the hurricane season tend to enhance hurricane activity. The SST relationship is based on data over the period 1920–1933. An early study by Chapel (1934) recognized the importance of the monsoon flow to hurricane formation across the western Caribbean. The confluence zone formed by the northeast trades from the Atlantic and the cross-equatorial flow from the Pacific produces favorable low-level conditions for storm formation over the western Caribbean Sea during October and November.

The importance of easterly waves as precursors to hurricane formation is recognized by Dunn (1940). His work provides evidence that easterly waves often originate

---

[1]The maintenance of disturbances can be considered separate from their intensification (see Chapter 2).

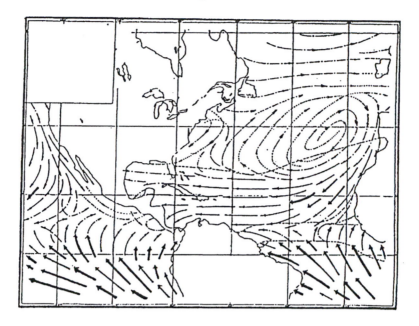

Fig. 13.1: Surface winds over the North Atlantic Ocean during the month of September. Direction lines are based on the dominant wind computed for each 5° latitude-longitude box. Arrow width is proportional to the relative wind steadiness. Reproduced from Dunn (1940).

in association with the inter-tropical convergence zone (ITCZ). The ITCZ forms at the confluence between the trade winds of both hemispheres. Dunn (1940) notes the development of an easterly wave along the ITCZ is more likely when the ITCZ reaches its northernmost position near 12°N latitude. Climatologically this occurs in September and is accompanied by southeast trades along the west African coast. The winds remain northeasterly in a region to the south of the islands of Cape Verde. Tropical easterly waves developing in this area during the Cape Verde season are the origin for some of the strongest hurricanes of the North Atlantic. The September surface wind circulation over the North Atlantic Ocean illustrated by Dunn (1940) is reproduced in Figure 13.1. The dominant circulation around the anticyclonic high pressure near the Azores is clearly indicated. The southwesterly monsoon flow (sea to land) off the west coast of Africa is evident. The air moving southward from the eastern North Atlantic on the eastern flank of the anticyclone converges with the southwesterly winds of the monsoon to enhance convergence. Significant convergence of the winds at low levels supports deep convection.

Dunn (1940) also remarks on the penetration of fronts into the tropics. Polar fronts from the middle latitudes tend to dive southward over the western North Atlantic and Gulf of Mexico inhibiting the formation of hurricanes. The importance of a geographic separation between middle-latitude westerly winds and tropical easterly winds on hurricane activity is mentioned in Riehl and Shafer (1944) and in Riehl and

Burgner (1950). These studies note that the development of a tropical cyclone can be enhanced under certain circumstances by vertical shear of the horizontal winds. It is concluded, however, that the intensification process typically requires a deep layer of easterly winds with little or no shear. This is consistent with the notion that shear acts to remove the column of deep convection from the low-level circulation. A decoupling of the upper-level flow from the low-level circulation is a signature that the storm will not intensify.

The association between seasonal hurricane activity and large-scale SLP fluctuations is revisited in Namias and Dunn (1955). They agree with early studies that increased North Atlantic hurricane frequency occurs when the subtropical high pressure over the eastern North Atlantic (Azores high) extends northeastward into Europe. They suggest, however, that the explanation lay in the penetration of middle-latitude cyclonic vorticity into the tropics rather than a strengthening of the trade winds. Low-level cyclonic vorticity is associated with a trough of low pressure, aligned north-south, along the west African coast. Dunn (1956) notes that disturbances originating over Africa pass through this trough before becoming hurricanes farther to the west. The work of Dunn and Miller (1964) is also in agreement with many of the earlier studies. They remark that persistent SLP anomalies over portions of the tropical Atlantic have a strong influence on hurricane frequency. Specifically, when the Azores high shifts northeastward, the Bermuda high is weakened and hurricane frequency increases over much of the region including the western North Atlantic and the Caribbean Sea. It is understood that higher sea-level pressures lead to more sinking air (subsidence), which leads to drying and the suppression of significant convection.

Riehl (1956) notes that middle-latitude weather systems can have a favorably influence on the formation and development of hurricanes. He argues that a large amplitude trough in the middle-latitude westerly wind field can initiate the development of a hurricane at lower latitudes. If the trough extends far enough south, it may split leaving a part of the circulation at low latitudes. As the circulation drifts westward it has the potential to develop into a hurricane under proper conditions. These conditions include weak shear and deep convection near the center of circulation. The incidence of baroclinically-enhanced hurricanes is recognized as an important climatological feature of North Atlantic tropical cyclone activity. He also comments on the possibility of time-lagged relationships between climate factors and hurricane activity. He notes there is a tendency for hurricane activity to be low if during late spring the subtropical anticyclone is stronger than normal and oriented east to west. The relationship holds because of the persistence of the anticyclone in producing favorable upper-level divergence patterns. This association is also mentioned in a later study by Miller (1958).

13.2 Composite Charts

Data on conditions of the atmosphere at upper levels were largely unavailable until the 1930s. Reliable climatological analyses of these data were not performed until the middle 1950s. A systematic study on the differences in atmospheric structure with respect to hurricane activity over a portion of the North Atlantic is performed in Namias (1955) using composites (superposed epochs). Comparisons are made between years of extreme hurricane activity. Heights on the pressure surface are averaged separately for years with few and years with many hurricanes. Differences in the composite charts reveal factors associated with large swings in hurricane activity. The work is important in highlighting a tool for understanding hurricane climate. Ballenzweig (1959) applies the technique to study total hurricane activity during the peak months of the season. Using data over the period 1933–1955, he contrasts two 700 mb height charts. One chart is the average departures from normal for the five seasons with the greatest number of tropical cyclones and the other is for the five seasons with the least number of storms. Note that the two charts produced by Ballenzweig (1959) are shown in Figures 13.2 and 13.3.

Seasons with significant tropical cyclone activity show a positive 700 mb height anomaly stretching across the North Atlantic basin near 40°N latitude. Largest anomalies are slightly farther south over the central Atlantic. Negative height anomalies are centered over Iceland and over portions of west Africa. Weaker negative anomalies are noted over the subtropical North Atlantic and Caribbean Sea. The wind field associated with this height anomaly field features westerly winds displaced farther north than usual. A northward displacement of the westerly wind pattern is accompanied by a northward shift of the subtropical anticyclones and their attendant zones of deep easterlies. This pattern favors an enlargement of the area favorable for hurricane origin and development. The minimum tropical-cyclone activity chart shows similar positive anomalies at 700 mb. However, the area of significant positive departures is smaller and the center is shifted south to 30°N latitude. Negative anomalies extend from the northern Great Lakes eastward to the central North Atlantic.[2] Weak negative anomalies remain over the Caribbean. The flow associated with this height pattern is characterized by a southward penetration of westerly winds. In general, the penetration of westerlies into the subtropics restricts hurricane development. Similar composite charts show well-defined circulation patterns associated with hurricane activity over limited areas of the North Atlantic. These features form the basis for statistical models used in the prediction of sub-basin hurricane activity (Chapter 15).

Based on composite maps, Ballenzweig (1959) discusses possible middle-latitude influences on annual hurricane activity. The strong negative height anomalies north of the positive departures over the central North Atlantic suggest above normal westerlies near 40°N. An intensified westerly wind flow is associated with stronger middle-latitude fronts capable of penetrating south over the subtropics. The fronts (baroclinic zones) and their attendant low-level vorticity provide a mechanism by which hurricanes can form. Development of baroclinically-enhanced hurricanes is more likely under conditions that keep the middle latitude trough progressing toward the east. In this case the trough is replaced by an upper-level ridge, which limits the amount of ver-

---

[2]Strong negative anomalies are also noted over western Europe.

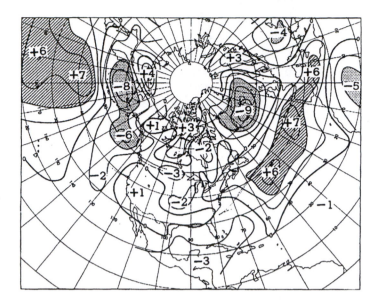

Fig. 13.2: Height field at 700 mb. Values are average departures from normal in tens of feet for the five years (August through October) of maximum tropical-cyclone incidence. Values are based on data over the period 1933–1955. Reproduced from Ballenzweig (1959).

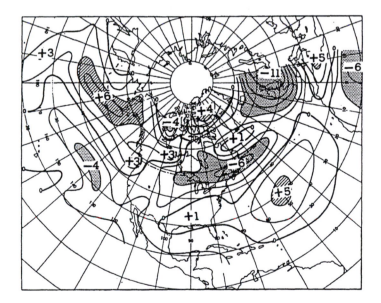

Fig. 13.3: Height field at 700 mb. Values are average departures from normal in tens of feet for the five years (August through October) of minimum tropical-cyclone incidence. Values are based on data over the period 1933–1955. Reproduced from Ballenzweig (1959).

tical shear over the developing system. Willett (1955) comments on the tendency for northerly occurring hurricanes because of the increased baroclinic patterns observed in the middle latitudes. These findings indicate that conditions hostile toward hurricane formation in general may be conducive to hurricane formation from baroclinic disturbances.

13.3   Satellites and Easterly Waves

Satellites ushered in a new era in meteorological observations.[3] Satellite pictures provide a comprehensive view of the tropics and pinpoint suspect areas of deep convection. From satellites it becomes apparent that most of the disturbed areas featuring shower and thunderstorm activity never develop and become hurricanes. The origin and movement of easterly waves are readily monitored from satellites. Satellite imagery shows that many easterly waves can be traced to the African continent. About half of the waves that exit the west coast dissipate before reaching the Antilles. Only a fraction of the waves develop into North Atlantic hurricanes. Simpson et al. (1969) discuss reasons for this observation. They note that as the developing wave travels westward, there is an increased likelihood that it will encounter unfavorable westerly winds penetrating southward from high latitudes. These winds disrupt the associated low-level wind patterns and break apart the deep convection. Based on an analysis of satellite pictures from the period 1967–1991, Avila and Pasch (1992) find that approximately 60 easterly waves per year can be expected to traverse the North Atlantic during the hurricane season. Some years feature fewer waves while others see quite a few more. However, the abundance of easterly waves is not a reliable indicator of seasonal hurricane activity as numerous other environmental conditions are needed for hurricane development.

Satellites can help distinguish between hurricanes originating from purely tropical factors from those that develop with the aid of middle latitude dynamics. Easterly wave development over the deep tropics provides the textbook example for hurricane formation. The fluxes of sensible and latent heat from the ocean surface provide energy for moist ascent through the center of the storm. According to Frank (1970b), tropical cyclones driven exclusively by latent heat originate from tropical depressions in the ITCZ or from easterly waves. In contrast, tropical cyclones driven only partially by latent heat are more likely the result of baroclinic disturbances. These disturbances include middle-latitude fronts and cutoff upper-level low-pressure centers, among others. Roughly half of all hurricanes over the North Atlantic basin form solely from easterly wave development. Avila (1991) notes that many eastern North Pacific hurricanes have origins that can be traced to African easterly waves. Easterly waves as seedlings for hurricanes are unique to the North Atlantic and eastern North Pacific tropical cyclone basins. The relative frequency of hurricanes originating from enhancing middle-latitude baroclinic factors may also be unique to the North Atlantic.

---

[3] See Chapter 3.

13.4   Statistical Relationships

With the availability of larger and more comprehensive data sets, the physical association outlined in earlier studies were tested rigorously with statistics. The additional
data also provided opportunities to uncover new relationships between hurricane activity and climate. Shapiro (1982) shows that the annual abundance of hurricanes is
significantly correlated with SLPs over portions of the North Atlantic. He also reveals
a statistical relationship between hurricane activity and SSTs in the Atlantic. This relationship is spatially incoherent, but holds for SSTs over limited areas of the North
Atlantic (Raper 1993). Significant relationships between North Atlantic hurricanes
and the quasi-biennial oscillation (QBO) are emphasized in Gray (1984a). He also un-
covers a hurricane relationship with SST anomalies in the eastern and central tropical
Pacific Ocean connected to the El Niño events. The QBO relationship is confirmed in
Shapiro (1989). Using 30 mb zonal winds[4] a statistically significant relationship to the
abundance of hurricanes is noted. In fact, the relationship is strong even when zonal
wind leads the hurricane activity by as much as 3 months. This lagged relationship is
exploited in early forecast models.

The QBO (see Chapter 10) is an oscillation of stratospheric winds with an average
period near 27 months. The largest amplitude of the oscillation is found in the tropics, where the winds fluctuate between strong easterlies and weak easterlies or weak
westerlies. El Niño or (warm ENSO) refers to the aperiodic warming of the ocean
waters over the equatorial eastern and central tropical Pacific. The warming occurs
roughly every 3 to 7 years, changing the pattern of deep convection over this region
of the world and affecting global weather patterns including upper-level tropical winds
(Arkin 1982). Seasonal variability of the tropical winds both at upper and lower levels
over the North Atlantic are strongly correlated with the ENSO cycle and can be used
as a gauge of North Atlantic hurricane frequency (Shapiro 1987). The relationship between North Atlantic hurricanes and El Niño events is confirmed using a low-resolution
general circulation model (GCM) (Wu and Lau 1992). Although the exact relationship
remains uncertain, it is speculated that both the QBO and ENSO are related to North
Atlantic hurricane activity through upper-tropospheric vertical shear of the horizontal
winds over the prime hurricane genesis regions. For example, it appears that when
the QBO is in its westerly phase, the vertical shear of horizontal winds is reduced and
hurricane formation is more likely (see Chapter 10).

The relationship of easterly waves to hurricane activity over the North Atlantic
is further substantiated in Gray (1990). He shows an increased incidence of major
hurricanes occurs in conjunction with periods of above-normal rainfall over parts of
western Africa. The annual frequency of intense North Atlantic hurricanes was greater
over the period 1947 through 1969 when rainfall in the western Sahel[5] was abundant.

---

[4]Zonal wind refers to the component of the horizontal wind oriented west to east.

[5]The Sahel is a narrow east-to-west zone of semiarid land over northern Africa, located between the
Sahara to the north and the rain forests and grasslands to the south. Rainfall is quite variable, often with the
potential for extended years of drought.

Table 13.1: Darwin pressures and Australian tropical cyclone activity. Values are correlations based on pressure anomalies averaged from June through August, and tropical cyclone frequency around Australia during the following tropical cyclone season. The period 1950–1958 represents the time before satellite observations. Values are from Nicholls (1979).

Time Interval $\rightarrow$	1950–1974	1950–1958	1959–1974
Oct–Dec	−0.48	−0.75	−0.62
Jan–Feb	−0.19	−0.74	−0.39
Mar–May	−0.15	−0.06	−0.30
Entire Season	−0.40	−0.79	−0.68

In contrast, during the period 1970–1987 drought conditions prevailed over the region and the abundance of intense hurricanes was considerably less.

The suggestion is made that these long-period changes in rainfall are linked to changes in global-scale ocean circulation processes. In particular, warm SST anomalies in the North Atlantic Ocean driven by a combination of a warm gulf-stream current and subsidence of cold, salty water at high latitudes can reverse and become cold anomalies. This happens as fresh (less dense) water from melting polar ice inhibits the high-latitude oceanic sinking (see Chapter 14). Cold North Atlantic SST anomalies are linked to a southward displacement of the ITCZ resulting in drought conditions over the Sahel. Dry conditions in the Sahel tend to increase the north-to-south thermal gradient in the lower troposphere thereby intensifying the middle-level (3 to 4 km) easterly jet over west Africa. These changes occur over time periods involving decades. Landsea and Gray (1992) show the association between western Sahel rainfall and North Atlantic hurricane frequency is strong. Speculation centers on the idea that an intensified low-level easterly jet more quickly advances the pre-hurricane cloud clusters, diminishing their chances for sustained organization. However, it is possible that cold North Atlantic SSTs directly inhibit the development of African tropical waves by suppressing convection. In this case, African rainfall serves as a proxy for SST anomalies with SSTs controlling both rainfall in western Africa and hurricane activity in the North Atlantic basin.

13.5   Early Prediction Models

A predictive relationship between large-scale pressure patterns near Australia during winter and tropical-cyclone frequency over the region during the following spring and summer was noted by Neville Nicholls of the Bureau of Meteorology Research Centre (BMRC) in Australia. He finds that when Darwin pressures are below average during the winter months, the likelihood that the tropical-cyclone season during the spring will be active increases. Table 13.1 shows the linear correlation between Darwin surface pressures and tropical-cyclone activity for different parts of the season based on the analysis in Nicholls (1979). The correlation is given for the time period before and after satellites. A relationship is noted over the entire season, but is strongest during October through December.

Darwin pressure is an indicator of the state of El Niño-Southern Oscillation (ENSO) and studies have verified the relationship between Australian tropical-cyclone activity and various indices of ENSO (Nicholls 1984, Solow and Nicholls 1990). A re-examination of this relationship through the 1990–1991 season indicates that the Southern Oscillation Index (SOI) and Australian tropical-cyclone numbers continues to be closely related (Nicholls 1992). However, there are secular changes in tropical-cyclone activity unrelated to changes in ENSO occurring since the 1986–1987 season, leading to a bias toward overestimation of the seasonal number of storms. As such, Nicholls (1992) suggests that the difference in tropical-cyclone activity from the previous season to the coming season be predicted based on observed changes in the SOI rather than actual activity. Interestingly, Basher and Zheng (1995) find an increased risk of tropical-cyclone activity during El Niño years for the southwest Pacific region from the Solomon Islands to French Polynesia. A suggestion is also made to forecast the change in hurricane activity from the previous season rather than forecast the level of activity directly.

The first prediction models for North Atlantic hurricane activity were developed by William M. Gray of Colorado State University.[6] The models used rules extracted from data over the period 1950–1982. A number indicating average activity is specified as a constant and correction factors (adjustment terms) are added or subtracted depending on whether the magnitude or sign of the factor (independent variable) is favorable or unfavorable for hurricane development. For instance, the total number of hurricanes and tropical storms expected in a season is expressed in terms of the season average as follows:

$$\text{Total Number} = \text{AVG} \pm \text{QBO} \pm \text{ENSO} \pm \text{SLPA}. \tag{13.1}$$

The QBO is the 30 mb zonal wind anomaly correction factor. If the anomaly is westerly add 1.5 to the seasonal average (AVG), but if it is easterly subtract the amount from the average. If the wind direction is changing from one phase to the other this term is set to zero. ENSO is the El Niño influence. If the season will occur during a warm phase subtract 2 storms for moderate conditions and 4 storms for strong conditions, otherwise add 0.7 to the average. SLPA is the sea-level pressure anomaly estimated from stations over the Caribbean region. If SLPA is less than -0.4 mb or less than -0.8 mb, add 1 or 2, respectively. If SLPA is between $+0.4$ and $+0.8$ mb or greater than $+0.8$ mb, substract 1 or 2, respectively. If SLPA is between -0.4 and $+0.4$, make no correction to the average number of storms. This is shown as follows:

$$\left\{ \begin{array}{ll} \text{SLPA} < -0.4\,\text{mb or SLPA} < -0.8\,\text{mb} & \text{add 1 or 2} \\ 0.4\,\text{mb} < \text{SLPA} < 0.8\,\text{mb or SLPA} > 0.8\,\text{mb} & \text{substract 1 or 2} \\ -0.4\,\text{mb} < \text{SLPA} < 0.4\,\text{mb} & \text{no correction} \end{array} \right\} \tag{13.2}$$

---

[6]The early models were based on three predictors including the phase of the stratospheric QBO, the occurrence and strength of El Niño, and mean sea-level pressure anomalies over the Caribbean basin (Gray 1984b).

Evaluation of the models using the same data set as used to develop the rules suggest the algorithms are skillful at predicting hurricane activity a few months in advance. The first official forecasts were issued for the 1984 hurricane season. The models provided guidance for estimating the abundance of tropical storms and hurricanes. The inaugural forecast was made on May 24 before the start of the official hurricane season. An update to the forecast was issued on July 30. Forecasts were made for four categories of activity including number of hurricanes, named storms, hurricane days, and named storm days (see Chapter 14). Seasonal forecasts from the CSU group have been issued routinely since 1984. A summary of the performance of the early forecasts is given in Hastenrath (1990). Overall the skill level has been impressive.

13.6   Modifications

Modifications and extensions to the original forecast models have been over the years. Following the discovery of the relationship between west African rainfall and hurricane activity, the forecast algorithms were modified to include rainfall over the Sahel and the Gulf of Guinea regions as predictor variables. When rainfall during the hurricane season is above average so is the potential for intense hurricane activity. This relationship is not surprising as the majority of intense hurricanes originate from disturbances initially over west Africa (Frank and Clark 1977). The models were also changed to include the dependent categories of intense hurricanes and intense hurricane days. Since low-latitude originating hurricanes remain over warm tropical waters for a substantial period of time, they have a greater opportunity to intensify. The strength and organization of the initial disturbance is linked to seasonal rainfall conditions over western Sahel. More rainfall occurs in association with healthier disturbances. Importantly, July rainfall provides a good indicator of August and September precipitation over these regions (Bunting et al. 1975). Therefore, early season rainfall over west Africa is an important variable for forecasting the character of the impending hurricane season. Moreover it is possible to extend the lead time of the forecasts based on the west African precipitation relationship. This is because precipitation anomalies over the Gulf of Guinea region are correlated between successive years. Dry years tend to follow other dry years and vice versa (Landsea and Gray 1992). Thus rainfall over portions of western Africa during September through November will portend rainfall over the same region 9 to 12 months later. Rainfall deficits accumulated over the last three months of the hurricane season can be used to forecast hurricane activity 9 to 11 months later. Another modification was a change from the rule-based prediction algorithms to multiple linear regression.

Building on the work of Gray et al. (1993) and considering the fact that not all North Atlantic basin hurricanes are the same with respect to origin and development mechanisms, Hess et al. (1995) show that the forecast models are predicting the tropical-only hurricane activity rather than all hurricanes. For example, the value of Gulf of Guinea rainfall (Figure 13.4) as a predictor for all hurricanes is compared with its value as a

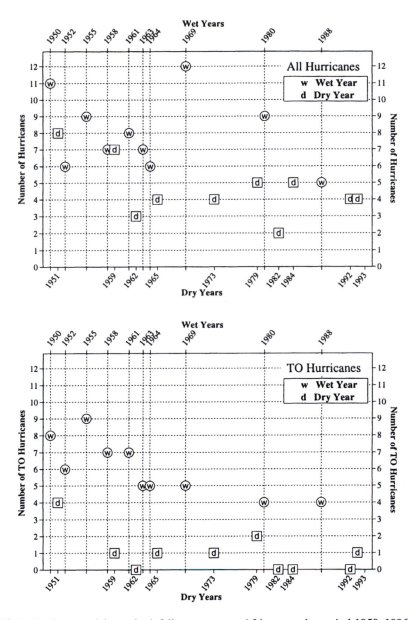

Fig. 13.4: Hurricane activity and rainfall over western Africa over the period 1950–1996. The relationship with all hurricanes is shown in the top panel and the relationship with tropical-only hurricanes is shown in the bottom panel. Abundance of hurricanes are noted for the ten very wet and ten very dry years over the Gulf of Guinea. The rainy season runs from September through November. Hurricane activity is for the following hurricane season.

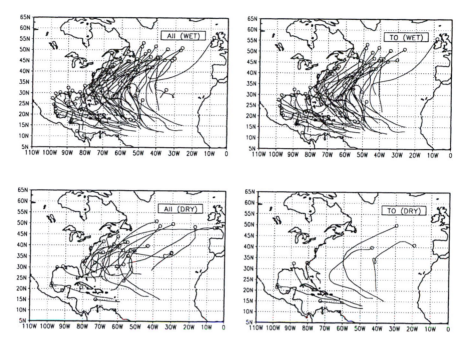

Fig. 13.5: Hurricane tracks during wet and dry years over the period 1950–1993. Years are based on the previous year's rainfall in the Gulf of Guinea region of western Africa. The 10 wettest and 10 driest years along with the number of hurricanes and the number of tropical-only (TO) hurricanes are given in Figure 13.4. The number of all hurricanes is 80 and 48 for wet and dry years, respectively. The number of tropical-only hurricanes is 60 and 10 for wet and dry years, respectively. After Hess and Elsner (1994b).

predictor for tropical-only hurricanes. The number of hurricanes during the year following the 10 wettest and 10 driest years over the Gulf of Guinea region of western Africa are shown. The annual number of hurricanes in chronological order is given in the top panel and the number of tropical-only hurricanes is given in the bottom panel. Wet years are associated with more North Atlantic hurricanes compared to dry years, but the association is particularly strong for the category of tropical-only hurricanes. This is to be expected since tropical-only hurricanes typically form from African easterly waves. The relationship appears to be less coherent with time. Hurricane tracks are plotted for the 10 wettest and 10 driest years with respect to both all and tropical-only hurricanes in Figure 13.5. Wet years feature a greater number of all and tropical-only hurricanes. Dry years feature fewer storms. However, the ratio of the number of hurricanes between wet and dry years is 6 to 1 for tropical-only hurricanes compared with less than 2 to 1 for all hurricanes. Moreover, when all five predictors of Gray et al. (1992) are used to forecast hurricane activity six to 11 months in advance, the linear correlation between actual and predicted values increases from 0.58 for all hurricanes to 0.82 for the subset of tropical-only hurricanes (Hess and Elsner 1994a). Apparently baroclinically-enhanced hurricanes are the random component driving down the correlation in the forecast models. This finding is consistent with earlier studies that pointed to the importance of middle-latitude influences in determining the character of a hurricane season. A closer look at forecast methodologies is presented next.

14

Seasonal Forecast Models

Forecasting is about making predictions of future events consistent with one's best judgments. Consequently, it is desirable to allow for consistent revisions of those judgments as more data become available. Statistical models are efficient means of logically understanding the data and should be used as a guide to forecasting. Their utility arises from the stochastic nature of climate. Hurricane climate prediction faces two sources of uncertainty. First, since climate is weather over longer time scales, the fundamental limitation of weather forecasts—resulting from the chaotic nature of the atmosphere—renders perfect climate predictions impossible. That is, even if knowledge of hurricanes is exact and comprehensive, the hurricane climate forecaster would still need to couch predictions in terms of probabilities. Second, knowledge about climate is neither exact nor complete, leading to another source of uncertainty. There is reason for hope because hurricane climate forecasts are about broad, general characteristics of activity rather than about the details of individual events; and there is no fundamental limitation to this kind of knowledge. Concern centers on whether or not Miami will get hit by a major hurricane during the next five years, not on what will happen to the next storm that develops.

Though climate is surely more than a number-generating machine, it is useful to consider it as such. In this way, each hurricane season represents the outcome of a stochastic process. The nature of stochastic outcomes can be modeled quite faithfully, despite the fact that physical processes operating to produce a hurricanes are complex, contingent, and only partially understood (Epstein 1985). Thus understanding stochastic processes are essential to the problem of hurricane climate prediction. Demonstrating the potential for prediction of hurricane activity based on conditions antecedent to the hurricane season is a significant breakthrough in the field of hurricane climate. The two strongest predictive signals are the El Niño-Southern Oscillation (ENSO) and the quasi-biennial oscillation (QBO) (see Chapter 10), both of which have been implicated in modulating seasonal hurricane activity, not only over the North Atlantic, but over other tropical cyclone basins as well (McBride 1995).

A variety of models for forecasting seasonal North Atlantic hurricane activity are presently in use. All of them rely on statistical relationships. Most models are directed at predicting the character of the next hurricane season as a whole. Predictions of the total number of hurricanes are often emphasized. Newer probabilistic models have been developed to forecast sub-basin activity. This chapter reviews some of the seasonal forecast models used to predict North Atlantic hurricane activity. In particular the track record of the Colorado State University group is noteworthy. Within the framework of *Bayes's theorem* it is possible to combine the forecasts of different groups to obtain a probabilistic model. It should be kept in mind that changes and updates to these models are frequently made.

14.1   Measures of Seasonal Activity

The character of a hurricane season can be measured in a variety of ways. The total number of hurricanes occurring in a season (H) is a popular metric of hurricane activity. The number of tropical cyclones or the number of intense hurricanes are also frequently considered. The number of tropical cyclones occurring in a season is denoted NS for named storms.[1] The number of intense hurricanes is denoted MH, for major hurricanes. The number of days reporting tropical cyclone activity provides another perspective on the character of a season. For example, *named storm days* (NSD), *hurricane days* (HD) and *major hurricane days* (MHD) are defined as the cumulative duration of days over all cyclones of a particular category. These quantities represent the sum over all 6-hour observations. The quantity called *net tropical cyclone activity* (NTC) is a measure of the general character of a season. NTC activity is defined as:

$$\text{NTC} \equiv (\%\text{NS} + \%\text{H} + \%\text{MH} + \%\text{NSD} + \%\text{HD} + \%\text{MHD})/6, \qquad (14.1)$$

where percentages refer to seasonal activity as a ratio of the long-term (usuallly 1950–1990) average. For example, %H for 1996 is 153% because there are nine hurricanes compared to the 1950–1990 average of 5.9. The %H for a season with six hurricanes is approximately 100%. NTC is biased toward major hurricane activity. This is because the averages are smaller for major hurricane activity. A season with 10 hurricanes, but no intense hurricanes will appear as below average, since NTC will be less than 100%. NTC is more sensitive to fluctuations in intense hurricane activity than it is to fluctuations in tropical storm activity.

Other measures of seasonal hurricane activity for the North Atlantic basin include *hurricane destructive potential* (HDP) and *maximum potential destruction* (MPD). The HDP is calculated as the square of the maximum sustained 1-minute near-surface wind for each 6-hour observation summed for each tropical cyclone and for all cyclones in a season. The MPD is the sum of the square of the absolute maximum (one value for each tropical cyclone) over all cyclones in a season. Both HDP and MPD are expressed using

---

[1]For tropical storms, the term *named storms* is used. This refers to the fact that, at this point in its development, it is given a name by the National Hurricane Center (see Chapter 1). Names were used for North Atlantic tropical cyclones beginning in 1950. Named storms (NS) refers to any tropical cyclone that reaches at least tropical storm intensity (18 m s^{-1}).

values that have been divided by 10^5. Destruction potential is closely related to kinetic energy, which is proportional to the square of the wind speed. Thus, two tropical cyclones each with maximum-sustained winds of 30 m s^{-1} striking the same location will theoretically do only half as much damage as a single hurricane with winds of 60 m s^{-1}. The different measures of activity provide slightly different perspectives on the nature of the hurricane season, but there is a strong degree of interdependence among them. Models built to predict the various components of activity will be similar.

14.2   Cross Validation

Data which describe relationships between climate and hurricanes are useful for building forecast models. Models are used to predict future levels of hurricane activity. Forecast models require an accurate estimate of skill. For statistical models this can be achieved through a procedure called *cross-validation*. Cross-validation procedures, including important issues related to proper implementation for regression models, are outlined in Michaelsen (1987). The effectiveness of cross-validation for estimating forecast skill is emphasized in Barnston and van den Dool (1993). Because statistical models see widespread use in forecasting climate, a sound understanding of validation procedures is important. Misapplication or misinterpretation can lead to over confidence about how well the model will perform operationally. Potential pitfalls to proper implementation of cross-validation,[2] are examined here.

In any set of data there is both useful information (signal) about the underlying processes being examined and extraneous information (noise) related to coincidences in the sampling process and measurement errors. For instance, we noted earlier that baroclinically-enhanced hurricane activity is the noise component in the known climate processes responsible for the abundance of hurricanes over the North Atlantic. The principal challenge to the researcher following a statistical approach to prediction is to devise an empirical method that strengthens signals in the available data and dampens noise. A desirable method for selecting a prediction rule from historical hurricane data is one that accurately captures the fundamental relationships between the variables, without counting too heavily on past coincidences. This is becaue these coincidences are unlikely to show up in future hurricane activity. Given the availability of computers, it is easy to build complex empirical rules from a limited data set that "predict" the past accurately. The central question is whether such rules reflect the underlying processes or whether they are simply exploiting random agreements. Rules that overemphasize coincidences are likely to perform poorly when the algorithm is used to make forecasts of future activity.

Cross-validation is an attempt, with a limited data sample, to simulate actual forecast situations and provide a truthful measure of an empirical procedure's ability to produce skillful prediction rules. Predictive skill is the skill expected when the prediction rule, chosen by the procedure, is used operationally to forecast the future. Predictive

---

[2]This information is originally put forth in Elsner and Schmertmann (1994).

skill estimated from historical data will, in general, be biased toward these particular historical values. Cross-validation works by developing a separate prediction rule for each observation in the data set based only on the remaining observations. The other observations represent a imaginary reordering of "history" from which to predict the omitted observation. The resulting "predictions" on past data are termed *hindcasts*. A successful cross-validation will remove the noise specific to each observation, and estimate how well the chosen procedure selects prediction rules when this coincidental information is unavailable.

Some formal notation is helpful. The presentation follows closely the work in Elsner and Schmertmann (1994). Let x (the independent variables) denote a vector of predictors, and let y (the dependent variable) denote an outcome. A prediction rule f is a repeatable mapping that produces a predicted outcome $\hat{y} = f(x)$ for any vector of predictors x. In this general composition, a prediction rule f might be a linear regression model with specific coefficients (e.g., $\hat{y} = 3 - 4x_1 + x_2$), but it might also be something more unusual, such as a neural network with a specific set of weights or a classification tree with a specific set of decision rules. Indeed, the need for proper cross-validation increase with model sophistication. Simple models are transparent in regards to their exploitation of coincidences. For these models it is easy to adjust the estimation of skill. More complex algorithms use coincidences in ways that make it appear as if the harmony is part of the signal. In these cases it is easy to be fooled by the spurious coherence.

Let F denote the set of all prediction rules under consideration by the researcher. For example, F could be the set of all linear combinations of x, or the set of all neural networks with a given architecture. The problem facing the researcher is to use the available data on x and y to choose a single prediction rule $f \in F$. Let A denote an algorithm that makes this choice. An algorithm A is a predefined procedure that takes as input a data set containing multiple observations on x and y values, and produces as output a single, "best" prediction rule f out of the set F. Standard ordinary-least-squares (OLS) regression is one such procedure (algorithm). The OLS algorithm selects $f(x)$ as shown:

$$f(x) = x^T[(X^TX)^{-1}X^Ty] \qquad (14.2)$$

from a set F of all prediction rules having the form

$$f(x) = x^T[\beta], \qquad (14.3)$$

where the superscripts T and -1 denote the matrix operations of transpose and inverse respectively, and where β is a vector of coefficients. However, the algorithm A could be greatly more complex. For example, the algorithm might first consist of choosing the number of independent variables used in the prediction rule. The algorithm continues with an OLS estimation of the coefficients. The quintessential point as argued is that cross-validation is performed on an algorithm (A) rather than on a particular prediction rule (f). Understanding this point is the key to proper application of cross-validation.

Table 14.1: Predictors used in the CSU forecast models. The predictors were used in forecasting seasonal hurricane activity during the 1996 hurricane season. A description of the variables is given in the text and listed in Table 14.2. Plus (+) indicates the variable is included in the model for the specified forecast initialization time. Note the upper atmospheric zonal winds (U50 and U30) are used in the models at all three forecast times.

	Forecast Initialization Time		
Predictors	Early Dec	Early Jun	Early Aug
RG–1	+	+	+
RS–1	+	+	.
RS	.	.	+
ΔP	.	+	.
ΔT	.	+	.
U50	+	+	+
U30	+	+	+
|U50-U30|	+	+	+
ZWA	.	+	+
SLPA	.	+	+
SSTA	.	+	+
ΔSSTA	.	+	.
SOI	.	+	+
ΔSOI	.	+	.

In short, there are three considerations in performing cross-validation for the purpose of accurately estimating forecast skill of a chosen prediction rule. These are:

• First, hindcasts must be performed on data that is not used in developing the algorithm (out-of-sample) data. In-sample hindcasts only reflect the degree to which the prediction rule chosen by A fits the data. It is a biased estimate of how well the rule will work operationally.

• Second, prediction rules for hindcasts should not be chosen based on decisions that require information from the entire data sample. This misapplication occurs, for example, when decisions are made to accept additional predictors based on out-of-sample correlation. If this condition is violated then the algorithm is not cross-validated.

• Third, the subsample from which a hindcast is generated must be independent of the omitted observations. If neighboring events in time are correlated then successively removing a single observation is not suitable since "future" information will be used in the algorithm's choice of a prediction rule.

These three considerations are all variations of the general necessity that the forecast target be independent of the development sample, where the development sample is defined as the section of the data from which the rule is derived, and the forecast target is the portion of the data used for predictions. More specifically, the forecast target must not be allowed to influence the development of the prediction rule in any way.

Table 14.2: Predictor variables for the CSU seasonal forecast models.

Variable	General Description
RG–1	observed rainfall in the Gulf of Guinea from the previous year during August to November, except in the early December forecast where it is the current year's rainfall
RS–1	observed rainfall in the West Sahel from the previous year during August to September, except in the early December forecast, where it is the current year's rainfall
RS	observed rainfall in the West Sahel during June-July prior to the early August forecast
ΔP and ΔT	west African February to May surface pressure and temperature gradients, respectively
U50 and U30	extrapolated values of the zonal (west to east) winds for the coming September at 30 and 50 mb near $10°N$ as a QBO index
$|U50-U30|$	magnitude of the extrapolated vertical shear of horizontal winds between 30 and 50 mb
ZWA	June-July anomalies of the zonal wind at 200 mb over selected stations (Kingston, Curacao, Barbados, and Trinidad) in the Caribbean basin
SLPA	June-July anomalies of sea-level pressures at selected stations in the Caribbean basin
SOI and SSTA	El Niño predictors, where SOI is the value of the Southern Oscillation Index for June-July, and SSTA is sea-surface temperature anomalies over the central Pacific Ocean during June-July for the early August forecast and during April-May for the early June forecast
ΔSSTA and ΔSOI	the change in SSTA and SOI values between February and April-May

14.3   CSU Regression Models

Starting in 1984, William M. Gray and colleagues at Colorado State University (CSU) began issuing operational forecasts of hurricane activity over the North Atlantic basin. Predictions were issued several times a year. For instance, forecasts for the 1996 hurricane season were issued in early December of 1995, early June of 1996 and early August of 1996. From Chapter 4 we saw that early August marks the start of the most active part of the North Atlantic hurricane season. Seasonal forecasts were made for various measures of tropical-cyclone activity outlined at the beginning of the chapter. The CSU group modifies their forecast models to incorporate new research findings. The general forecast strategies are outlined here. Details concerning their early December models are described in Gray et al. (1992). Their early June models are described in Gray et al. (1994), and their early August models are depicted in Gray et al. (1993). The various predictors are displayed in Table 14.1. A brief description of the predictors are given in Table 14.2.

Table 14.3: Predictions of North Atlantic hurricane activity. Values are the forecasts made by the CSU group over the period 1984–1993. Values from the early December, early June, and early August forecasts are given along with the observed level of hurricane activity. Forecasts based on climatological averages over the same period have a mean-absolute error (MAE) of 1.6 for the number of hurricanes (H) and 8.8 for hurricane days (HD). Adapted from Hess and Elsner (1994b).

	Forecast Initialization Time			
	Early Dec	Early Jun	Early Aug	Observed
Year	H/HD	H/HD	H/HD	H/HD
1984	.	7/30	7/30	5/18
1985	.	8/35	7/30	7/21
1986	.	4/15	4/10	4/10
1987	.	5/20	4/15	3/5
1988	.	7/30	7/30	5/24
1989	.	4/15	4/15	7/32
1990	.	7/30	6/25	8/27
1991	.	4/15	3/10	4/8
1992	4/15	4/15	4/15	4/16
1993	6/25	7/25	6/25	4/10
MAE	1.0/8.0	1.4/9.5	1.3/7.4	.

Data values for RG, RS, and SOI are expressed in terms of standardized anomalies. Values of SLPA, ZWA, and SSTA are deviations from average based on the period 1950–1990. The method of least-absolute-deviation (LAD) regression is used to determine the coefficients in all the models. LAD regression is preferred to traditional ordinary-least-squares (OLS) regression because it is less sensitive to outliers in the data. Outliers are values that depart significantly from the mean. The method of LAD was introduced in by Roger Joseph Boscovich in 1757, 50 years before the introduction of the more popular method of OLS. The prevalence of OLS over LAD is due to its computational simplicity. Another factor is the theory behind OLS was developed earlier than the theory behind LAD. Computation is no longer a significant issue and LAD is available with most statistical software.

Statistical models are used only as guidance in making seasonal hurricane forecasts.[3] Often it is desirable to hedge one way or the other after reviewing the model forecasts. For example, if the models are predicting seasonal activity to be slightly above the climatological average, the official forecast might call for activity to be much above average based on additional information not available to the model. The track-record of the CSU group in making seasonal hurricane forecasts is shown in Table 14.3. The values are based on forecasts made over the period from 1984 through 1993. Values of the number of hurricanes (H) and number of hurricane days (HD) represent their official forecast rather than the output from their statistical models. For most years the

---

[3]Output from the models are checked for consistency.

predicted number of hurricanes is within one or two storms of the actual. In particular, forecasts of below average activity for 1986 and 1992 were excellent. Overall fore- cast performance, as indicated by the mean-absolute error (MAE),[4] over the 10 years from 1984 through 1993 is respectable. It should be noted that the first early-December forecast was issued in 1991 for the 1992 season.

The early August hurricane forecasts made over the longer period 1984–1996 are shown in Figure 14.1. The top panel shows the forecast number of hurricanes versus actual number for each year and the bottom panel shows the regression of actual versus predicted. The lull in North Atlantic hurricane activity during the early 1990s is well predicted as is the increase in activity during 1995 and 1996. The magnitude of the increase was underestimated. There is no significant bias in the forecast record. The 1989 season has the largest outlier value. The correlation coefficient between actual and predicted values is 0.72.

14.4   FSU Regression Models

Our group at Florida State University (FSU) has been issuing hurricane climate fore- casts since 1993. The models produced at FSU differ from the CSU models in their emphasis on probability statements. Models are also developed to forecast sub-basin hurricane activity (see Chapter 15) and to forecast several seasons in advance (see Chapter 16). The two components of basin-wide activity that we examine are the sea- sonal number of hurricanes (H) and the seasonal number of major hurricanes (MH). To predict the number of hurricanes we use an OLS linear regression to estimate the number of tropical-only hurricanes ($\widehat{H}_T$), to which we add an average number of baroclinically-enhanced hurricanes ($\bar{H}_B$). The average is adjusted conditionally on the number of tropical-only hurricanes as the two components are negatively corre- lated. The conditional adjustment amounts to removing the variance explained by the tropical-only component of hurricane activity. The prediction model can be expressed as:

$$\widehat{H} = \bar{H}_B + \widehat{H}_T, \tag{14.4}$$

where

$$\widehat{H}_T = \beta_0 + \sum_{i=1}^{3} \beta_i x_i, \tag{14.5}$$

and where the β_is are coefficients on the predictors. The three predictors (x_1, x_2, and, x_3) include U30, RS, and RG−1 as identified by the CSU group (see Table 14.2). Positive correlation is noted with both rainfall and upper-level wind anomalies.

To predict the number of intense hurricanes the FSU group uses a Poisson regres- sion.[5] The assumption that the occurrence of major hurricanes over the North Atlantic

---

[4]The mean-absolute error is computed as $\frac{1}{N}\sum_{n=1}^{N}|f_n - o_n|$, where f_n and o_n are the forecast and observed values for the nth season, respectively.

[5]After the French mathematician Siméon Denis Poisson (1781–1840).

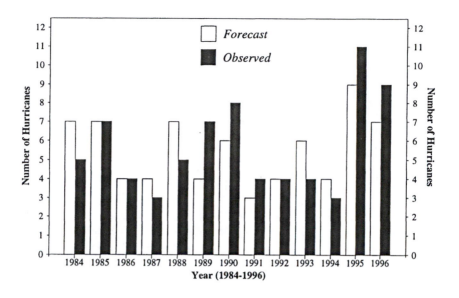

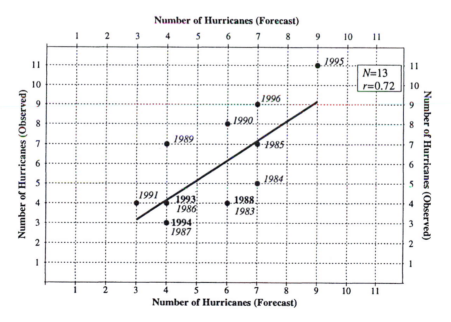

Fig. 14.1: Predictions of North Atlantic hurricanes. Values are the early August forecasts of hurricane frequency made by the CSU group over the period 1984–1996. The top panel shows the actual versus predicted number of hurricanes for each year. The bottom panel shows the scatter plot of actual versus predicted number of hurricanes, with the years marked for individual seasons. The solid line indicates the linear regression line for the scatter of points. The correlation coefficient between actual and predicted values is a respectable 0.72 over the 13 years.

Table 14.4: Predictions of North Atlantic intense hurricane activity. Values are the forecasts made by the FSU model over the period 1993–1996. Values are probabilities in percent for the early December initialization time. Forecast lead time is 6 months. Bold numbers indicate the verification. The forecast for 1996 was poor.

Year	Forecast Number Of Major Hurricanes (MH)					
	0	1	2	3	4	5+
1993	16.8	**29.9**	26.7	15.9	7.1	3.3
1994	**15.9**	29.2	26.9	16.5	7.6	3.7
1995	4.7	14.4	22.0	22.4	17.1	**19.4**
1996	25.9	35.0	23.6	10.7	3.6	**1.2**

can be described using a Poisson distribution is suggested in Xu and Neumann (1984). The Poisson model for seasonal number of major hurricanes is expressed as

$$MH = \exp(\gamma_0 + \sum_{i=1}^{4} \gamma_i x_i). \tag{14.6}$$

The predictors are the same as those used in the model for tropical-only hurricanes, except that U50 is added to provide a more complete description of the QBO. Maximum likelihood criterion are used to estimate the predictor coefficients. Details of the model are given in Elsner and Schmertmann (1993). Both the hurricane and major hurricane models are skill-evaluated using cross-validation. The linear correlation between actual and predicted numbers of tropical-only hurricanes is 0.61, with a MAE of 0.88 storms. The model of intense hurricanes produces a correlation of 0.84 and a MAE of 0.74. The track record of the FSU intense hurricane model is shown in Table 14.4. The forecast number of major hurricanes is expressed as a probability distribution. The peak in the distribution suggests provides a reasonable forecast of the expected number of intense hurricanes. The probabilities in bold represent the actual incidence. For example, in 1993, the preseason probabilities peaked at one intense hurricane. The actual number for 1993 was one. Overall the forecast model has performed reasonably well. In particular, the heightened activity during 1995 was foreshadowed by a low probability of fewer than two intense hurricanes. Since both the CSU and FSU group base their predictions on regression-type models that use nearly the same predictors, forecasts for individual years are not expected to differ by too much.[6]

14.5   Poisson Random Process

The Poisson approximation is useful for estimating probabilities associated with rare or extreme events. Its role in rare-event analysis is as universal as the role of normal

---

[6]Both the FSU and CSU models performed poorly on the 1998 North Atlantic hurricane season. Interannual variability in seasonal activity during the 1990s has been exceptionally large (see Chapter 20).

approximation in the analysis of expected values. A Poisson distribution describes the number of times that some event occurs as a function of time, where the events occur at random times. The event might be the occurrence of an intense hurricane in the North Atlantic basin, the occurrence of 50 m s^{-1} wind in a particular county, or the development of a tropical-only hurricane in the Caribbean Sea. In these cases, a single event occurs at a random time and the process amounts to counting the number of occurrences over some period. Hence the term *Poisson counting process.*

The Poisson distribution is an approximation to the binomial distribution when the probability of occurrence is small, but the number of occasions on which it can occur is large. In fact one of the oldest limit theorems of probability is the Poisson "law of small numbers." In its basic form, the law of small numbers states that the binomial distribution $Bi(n, p)$ converges to the Poisson distribution $P_o(\lambda)$ as $n \to \infty$ if $p = \lambda/n$ for some $\lambda > 0$.[7] The usefulness of the binomial distribution for modeling seasonal hurricane occurrences is limited. Although the mean is well defined by the ratio of the total number of storms to the number of years, one cannot say anything about the probability of a major hurricane occurring at a particular time during the season. In such cases the Poisson distribution has more importance than just as an approximation to the binomial distribution (Porkess 1991).

The two conditions for a Poisson random process are (1) events must occur only once in any increasingly small interval of time, and (2) the interarrival times be statistically independent. The number of events occurring in any time interval is independent of the number occurring in any other non-overlapping interval. A consequence of these two conditions is that the number of events in any finite time interval is described by the Poisson distribution. The average rate of occurrences in the time interval is the Poisson intensity denoted as λ. Let $X(t)$ be the number of events as a function of time. Then the probability of exactly k occurrences over a time interval $(0, t)$ is given by the cumulative marginal probability distribution function

$$F_X(k) = P[X(t) = k] = \frac{(\lambda t)^k e^{-\lambda t}}{k!}, \quad k = 0, 1, 2, \ldots, \infty, \quad (14.7)$$

where factorial k is denoted by $k!$ and defined as $k! = k(k-1)(k-2) \cdots 1$. It is convenient to define $0! = 1$. Assume the occurrence of intense North Atlantic hurricanes follows a Poisson random process, then the distribution is used to answer the following question. What is the probability of observing two, three, or four in a given season? In this case $t = 1$ year (season) and the average number of intense hurricanes in a season is 1.9 (see Chapter 7). Thus,

$$P[X = 2, 3, \text{ or } 4] = \frac{1.9^2 e^{-1.9}}{2(1)} + \frac{1.9^3 e^{-1.9}}{3(2)(1)} + \frac{1.9^4 e^{-1.9}}{4(3)(2)(1)} = 0.522. \quad (14.8)$$

Under these conditions there is a better than 52% chance of observing either two, three, or four major hurricanes each season.

---

[7]This was proved by Poisson in 1837.

Sometimes it is desirable to know the conditional and joint probability distribution functions. For example, what is the probability of 20 major hurricanes in a 5-year period, given that there has been eleven intense storms in the first two years of the period? The conditional probability of k_2 occurrences over the time interval $(0, t_2)$, given that k_1 events occurred over the interval $(0, t_1)$, is the probability that $k_2 - k_1$ events occur over (t_1, t_2). This is given by

$$P[X(t_2) = k_2 | X(t_1) = k_1] = \frac{[\lambda(t_2 - t_1)]^{k_2 - k_1} e^{-\lambda(t_2 - t_1)}}{(k_2 - k_1)!}, \qquad \text{for} \quad k_2 \geq k_1.$$
$$(14.9)$$

The joint probability of k_2 occurrences at time t_2 and k_1 occurrences at time t_1 is the product of the marginal and conditional distributions

$$P(k_1, k_2) \quad = \quad P[X(t_2) = k_2 | X(t_1) = k_1] P[X(t_1) = k_1], \qquad (14.10)$$

$$P(k_1, k_2) \quad = \quad \frac{(\lambda t_1)^{k_1} [\lambda(t_2 - t_1)]^{k_2 - k_1} e^{-\lambda t_2}}{k_1! (k_2 - k_1)!}, \qquad \text{for} \quad k_2 \geq k_1. \quad (14.11)$$

Thus the joint probability of six intense hurricanes in 3 years and 10 intense hurricanes in 5 years is 3.1%. The Poisson distribution allows one to make a variety of probability statements about the future occurrence of discrete random events.

14.6 Probabilistic Prediction

With different groups approaching the problem of seasonal hurricane prediction from different angles, it is useful to consider the prospect of probabilistic modeling that combines information from disparate sources. A statistical model of seasonal hurricane numbers is an oversimplification of the climate system involving complicated physics and dynamics. It is clearly a mistake to assume that any model is a true representation of the underlying processes. Moreover, often information for making a particular forecast is available from different sources. An approach to probabilistic (or Bayesian) prediction of annual hurricane activity is to first assume some prior information about what to expect for the upcoming season and then calculate the posterior distribution of this expectation. This strategy produces a predictive distribution for hurricane activity. Here we describe and apply a Bayesian prediction model as an example of probabilistic forecasting. The original description was given in Elsner (1996). The use of Bayesian statistics in studies of weather and climate are discussed in Epstein (1985). The issue of detecting climate change from historical records using a Bayesian approach[8] is presented in Solow (1988).

14.6.1 A Simple Example

Let $\Theta = \{\theta_0, \theta_1, \theta_2, \ldots, \}$ be the set of all possible numbers of intense hurricanes in a given year, where θ_0 denotes the occurrence of no intense hurricanes, θ_1, the occurrence

---

[8] After the Rev. Thomas Bayes (1744–1809).

of one intense hurricane, and so on. There are two pieces of information available to the forecaster. There are estimates $\pi_1(\theta_i)$ from an expert, where $\pi_1(\theta_i)$ is the forecast probability of θ_i for a chosen year. There is also the Poisson regression model, where we let π_2 be the number of major hurricanes (MH). The problem amounts to combining these two pieces of information to make the best possible forecast of intense hurricane activity. To construct an overall probability model for this situation, it makes sense to focus on modeling the accuracy of the experts. Thus it is necessary to determine densities $f(p|\theta)$ reflecting the probability p that an expert would likely provide under each situation. For example, suppose we review past predictions made by the CSU group and find that when the annual number of intense hurricanes is θ, predictions p follow a distribution $f(p|\theta)$. Once $f(p|\theta)$ is specified, Bayes's theorem is applied directly to obtain

$$\pi(\theta_0|p) = \frac{f(p|\theta_0)\pi_2(\theta_0)}{[f(p|\theta_0)\pi_2(\theta_0) + f(p|\theta_1)\pi_2(\theta_1) + \cdots + f(p|\theta_n)\pi_2(\theta_n)]}, \tag{14.12}$$

or more generally,

$$\pi(\theta_0|p) = \frac{f(p|\theta_0)\pi_2(\theta_0)}{\sum_{i=0}^{n} f(p|\theta_i)\pi_2(\theta_i)}. \tag{14.13}$$

Similarly, we can write

$$\pi(\theta_1|p) = \frac{f(p|\theta_1)\pi_2(\theta_1)}{\sum_{i=0}^{n} f(p|\theta_i)\pi_2(\theta_i)}, \tag{14.14}$$

or more generally,

$$\pi(\theta_j|p) = \frac{f(p|\theta_j)\pi_2(\theta_j)}{\sum_{i=0}^{n} f(p|\theta_i)\pi_2(\theta_i)}. \tag{14.15}$$

Note that in Eq. 14.15, $j = 0, 1, 2, \cdots, n$. This probabilistic modeling process is a sound way to proceed. The critical factor in evaluating any group's predictions is the skill of their previous predictions, and anything short of probabilistic modeling of this skill is likely to be inadequate (see Berger 1985).

As an example, Table 14.5 gives the performance of the CSU forecasts over the years 1990–1995. From this we sketch a prior distribution $f(p|\theta)$, where

$$f(p|\theta) = \begin{cases} 2, & \text{if } \theta=0; \\ 2.33, & \text{if } \theta=1; \\ 1, & \text{if } \theta=2; \\ 2, & \text{if } \theta=3; \\ 3, & \text{if } \theta=4. \end{cases} \tag{14.16}$$

Note that the prior distribution could be determined differently. The estimated probabilities based on a Poisson regression $[\pi_2(\theta_i)]$ for each possible number of intense hurricane are given in Table 14.6. From the two sets of information, Bayes's theorem is applied in a straightforward manner.

Table 14.5: Intense North Atlantic hurricanes over the period 1990–1995. Values are the predicted and observed numbers of intense hurricanes. Predictions are the early August forecasts made by the CSU group. Here θ denotes the number of intense hurricanes. The difference is the predicted minus the observed.

Year	Predicted θ	Observed θ	Difference
1990	3	1	+2
1991	1	2	−1
1992	1	1	0
1993	3	1	+2
1994	2	0	+2
1995	3	5	−2

The numerator of Eq. 14.15 is computed from Eq. 14.16 and Table 14.5. The denominator is calculated as

$$\sum_{i=0}^{4} f(p|\theta_i)\pi_2(\theta_i) \;=\; 2(0.259) + 2.33(0.350) + 1(0.236)$$

$$+ \; 2(0.107) + 3(0.036). \tag{14.17}$$

The above procedure produces a posterior probability distribution for θ. Under very general conditions the mean of the posterior distribution minimizes the Bayes risk when the loss function is quadratic (squared difference). The posterior probabilities are a combination of prior knowledge and statistical evidence (Epstein 1985). In the situation where the prior knowledge is useless, the statistical evidence remains uninfluenced. The underlying philosophy here can be extended by noting that π_2 depends solely on the intensity of the Poisson random process (λ). Accordingly, it is consistent to assume that λ is known only in terms of probability statements so that we can perform a Bayesian estimation to determine π_2 by first assuming a prior distribution for λ. For the Poisson process there exists identical formulas (conjugate distributions) for expressing judgments about λ before and after reviewing the data (see Epstein 1985).

14.6.2   Bayesian Estimation

In Bayesian estimation, the model parameters are considered random variables that have *prior* distributions reflecting either the strength of one's belief about possible values or other collateral information. Consider the Poisson model parameter λ. The main focus of Bayesian estimation is that of combining prior information about the parameter with direct sample evidence. This is accomplished by determining $\phi(\lambda|x)$, where $\phi(\lambda|x)$ is the conditional density of λ given the sample value taken on by x. In contrast to the *prior* distribution of λ, the conditional distribution, which also reflects the direct sample evidence, is called the *posterior* distribution of λ.

Table 14.6: Forecast distribution of intense North Atlantic hurricanes for 1996. Values are based on the Poisson model (Eq. 14.6) with predictor data from early December of 1995.

θ_i Value	0	1	2	3	4
$\pi_2(\theta_i)$ Value	0.259	0.350	0.236	0.107	0.036

If $h(\lambda)$ is a value of the prior distribution of parameter λ, and we wish to combine this information with direct sample evidence about λ, say, the value of x, we determine the posterior distribution of λ by

$$\phi(\lambda|x) = \frac{f(\lambda, x)}{g(x)} = \frac{h(\lambda)f(x|\lambda)}{g(x)}. \tag{14.18}$$

Here $f(x|\lambda)$ is a value of the sampling distribution of the statistic x for a given value of the parameter λ, $f(\lambda, x)$ is a value of the joint distribution of λ and x, and $g(x)$ is a value of the marginal distribution of x. Note, once the posterior distribution of a parameter is available, it can be used to make point estimates or it can be used to make probability statements about the parameter.

For instance, let $x = \theta$, where again θ is the number of intense North Atlantic hurricanes in a given season. If θ has a Poisson distribution and its parameter λ has, as its prior distribution, the *gamma* distribution, then the posterior distribution of λ given an observation of θ is a gamma distribution with the parameters $\alpha + \theta$ and $\beta/(\beta + 1)$ (instead of α and β). And the mean of this posterior distribution of λ is

$$\mu = \frac{\beta(\alpha + \theta)}{\beta + 1}. \tag{14.19}$$

Suppose a forecaster knows that the annual number of intense hurricanes for the North Atlantic is a random variable having a Poisson distribution, whose parameter (λ) has a prior gamma distribution with $\alpha = 11$ and $\beta = \frac{1}{5}$. Using a Poisson regression, the model predicts three intense hurricanes for the upcoming season, what would the forecaster's estimate of the hurricane season be if only direct information was used? In this case the model contains only the direct information, so the forecaster predicts three intense hurricanes. In contrast, if *only* prior information is used the forecaster relies on the equation $\mu = \alpha\beta = 11(\frac{1}{5})$ and predicts 2.2 intense hurricanes.[9] If both pieces of information are combined the answer is 2.3 intense hurricanes based on the equation

$$\mu = \frac{\beta(\alpha + \theta)}{\beta + 1} = \frac{\frac{1}{5}(11 + 3)}{\frac{1}{5} + 1} = \frac{2.8}{1.2} = 2.3. \tag{14.20}$$

Notice that the forecaster could provide the probability that the forecast value of θ is actually between two and four. Again, the usefulness of the Bayesian approach in

---

[9]The mean and variance of the Poisson distribution are $\mu = \lambda$ and $\sigma^2 = \lambda$, while the mean and variance of the gamma distribution are $\mu = \alpha\beta$ and $\sigma^2 = \beta^2\alpha$.

the context of seasonal hurricane forecasts is the ability to combine subjective prior information with statistics in a formal and rigorous way. In the end this produces the best possible forecast consistent with one's best judgments. The ideas presented in this chapter provide a guide to developing climatological forecast models of hurricane activity.

15

Sub Basin Forecast Models

Operational tropical cyclone prediction is established in the North Atlantic, the western North Pacific, and the Australian region. Forecasts that describe the general character of the upcoming season are routinely made. Emphasis is on prediction of the abundance of storms over the entire basin. The nature of basin-wide forecasts render them somewhat intangible for general use. In other words, basin-wide seasonal forecasts fail to provide enough detail for significant utility. For instance, suppose that several months in advance of the hurricane season, a prediction of three hurricanes and one intense hurricane is offered. In essence the forecast calls for a below-average season. Even if the forecast verifies, coastal dwellers may be lured into a false sense of security due to the prediction of below-average activity. Since it takes only a single hurricane to cause catastrophic losses, information on the frequency of storms for next year has limited value. Hurricane *Alicia* in 1983 ranks as the eleventh costliest hurricane on record for the United States,[1] but was one of only three hurricanes during that year. Hurricane *Andrew* in 1992 also arrived in a year with little activity.

The failure to specify location reduces the usefulness of seasonal forecast models. Accurate prediction of an active season may not prove beneficial to coastal residents if the location of the hurricane activity can not be identified. Despite a relationship between basin-wide hurricane activity and the number of U.S. hurricanes, many active hurricane seasons occur with only few, if any direct hits (e.g., 1981). Conversely, an inactive season, such as 1983, can prove to be quite damaging if the path of one hurricane crosses a vulnerable area. Table 15.1 lists the percentage of U.S. hurricanes to total hurricanes for each year over the period 1886–1996. Largest percentages are noted for years before 1920. This is due to the detection bias of storms at sea during the years before aircraft. All four of the storms during 1909 made landfall as hurricanes in the United States. Consistently large percentages are noted in years during the 1940s. Eighty-six percent (6) of the hurricanes during 1985 made landfall. Only 20% of the hurricanes of the active 1995 and 1996 seasons made U.S. landfall. As seen from Table 15.1, there are only two years (1907 and 1914) with no North Atlantic hurricanes.

---

[1]See Chapter 18.

Table 15.1: Ratios of U.S. to all hurricanes. Values are whole percentages of the annual number of U.S. hurricanes to total number of North Atlantic hurricanes over the period 1886–1996. Zero indicates no U.S. hurricanes and X indicates no North Atlantic hurricanes. Y denotes the last digit of the year. Average percentage in each decade is provided in the bottom row of each of the columns.

Y	188	189	190	191	192	193	194	195	196	197	198	199
0	.	0	33	67	50	0	50	27	50	20	11	0
1	.	12	67	67	50	0	50	0	12	50	0	25
2	.	0	0	50	0	33	50	17	0	33	0	25
3	.	50	25	67	33	50	20	50	14	0	33	25
4	.	40	50	X	40	33	43	25	67	25	20	0
5	.	0	0	67	100	40	60	33	25	17	86	18
6	88	50	67	55	38	43	33	25	29	17	50	22
7	30	100	X	50	0	0	60	33	17	20	33	.
8	60	75	20	33	50	67	50	0	20	0	20	.
9	20	40	100	100	67	33	43	43	17	60	43	.
	50	37	40	62	43	30	46	25	25	24	30	16

Our group at FSU has made an effort to improve the specificity of the season forecasts. In particular we have developed models to forecast the occurrence of hurricanes in separate sub-basins of the North Atlantic. The models, which are categorical, are developed in Lehmiller et al. (1997). They forecast the probability of hurricanes and intense hurricanes over the Caribbean, Gulf of Mexico, and along the southeast coast of the United States with lead times up to 6 months ahead. The probabilities are determined from discriminant analysis using an assortment of predictors. This chapter examines these models in detail.

15.1 Sub Basins and Coastal Regions

The North Atlantic can be divided into regional sub-basins. Several regions are chosen a priori based on precedent and geography rather than on hurricane climatology or model optimization. Because of the problems associated with a proper cross-validation of model performance (see Chapter 14), it is unwise to choose the basins based on model performance. The regions include the Caribbean Sea, the Gulf of Mexico, the southeast U.S. coast, and the northeast U.S. coast (Figure 15.1). The Caribbean Sea and the Gulf of Mexico are natural geographic sub-basins of the North Atlantic. The two east-coast regions are defined as coastal segments chosen to separate the more active southeast from the less active northeast. The Caribbean region is enclosed by a line running through the spine of the Greater Antilles and a line connecting the islands of the Lesser Antilles. The Gulf of Mexico is enclosed by a line from Key Largo to the northeast tip of Cuba and by a line from Cuba to northerneastern Yucatan.

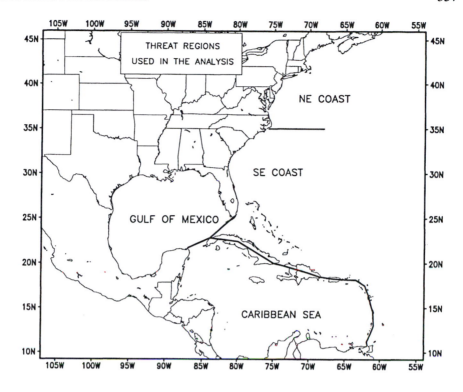

Fig. 15.1: Sub-basins of the North Atlantic. The east coast is sub-divided into two coastal regions [southeast (SE) and northeast (NE)] at Cape Hatteras, NC. The coast is divided at the eastern tip of North Carolina's Outer Banks. For the sub-basins it is enough if a hurricane enters or forms in the region, while for the two coastal regions a hurricane must make landfall to be counted as an occurrence. After Lehmiller et al. (1997).

Necessity requires that the east-coast regions consist only of landfalling hurricanes, as no natural markers exist to allocate geographic sub-basins along the coast. Furthermore, an attempt to create east-coast regions would necessarily be arbitrary. Model accuracy could be improved by rearranging the regions. A further justification for dividing the North Atlantic into both sub-basins and landfalling regions is that tropical cyclones of hurricane intensity in both the Gulf of Mexico or Caribbean Sea rarely fail to hit a populated area (e.g., only four hurricanes in the Gulf of Mexico failed to make landfall over the period 1950–1996). On the other hand, many storms curve quite close to the U.S. east coast but never make landfall.

A climatology of hurricanes is determined and discriminant statistical models developed to make long-lead forecasts within each region. Statistics of hurricane incidence are based on the period 1950–1996. This corresponds to the period over which data were available for the predictors. Predictors include the seasonal activity parameters discussed in Chapter 14. Additional predictors are included to forecast landfall occurrence along the U.S. east coast. Using an algorithm that chooses a subset of available predictors, significant and skillful models are found for three out of the four regions. No skillful model is found for the northeast coastline.

15.2 Climate Factors

Climate factors that control the likelihood of hurricane formation (e.g., SLPs) and hurricane development within specific geographic regions of the North Atlantic are not well understood. Large-scale circulation patterns that exert a control on the eventual paths of hurricanes have also not been extensively studied. Hurricane steering winds are controlled to some extent by the dominant pressure systems, which often display some persistence throughout a season. According to Ballenzweig (1959), atmospheric conditions that dictate the likelihood of hurricane formation and growth are unique to individual sub-basins. For instance, anomalous easterly flow in the middle and upper troposphere supports increased tropical-cyclone formation and conditions capable of spawning hurricane activity in the eastern North Atlantic and the Gulf of Mexico. On the other hand, an extension of the polar trough into the western Caribbean Sea often has a substantial impact on the frequency of hurricane development during the later part of the hurricane season.

Factors related to hurricane development over the North Atlantic as a whole are outlined in Gray (1984a) and Shapiro (1989), but a sharper focus on conditions conducive to the formation of hurricanes over individual sub-basins is still needed. Some progress has been made. Landsea et al. (1992) are able to link western Sahelian monsoon rainfall to the number of intense landfalling hurricanes in the United States. Over the Gulf of Mexico, hurricane activity is largely confined to the period from the middle of August to the middle of October, coincident with the climatological peak of the North Atlantic hurricane season. This includes hurricanes entering the Gulf and ones originating there. Hurricane incidence outside this window is limited and episodic. In fact, it is rather rare for a Gulf of Mexico hurricane during June or July, or after the middle part of October, but it is not without precedent. In fact, compared to the rest of the North Atlantic, the Gulf of Mexico is relatively more active outside the peak months, particularly during June. Hurricane *Audrey* in 1957 was a Gulf of Mexico intense hurricane that formed before the start of August.

A total of 56 hurricanes occurred over the Caribbean Sea during the period 1950 through 1996. This makes in an average of 1.2 hurricanes per season. Again, this includes hurricanes forming in the region and those passing into it. The 1996 season featured a record six hurricanes over the Caribbean. Caribbean hurricanes were more frequent during the 1950s and 1960s compared to the 1980s and 1990s. The 1950 season featured 4 Caribbean hurricanes. Only a few years during the active 1950s and 1960s are without a hurricane either forming or passing through the Caribbean Sea. Since the middle 1970s, a substantial reduction of hurricane activity over this region of the North Atlantic is evident.

Given that moderate to strong El Niño (warm ENSO) events inhibit tropical-only hurricane activity, the extended episodes of El Niño over this period may be responsible for the reduction in frequency of Caribbean hurricanes. Rainfall deficits in western African rainfall might also be linked to the dearth of low-latitude hurricanes. No Caribbean hurricane occurred from 1982 through 1986 and again from 1990 through 1994. The relative absence of hurricanes in the Caribbean Sea after about 1970 is nearly coincident with the reduction in tropical-only hurricane formations. The incidence of hurricanes threatening Cuba shows a decreasing trend since 1900. Cuba was last hit by a major hurricane in 1952. The decrease in hurricane activity over the northern Caribbean Sea is commensurate with a substantial increase in north-to-south surface-pressure gradient and a corresponding increase in the mean easterly flow over Cuba since the 1940s (Naranjo-Diaz and Centella 1998). Despite this general lack of Caribbean hurricanes during the 1980s, Reading (1989) indicates that activity is above what it was during the decades of the 1870s, 1910s, and 1920s. The 1995 and 1996 North Atlantic hurricane seasons featured a return of Caribbean hurricane activity, although this activity was focused over the eastern portion of the basin.

The U.S. east coast from Florida to the Outer Banks of North Carolina is particularly vulnerable to hurricane strikes. A total of 28 southeast hurricanes occurred over the 47-year period from 1950 through 1996. Most of the strikes (25) occurred between August and October. For the northeast U.S. coast, Kocin and Keller (1991) indicate that hurricane visits show a tendency for alternating periods of activity. For instance, the 1890s, 1950s, and early 1960s are relatively active. Periods of inactivity include the years from 1900 through 1930 and the middle 1960s through the early 1990s. Overall, hurricane landfalls along the northeast coast are rare. In fact, only eleven hurricanes made landfall in the northeast during the period 1950 through 1996. Consequently, no forecast model is found that performs better than climatology in this region. The annual probability of a northeast hurricane is low. The base rate swamps the success of even a good forecast model. In other words, the climatological prediction of no northeast hurricane is quite accurate year in and year out. On the other hand, one can argue that the utility of a forecast scheme based only on climatology is less than one that occasionally forecasts a rare strike correctly.

15.3 Base Rates and Forecast Accuracy

Base rates are known as climatology in the meteorological literature. It is important to understand base rates when evaluating predictive skill for models that forecast relatively rare events. Even a model that predicts landfalls with 80% accuracy, will correctly predict a strike only 31% of the time if the base rate of hits is only 10%. This counterintuitive fact can be seen with the aid of a contingency table (Table 15.2). Suppose that 100 predictions are made with a model having an accuracy of 80%. Suppose further that these 100 forecasts are made for 100 years over which only 10 years have a hurricane strike. Since the overall accuracy is 80% it is 80% for both the hit and no hit categories. With a base rate (climatology) of 10%, correctly forecasting a hit when a hit

Table 15.2: Forecast contingency table. Values are the number of years in each category. The forecast model has 80% accuracy against a base rate (climatology) of 10%. Based on 100 years data 10 years have one hurricane landfall. The number of landfall years forecast by the model is 26. The symbol Y_{11} refers to the number of years in which a forecast of no landfall is issued when no landfall occurs, etc.

	Forecast No Hit	Forecast a Hit	Total
No Hit Occurs	72 (Y_{11})	18 (Y_{12})	90
Hit Occurs	2 (Y_{21})	8 (Y_{22})	10
Total	74	26	100

occurs happens eight times, but incorrectly predicting a hit when no hit occurs happens eighteen times for an accuracy of only 31% (8/26). No forecast algorithm can escape this fundamental analysis. Ignoring base rates can lead to overconfidence in expressing the skill level of forecast models. Using the above numbers, a forecast model would need to be accurate better than 90% of the time in order for landfall predictions to be more accurate than guessing the outcome of a coin flip. The high base rate associated with the probability of no hurricanes along the northeast coast is the reason no skillful models have been found.

Various measures are available for expressing the information content of categorical forecasts (forecasts of discrete events). Most measures take into account in some form or another the base rates. Table 15.2 also shows a two-way contingency table for forecasts of hurricane landfall, where Y_{ii} is the number of samples within the category. Forecasts are perfectly accurate for $Y_{12} = Y_{21} = 0$. In reality, forecasts are rarely perfect and measures are necessary to evaluate the level of accuracy.[2]

One example is the probability of detection (POD) given as:

$$\text{POD} = \frac{Y_{22}}{Y_{21} + Y_{22}}. \tag{15.1}$$

In words, POD is the fraction of those occasions when the event occurred on which it was also forecast to occur. It represents the conditional probability of the event being forecast given that it actually happened. Values for POD range from zero for the worst case to one for perfect agreement between forecasts and observations. Note that POD does not accurately represent the level of skill of a prediction model since it fails to account for the false-alarm ratio given as

$$\text{FAR} = \frac{Y_{12}}{Y_{11} + Y_{12}}. \tag{15.2}$$

The FAR is that proportion of events that are forecast to occur but never do.

---

[2]Most common measures of forecast performance are scalar and nonprobabilistic.

Another scalar metric of forecast accuracy is the percentage of correct estimate (PCE) expressed as:

$$PCE = \frac{Y_{11} + Y_{22}}{Y_{11} + Y_{12} + Y_{21} + Y_{22}}. \tag{15.3}$$

Values of PCE (or hit rate) range from zero for worst case to one for perfect forecasts and credits correct forecasts equally whether the model is correct in predicting no hits or correct in forecasting a hit. In cases where one category is considerably more prevalent than the other (years with no hits along the northeast coast), this equivalence is not necessarily desirable as the previous argument on base rates suggests.

A common alternative measure in these cases is the *threat score* (TS) or sometimes called the *critical success index* (CSI) which is the same as the PCE if the category Y_{11} is eliminated. Formally the threat score is computed as

$$TS = \frac{Y_{22}}{Y_{12} + Y_{21} + Y_{22}}. \tag{15.4}$$

As with the other measures, values for TS range from zero to one. Using the data from Table 15.2 we find that the value for PCE is 0.80 suggesting a good forecast model, yet due to the relative infrequency of hurricane strikes the TS is only 0.29.

The forecast bias, computed as

$$B = \frac{X_{22} + X_{12}}{X_{22} + X_{21}}, \tag{15.5}$$

is a comparison of the average forecast with the average observation from categorical forecasts (see Wilks 1995). Unbiased models have B = 1 indicating an equality between the number of times the event was observed and the number of times the event was forecast, regardless of whether the forecasts and observations actually matched on particular occasions. A bias value greater than one indicates that the model over forecasts the event to occur, while a bias value less than one indicates an under forecast.

The most widely used scalar metric of forecast accuracy for categorical forecasts is the Heidke skill score given as

$$HSS = \frac{2(Y_{11}Y_{22} - Y_{12}Y_{21})}{Y_{12}^2 + Y_{21}^2 + 2Y_{11}Y_{22} + (Y_{12} + Y_{21})(Y_{11} + Y_{22})}. \tag{15.6}$$

The HSS is rooted in a reference to a fictitious model that generates random forecasts based on the base rates. A perfect model would give a value of one to HSS, however, a forecast model that performs worse than random guessing would have a negative HSS. For the case above, the HSS is 0.35, which is consistent with a measure of skill that includes the base rates.

Another measure of forecast skill which also includes a comparison with climatology is based on the normal approximation. Let $N = Y_{11} + Y_{12} + Y_{21} + Y_{22}$ be the total number of forecasts made, p_m be the PCE of the proposed model, and p_c be the maximum of $(Y_{21} + Y_{22})/N$ and $(Y_{11} + Y_{12})/N$ (i.e., the best skill available from a climatological forecast). Then the test statistic z is given by

$$z = \frac{\sqrt{N}(p_m - p_c)}{\sqrt{(p_c - p_c^2)}}. \tag{15.7}$$

It is desirable because it allows direct tests of statistical significance against the base rates (Devore 1991). Using normal approximation theory, the z-statistic can be converted to a probability statement from the corresponding p-value. This is the approach used by the FSU group in evaluating the skill of categorical forecast models of hurricane location. In general, the most useful metrics of forecast skill are probabilistic (Marzban 1998) or multi-dimensional like the reliability diagrams examined in Wilks (1995) and Hamill (1997).

Verification of probabilistic forecasts is more subtle. Unless probabilities are 0 or 1, forecasts of probability are neither right or wrong. A common measure used in verify probability forecasts is the Brier score (BS). Consider a forecast algorithm that predicts the probability of a discrete event, then the BS is given by

$$\text{BS} = \frac{1}{N} \sum_{n=1}^{N} (f_n - o_n)^2, \tag{15.8}$$

where f_n are the predicted probabilities, and o_n is 1 if the event occurs, and 0 if it does not occur. The BS gives the mean-square error of the probability forecasts.

15.4   Discriminant Analysis

The models developed by the FSU group to forecast sub-basin hurricane occurrences are based on multivariate linear discriminant analysis (LDA). These models use a categorical response variable[3] consisting of two distinct groups (i.e., hurricane year versus no-hurricane year), where a hurricane year is defined as a year during which at least one hurricane occurs in the basin. For the east-coast regions it is defined as a year during which at least one landfall occurs. For each model there are several potential predictors. Linear discriminant analysis is a statistical method that seeks to classify categorical data as a linear function of its predictors. It is an exact analogue of linear regression analysis except that the dependent variable is discrete rather than continuous.

Linear discriminant analysis works by creating a linear function of the predictors for each group. Consider the case where we have two groups (1 and 2) and four predictors. The methodology works by using the data in a sample to estimate linear functions for each group. Using the notation a_{ij} to denote the estimated linear coefficient for the

---

[3] A categorical variable is one for which the measurement scale consists of a set of categories. For instance, a hurricane is cataloged as either tropical-only (TO) or baroclinically-enhanced (BE).

ith group and jth predictor, the method yields

$$\text{Group 1:} \quad s_1 \;=\; a_{10} + a_{11}x_1 + a_{12}x_2 + a_{13}x_3 + a_{14}x_4 \qquad (15.9)$$

$$\text{Group 2:} \quad s_2 \;=\; a_{20} + a_{21}x_1 + a_{22}x_2 + a_{23}x_3 + a_{24}x_4 \qquad (15.10)$$

An observation is then classified into either group 1 or group 2 if the corresponding value of score s_i is the largest of the two values. Given new observations of the predictors, linear discriminant functions are used to predict its classification. If the groups lag the predictors in time, then the prediction is a forecast of the future.

A major issue concerning the procedure is the choice of an optimal method for determining the linear coefficients (see Mardia et al. 1979). Discriminant methodology is technically a Bayesian classifier, so the choice of the optimal method should seek to maximize the associated Bayesian classification rule. For the case of only two categories, the classification method developed by R. A. Fisher[4] asymptotically maximizes the Bayes classification efficiency regardless of predictor distributions, and the method in itself represents an effective classification method, regardless of statistical considerations provided that the two groups have the same population covariance matrices (Hand 1981).

Suppose there are n_1 observations of the multivariate random variable $\mathbf{X}_1$ from group 1 and n_2 observations of the random variable $\mathbf{X}_2$ from group 2, where

$$\mathbf{X}_1 = \begin{bmatrix} x_{11} \\ x_{12} \\ \vdots \\ x_{1n_1} \end{bmatrix}_{n_1 \times 1} \qquad \mathbf{X}_2 = \begin{bmatrix} x_{21} \\ x_{22} \\ \vdots \\ x_{2n_1} \end{bmatrix}_{n_2 \times 1}$$

From these data matrices, the sample means $\overline{\mathbf{x}}_1$ and $\overline{\mathbf{x}}_2$ are $p \times 1$ vectors given as

$$\overline{\mathbf{x}}_1 \;=\; \frac{1}{n_1} \sum_{j=1}^{n_1} \mathbf{x}_{1j}, \qquad (15.11)$$

$$\overline{\mathbf{x}}_2 \;=\; \frac{1}{n_2} \sum_{j=1}^{n_2} \mathbf{x}_{2j}. \qquad (15.12)$$

And the sample covariance matrices are as follows:

$$\mathbf{S}_1 \;=\; \frac{1}{n_1 - 1} \sum_{j=1}^{n_1} (\mathbf{x}_{1j} - \overline{\mathbf{x}}_1)(\mathbf{x}_{1j} - \overline{\mathbf{x}}_1)^{\mathrm{T}}, \qquad (15.13)$$

$$\mathbf{S}_2 \;=\; \frac{1}{n_2 - 1} \sum_{j=1}^{n_2} (\mathbf{x}_{2j} - \overline{\mathbf{x}}_2)(\mathbf{x}_{2j} - \overline{\mathbf{x}}_2)^{\mathrm{T}}. \qquad (15.14)$$

Assuming that each group has the same covariance structure, a $p \times p$ pooled sample covariance matrix ($\mathbf{S}$) is

$$\mathbf{S} = \frac{(n_1 - 1)\mathbf{S}_1 + (n_2 - 1)\mathbf{S}_2}{(n_1 + n_2 - 2)}. \qquad (15.15)$$

---

[4]Ronald A. Fisher was a founder of modern statistical design.

The pooled covariance matrix (**S**) is a linear combination of $\mathbf{S}_1$ and $\mathbf{S}_2$. It is an unbiased estimator of the population covariance matrix. The linear coefficients in Eqs. 15.9 and 15.10 are determined by *Fisher's method*. This is done by considering the distance between a sample observation and the centroid of all sample observations for that group. The distance metric in this case is the so-called *Mahalanobis distance function*. It adjusts the distance in each predictor dimension according to the variance of that predictor, so that the measure is scale invariant (Mardia et al. 1979). This distance measure D is

$$D^2 = (\overline{\mathbf{x}}_1 - \overline{\mathbf{x}}_2)^{\mathrm{T}}\, \mathbf{S}^{-1}\, (\overline{\mathbf{x}}_1 - \overline{\mathbf{x}}_2), \tag{15.16}$$

where an observation is classified to group i if the distance function is minimized for group i.

As an example of discriminant analysis, refer to Figure 15.2. For simplicity, the graph shows only the two-predictor case in the model for predicting the occurrence of a Caribbean hurricane initialized on early December data. The period of record is from 1950 through 1995. The linear discriminant methodology partitions the plane by a line and assigns observations to the two groups according to which side of the line the observation lies. Here, observations above and to the right of the line are allocated as years for which a Caribbean hurricane will occur (Y), while observations below the line are allocated as years for which a Caribbean hurricane will not occur (N). For this classification, the in-sample error rate, as can be determined by the graph, is 0.2 (9 out of 46 are wrong, giving an accuracy of 80.4%). Results are somewhat improved by incorporating additional predictors.

For this method to be valid, an assumption requires each group to have identical true covariance matrices. In the case of multivariate normal data, this method is also optimal. Since multivariate normal data rarely occur in practice, an alternative distance function can prove more accurate in certain cases; however, the above method works quite well even when the data are far from normal. More importantly it is necessary to evaluate how well the discriminant models classify the hurricane years in order to determine the model's predictive abilities. The simple approach is to evaluate the in-sample classification accuracy. Such estimates are biased low, providing too much confidence. The in-sample classification accuracy increases monotonically as additional predictors are added. Cross-validation techniques (see Chapter 14) are required to obtain nearly unbiased error estimates (Hand 1981). The above models are evaluated for skill using cross validation and the z-statistic.

15.5 Significant Models

The model selection algorithm is applied to each sub-basin for the occurrence of hurricanes and major hurricanes. The algorithm is applied at two initialization times: early December and early August (see Chapter 14). The available predictors for selection in the December model are the parameters used in the basin-wide models. The predictors for the August models included additional variables along with the basin-wide indices.

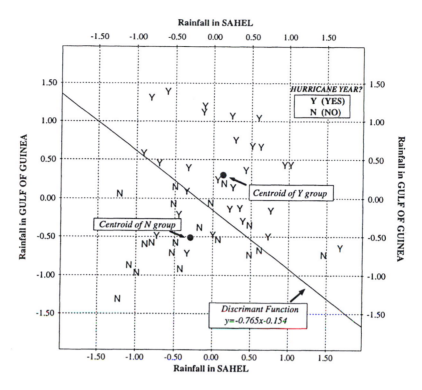

Fig. 15.2: Discriminant analysis for forecasting the occurrence of a Caribbean hurricane 6 months in advance. The predictor values are in standard deviation. Note that years with at least one Caribbean hurricane tend to occur when rainfall in the Gulf of Guinea and the Sahel regions of west Africa are above average. Adapted from Lehmiller et al. (1997).

The basin-wide predictors provide a synopsis of the large-scale atmospheric patterns associated with regional hurricane activity, while the extra predictors focus specifically on localized conditions. Models more skillful than climatology can not be developed for hurricane activity in the Gulf of Mexico or for landfalling hurricanes along the northeastern coast of the United States. The Gulf of Mexico sees hurricanes nearly every year, while landfalling storms in the northeastern are relatively rare. In both cases base rates are too high.

For the early December initialization, the algorithm identifies two predictors of Caribbean hurricane activity. These include the two African rainfall estimates (RG and RS). The model correctly predicts 37 of the 46 years (80.4%) in a cross-validation strategy. Statistical significance tests yield a bootstrap[5] p-value < 0.001 for the cross-validation accuracy and a normal approximation p-value of 0.002 when compared to the climatological accuracy of 27 of the 46 years (58.7%). The validation over all years

---

[5]Bootstrap refers to a statistical re-sampling procedure. Additional bootstrap samples are generated from the original sample by random selection with replacements.

in the period 1950–1995 indicates no error bias, and forecast errors show no consistent pattern. For this model, the cross-validated hindcast error is identical to the in-sample error estimate.

No other successful forecast models could be initialized by early December. However, three useful models using data through July are found. In particular, skillful models are possible for the Caribbean Sea and the Gulf of Mexico for predicting the occurrence of intense hurricanes. Although lead time is significantly reduced these models can still provide good information for the remainder of the season as the vast majority of intense North Atlantic hurricanes occur after July, and nearly all southeast hurricanes occur after the beginning of August. For predicting intense hurricanes in the Gulf of Mexico, the algorithm identifies a model that has a cross-validated hindcast accuracy of 36/46 or 78.3%.[6] The best strategy using climatology is to predict no intense hurricane for each year. That is, the occurrence of a major Gulf hurricane is less likely than not. The climatology yields an accuracy of 24/46 or 52.2%. Bootstrap p-values of less than or equal to 0.001 are obtained for this model, while the approximate z-value versus climatology is 3.54 (p-value < 0.001). No notable prediction bias exists. Slightly more significant results are obtained for predicting intense hurricanes within the Caribbean basin by late July. The algorithm selects a three-predictor model achieving a cross-validated hindcast accuracy of 37/46 or 80.4% using U30, RS, and ZWA. Climatological prediction accuracy for intense Caribbean hurricanes is identical to the accuracy for intense Gulf hurricanes. A prediction of no intense hurricanes gives an accuracy of 25/46 or 54.3%. The bootstrap p-value is < 0.001, while the normal approximation z-value is 3.55 (p-value < 0.001). No prediction bias is noted in the cross-validated hindcasts for this model.

No skillful model for predicting intense landfalling hurricanes along the southeastern United States is found; however, a significant model for landfalling hurricanes is possible. The model selects five-predictors including RS, UDIFF, the 700 to 200 mb vertical shear in the Miami/West Palm Beach area (VS-MIA/PBI), the July monthly sea-level pressure in Cape Hatteras (SLP-HAT), and the July monthly east coast sea-level pressure average (JCSLP). Cross-validated hindcast model accuracy is 81.4%, which exceeds a climatological prediction accuracy of 58.1%. Statistical significance tests of the model yields a bootstrap p-value of 0.0005 and a normal approximation z-score versus climatology of 3.09 (p-value = 0.001). To illustrate the significance of this model, Lehmiller et al. (1997) show the simulated bootstrap distribution (based on 4000 bootstrap samples) of the hindcast accuracy. The model value occurs in the extreme tail of the bootstrap distribution. As is the case for the other three models, no prediction bias exists with this model. A summary of results is given in Table 15.3. Four skillful models are obtained for forecasting sub-basin hurricane activity. Only one model is available for extended-range forecasts. Three of the four models use only the predictors identified as having an influence on basin-wide activity. The prediction

---

[6]In achieving this accuracy, the algorithm selects three predictors for the discriminant model: U50, U30, and SOI.

Table 15.3: Significant prediction models for sub-basin hurricane activity. All models are significant with a p-value < 0.005. H stands for hurricanes and IH for intense hurricanes. Southeast
stands for landfall along the southeast U.S. coast. Model initialization time (Time of Issue) for
each forecast region is the beginning of the month (e.g., Early December). Abbreviations for the
predictors are explained in the text.

Forecast Region	Time of Issue	Type	Accuracy (%) Model	Climate	Predictors Used
Caribbean	Dec	H	80.4	58.7	RG, RS
Caribbean	Aug	IH	78.3	52.2	U50, U30, SOI
Gulf	Aug	IH	80.4	54.3	U30, RS, ZWA
Southeast	Aug	H	81.4	58.1	RS, UDIFF, VS, SLP, JCSLP

model for landfalling hurricanes along the southeast coast includes additional predictors related to vertical wind shear and SLPs.

15.6 Physical Linkages

The early December prediction model for Caribbean Sea hurricanes is linked solely
to African rainfall. As African rainfall increases relative to the long-term average, the
probability of a Caribbean hurricane rises. The result is intuitive, as the great majority
of hurricanes that form in or track over the Caribbean Sea are tropical-only. Hence, an
increase in African rainfall drives an increase in tropical-only hurricanes that form from
easterly waves at low latitudes. Low-latitude originating tropical cyclones are more
likely to cross the Caribbean. Waves that fail to develop over the central Atlantic have
an opportunity to grow over the western Caribbean under conditions that are generally
favorable.

In contrast, the prediction models of major hurricane activity depend more strongly
upon variables measuring the vertical structure of the wind. While the Caribbean intense hurricane activity model also uses an African rainfall parameter, due most likely
to the tropical-only nature of Caribbean hurricanes, the other two predictors measure
the wind environment available to tropical cyclones occurring over the Caribbean region. The models appear to distinguish between necessary conditions of hurricane
formation. The condition of an initial disturbance is determined by African rainfall
(RS) and the condition of storm environment is determined by upper-level wind patterns (U30 and ZWA). A suggested physical mechanism relating the upper-level wind
phases (QBO) to hurricane development is discussed in Shapiro (1989) and in Gray et
al. (1992). A likely explanation involves the extent to which the troposphere couples
with the lower stratosphere, though debate remains as to the exact nature of the linkage.
Nevertheless, the effect upon intense hurricanes is clear. The QBO relates to the formation of tropical-only hurricanes that generate intense hurricanes. The ZWA parameter
relates to the vertical wind shear environment that is felt by a hurricane; obviously, the

more favorable the environment, the more likely it is that the hurricane will become intense.

Upper-level winds determine the probability of an intense hurricane over the Gulf of Mexico. All three predictor variables index the upper-level wind structure, though the relationship of the SOI to winds over the Atlantic deserves some comment. Warm ENSO conditions in the equatorial Pacific Ocean enhance convection resulting in strong upper tropospheric jets which traverse the North Atlantic. These jets enhance vertical shear of horizontal winds. Strong wind shear is an inhibiting factor for hurricane formation. The two other variables, U50 and ZWA, also pertain to vertical wind shear. Hurricanes occur nearly every year in the Gulf of Mexico. The Gulf sees both baroclinically-enhanced and tropical-only hurricanes. Thus it is suggested that a predictable criterion for the occurrence of an intense hurricane in this region should be related to the storm environment. The identification of vertical shear as an important predictor for sub-basin hurricane activity has legitimate physical justifications as weak shear plays an integral role in producing a favorable hurricane environment.

Interpretation of model results is more difficult for the southeastern coastline. The presence of African rainfall and wind shear in the model suggest that the usual tropical-only component of North Atlantic hurricane activity plays a part. Increased rainfall and weak shear lead to more deep tropical hurricanes that track northwestward toward the coast. However, the role of the east coast sea-level pressures and the 700 to 200 mb vertical shear over south Florida is more difficult to ascertain. It is possible that sea-level pressures could be a measure of tropical-cyclone steering mechanisms, provided that climatological persistence occurs. Lehmiller et al. (1997) speculate that July monthly vertical shear over south Florida indicates the extent of the intrusion of middle-latitude synoptic-scale features into the subtropics along the southeast coast. Another possibility is that this shear is measuring the presence or absence of subtropical (or polar) jets within the region. In any case, these predictors do not lend themselves to straightforward interpretation and additional research is needed. Availability of skillful models for sub-basin hurricane activity is a step closer to more useful long-range hurricane climate forecasts.

16

Prospects for Extended Range Outlooks

Forecast utility is a function of both skill and lead time. Seasonal hurricane forecasts of activity several years in advance have the potential for greater utility. Presently, lead time is about 9 months for seasonal forecasts issued in early December. However, as discussed earlier, annual North Atlantic hurricane frequencies show preferred periods of oscillation on times ranging from months to years. This offers the possibility for extended-range seasonal forecasts. Assuming the past is a good indicator of the future, regularities in the time series can be exploited to make multi-season prognostications.

This chapter examines two algorithms for doing this. Both approaches are based on extending the singular spectrum analysis (SSA) of Chapter 10. The first algorithm makes use of autoregressive-moving average (ARMA) models and a model selection based on Bayesian criterion. The second algorithm is a direct iterative approach. It is expected that these forecast algorithms, when used operationally, will yield only marginally skillful predictions against climatology. They could, however, offer considerable value as realized by benefits to decisionmakers in government and industry. The chapter ends with a discussion of the prospect for multi-decadal hurricane outlooks. Some of the presentation follows the work in Elsner et al. (1998).

16.1 The ARMA Approach

The singular spectrum analysis (SSA) is a method for decomposing a time series into temporal principal components. The principal components are used to create reconstructed components, which when summed give back the original record. With reference to the analyses performed in Chapter 10, the three dominant reconstructed components of the annual North Atlantic hurricane frequency record are computed as

$$
\begin{aligned}
y_1(i + j - 1) &= a_i^1 e_j^1 + a_i^2 e_j^2, \\
y_2(i + j - 1) &= a_i^3 e_j^3 + a_i^4 e_j^4, \\
y_3(i + j - 1) &= a_i^5 e_j^5 + a_i^6 e_j^6,
\end{aligned}
\tag{16.1}
$$

where the a_i's are the principal components ($i = 1, 2, \ldots, N_t - m + 1$) and the e_j's are the eigenvectors ($j = 1, 2, \ldots, m$), and where N_t is the length of the time series (in years) and m is the number of lags (also in years). The y_k's are called the reconstructed components. In contrast to Fourier spectral methods this approach—involving the singular-value decomposition (SVD) of the lagged-correlation matrix—does not assume periodic behavior in the time series. Classical time-series methods have problems in forecasting a low-pass filtered version of the data due to tapering at the ends of the record. The tapering is needed to avoid sudden jumps caused by the implied periodicity. Traditional methods, including (ARMA) models, can be applied to the reconstructed components because tapering is not an issue. For example, a maximum likelihood estimator can be used to determine the ARMA coefficients for each of the reconstructed components separately. These coefficients are employed to forecast hurricane activity several years into the future.

Suppose for example that the three reconstructed components are modeled with ARMA(1,0) processes (one autoregression (AR) parameter and no moving-average (MA) parameters) with coefficients α_k and σ_k. Then a prediction of the reconstructed component y_k is obtained by

$$\widehat{y}_k(N_t + 1) = \alpha_k y_k(N_t) + \sigma_k, \tag{16.2}$$

where $\widehat{y}_k(N_t + 1)$ is the predicted value. Note that by using $\widehat{y}_k(N_t + 1)$ as an observed value for record k we can get a two-time-step prediction $\widehat{y}_k(N_t + 2)$, which in turn can be used to get a three-time-step prediction and so forth for each reconstructed component. A prediction for the observed time series is issued by summing forecasts made separately on each of the components using

$$\widehat{y}(N_t + 1) = \sum_{k=1}^{3} \widehat{y}_k(N_t + 1). \tag{16.3}$$

More realistically, let separate ARMA models be built for each reconstructed components. In general, a time series $\{y(t)\}$ is modeled as an ARMA(p, q) process of the form:

$$\begin{aligned} y(t) - \phi_1 y(t-1) - \cdots - \phi_p y(t-p) &= \mu + \epsilon(t) - \theta_1 \epsilon(t-1) \\ &\quad - \cdots - \theta_q \epsilon(t-q), \end{aligned} \tag{16.4}$$

where the $\epsilon(t)$'s are assumed to be independent and normally distributed with mean zero and variance σ^2. In the model, p is the order of the autoregression (AR) term, and q is the order of the moving-average (MA) term. The AR and MA coefficients are denoted by $\phi = (\phi_1, \ldots, \phi_p)'$ and $\theta = (\theta_1, \ldots, \theta_q)'$, respectively. If some coefficients in the model are zeros, i.e., $\phi_i = 0$ for $i < p$ or $\theta_j = 0$ for $j < q$, then the model is parsimonious.

The first step in modeling a time series is to identify a tentative model for the data. Numerous criteria have been proposed for model selection in time-series literature (see,

Table 16.1: Time-series model of the first reconstructed component. Values are estimated coefficients, estimated standard errors, and t ratios for the model coefficients. The model is based on a detrended record of the annual hurricane frequency over the period 1901–1991. The proportion of the total variation explained by the model (r^2) is 0.81.

Coefficient	Estimate	Std. Error	t-Ratio
ϕ_1	0.372	0.082	4.54
ϕ_2	−0.508	0.081	−6.27
ϕ_3	−0.446	0.102	−4.37
ϕ_6	−0.325	0.109	−2.98
ϕ_9	−0.246	0.104	−2.37
ϕ_{10}	0.371	0.092	4.03
ϕ_{11}	−0.282	0.093	−3.03
ϕ_{13}	−0.252	0.098	−2.58
ϕ_{15}	−0.300	0.082	−3.66

e.g., Box and Jenkins 1976, Brockwell and Davis 1991). Schwartz (1978) suggests a Bayesian criterion (SBC), which is similar to Akaike's Bayesian criterion (BIC). Specifically, assume that an ARMA(p, q) model of M parameters is fit to a time record. Then the SBC for the fitted model is defined as

$$\text{SBC}(M) = -2\log[L(\widehat{\phi}, \widehat{\theta}, \widehat{\sigma}^2)] + M\log(N_t), \tag{16.5}$$

where $L(\widehat{\phi}, \widehat{\theta}, \widehat{\sigma}^2)$ is the maximum of the likelihood function for the parameters, and N_t is the record length. Among a group of adequate models to fit the time series, the one with the smallest SBC value will be selected as the best choice.

An ARMA(p, q) model is built for each of the series in Eq. 16.1 using observations over the period 1901–1991. The first 15 values of each reconstructed component (1886–1900) are ignored due to limitations in the algorithm. The last five values are withheld for the purpose of model validation. The Schwartz Bayesian criterion is used for model selection and the maximum likelihood method is used to estimate parameters in the chosen model. The final chosen model for $\{y_1(t)\}$, which is the dominant reconstructed component of the detrended hurricane series, is a parsimonious AR(15) model with the form:

$$
\begin{aligned}
y_1(t) &= \phi_1 y_1(t-1) + \phi_2 y_1(t-2) + \phi_3 y_1(t-3) + \phi_6 y_1(t-6) \\
&+ \phi_9 y_1(t-9) + \phi_{10} y_1(t-10) + \phi_{11} y_1(t-11) \\
&+ \phi_{13} y_1(t-13) + \phi_{15} y_1(t-15) + \epsilon(t).
\end{aligned} \tag{16.6}
$$

The estimated parameters along with their estimated standard errors are listed in Table 16.1. In the table, the r^2 for the fitted model is calculated using

$$r^2 = 1 - \frac{\sum_{t=1}^{n}[y_1(t) - \widehat{y}_1(t)]^2}{\sum_{t=1}^{n}[y_1(t) - \bar{y}_1]^2},$$

Table 16.2: Time-series model of the second reconstructed component. Values are estimated coefficients, estimated standard errors, and t ratios for the model coefficients. The model is based on a detrended record of the annual hurricane frequency over the period 1901–1991. The proportion of the total variation explained by the model (r^2) is 0.72.

Coefficient	Estimate	Std. Error	t-Ratio
ϕ_1	−0.639	0.077	−8.30
ϕ_2	−0.312	0.085	−3.67
ϕ_4	−0.421	0.068	−6.19
ϕ_{10}	0.355	0.087	4.08
ϕ_{13}	−0.227	0.065	−3.49
ϕ_{15}	−0.253	0.075	3.37
θ_2	0.487	0.110	4.43

where $\{\hat{y}_1(t), t = 1, \ldots, n\}$ are the fitted values based on the model. The value of $r^2 = 0.81$ means that the parsimonious AR(15) model explains about 81% of the total variation of the series $\{y_1(t)\}$.

A description for the second dominant reconstructed component $\{y_2(t)\}$ is the following parsimonious ARMA(15, 2) model of the form:

$$
\begin{aligned}
y_2(t) \;=\; & \phi_1 y_2(t-1) + \phi_2 y_2(t-2) + \phi_4 y_1(t-4) + \phi_{10} y_1(t-10) \\
+ \; & \phi_{13} y_1(t-13) + \phi_{15} y_1(t-15) + \epsilon(t) - \theta_2 \epsilon(t-2).
\end{aligned} \tag{16.7}
$$

The estimated parameters and standard errors are listed in Table 16.2. An $r^2 = 0.72$ indicates that about 72% of the total variation of the second dominant component of the detrended hurricane series is explained by the fitted model. The third dominant reconstructed component $\{y_3(t)\}$ has much smaller variations from year to year. A description for this series is an AR(8) model of the form:

$$
y_3(t) = \sum_{i=1}^{8} \phi_i y_3(t-i) + \epsilon(t). \tag{16.8}
$$

The estimated coefficients in this model are given in Table 16.3. About 83% of the total variation is explained by the fitted AR(8) model. Although r^2 is largest for this component, the variability is smallest. Only the model for the second component contains a moving-average term.

For comparison, a time-series model is also fitted to $\{x(t)\}$ by using 91 observations of the detrended hurricane record over the period 1901 through 1991. A description of this series is given by the following parsimonious AR(17) model:

$$
x(t) = \phi_{10} x(t-10) + \phi_{17} x(t-17) + \epsilon(t). \tag{16.9}
$$

The estimated coefficients are $\hat{\phi}_{10} = 0.238$ and $\hat{\phi}_{17} = 0.276$ with estimated standard errors of $\mathrm{se}(\hat{\phi}_{10}) = 0.101$ and $\mathrm{se}(\hat{\phi}_{17}) = 0.104$, respectively. The r^2 for this model is 0.10 (only 10% of variance is explained by this model).

Table 16.3: Time-series model of the third reconstructed component. Values are estimated coefficients, estimated standard errors, and t ratios for the model coefficients.. The model is based on a detrended record of the annual hurricane frequency over the period 1901–1991. The proportion of the total variation explained by the model (r^2) is 0.83.

Coefficient	Estimate	Std. Error	t-Ratio
ϕ_1	1.023	0.104	9.85
ϕ_2	-1.099	0.148	-7.43
ϕ_3	0.425	0.173	2.46
ϕ_4	-0.836	0.165	-5.07
ϕ_5	0.578	0.163	3.55
ϕ_6	-0.834	0.167	-4.99
ϕ_7	0.367	0.152	2.41
ϕ_8	-0.273	0.112	-2.44

Based on the r^2 values, it is clear that each of the three dominant reconstructed components are well fitted by an ARMA(p, q) model, but not the detrended hurricane series itself. This suggests that without prefiltering classical time-series analysis is of limited use in forecasting future values of the hurricane frequency record. The fitted models for the three components are used to forecast (hindcast) the last 5 years (1992–1996) in the record. The 5-year detrended observed and forecast values of the three dominant components are listed in Table 16.4 along with lower and upper bounds of the 95% confidence intervals of the real values. As seen in the table all values are within their respective 95% confidence intervals. Except for a few extreme cases such as the second component for 1994 and the first component for 1995, the forecast values are quite close to their corresponding actual values (detrended). The third dominant component appears to be more accurately predicted by its fitted model than other two components. This is partially due to the smaller variability of this series.

16.2 The Iterative Approach

The iterative approach is straightforward. To obtain a prediction for the next (future) value of the time series an initial guess $\widehat{y}_0$ is added to the observed time series. The lagged-correlation matrix is formed from the augmented time series having $N_t + 1$ values and a SVD of the matrix is performed. A reconstruction of the augmented time series is then made by discarding noise components and an updated prediction of the future value is made by letting

$$\widehat{y}_{(1)} = y'(N_t + 1) = \sum_{k=1}^{6} a_N^k e_m^k, \tag{16.10}$$

where a_N^k is the Nth value (in this case $N = N_t - m + 2$) of the kth principal component and e_m^k is the mth component of the kth eigenvector of the augmented trajectory

Table 16.4: Hindcasts of the three reconstructed components. The components are obtained from the detrended annual hurricane record. Values include normalized detrended frequencies, hindcasts, as well as lower and upper bounds on the 95% confidence intervals on the hindcasts.

Year	Normalized	Hindcast	Lower Bound	Upper Bound
		First Dominant Component		
1992	−1.156	−0.835	−1.362	−0.308
1993	−0.480	−0.457	−1.020	0.105
1994	0.353	0.138	−0.457	0.733
1995	1.739	1.102	0.381	1.823
1996	0.565	0.384	−0.350	1.119
		Second Dominant Component		
1992	0.243	0.232	−0.183	0.646
1993	−0.053	−0.208	−0.700	0.284
1994	−0.501	−0.101	−0.619	0.416
1995	0.323	0.110	−0.440	0.660
1996	−0.178	−0.254	−0.855	0.348
		Third Dominant Component		
1992	−0.236	−0.288	−0.404	−0.172
1993	−0.212	−0.297	−0.464	−0.130
1994	−0.136	−0.166	−0.333	0.001
1995	−0.028	−0.034	−0.223	0.154
1996	0.051	0.112	−0.117	0.341

matrix.[1] The updated prediction $\widehat{y}_{(1)}$ is then used to replace $\widehat{y}_{(0)}$ in the augmented record and the process is iterated until

$$|\widehat{y}_{(q)} - \widehat{y}_{(q-1)}| < \eta \quad \text{for} \quad q = 1, 2, \ldots \tag{16.11}$$

The convergence tolerance η is set to an arbitrary small value. Upon convergence $\widehat{y}_{(q)}$ is accepted as the predicted value for $y(N_t + 1)$.

Both the ARMA and iterative approaches produce forecasts at extended lead times. To obtain an accurate estimate of the algorithm's ability to forecast the future, the procedures are implemented on a subset of the time series where hindcasts are made on known "future" values. For example, let $N_t = 100$ years, but use only the first $N_T = 75$ years to build the algorithm making predictions on the 25 remaining years. Forecast values are collected from predictions made on the record from time $l = N_T + 1, N_T + 2, \ldots, N_t$. A new model is estimated based on data from $i = 2$ to $N_T + 1$. This model is used in predicting the record from time $l = N_T + 2, N_T + 3, \ldots, N_t$. Continuing this until the end of the record provides a sample distribution of forecasts for each lead time. The sample sizes range between $N_t - N_T$ for a one-year forecasts to one for forecasts made with a lead time of $N_t - N_T$. It is important that none of the withheld values are used in implementing the hindcast strategy. This no "look-ahead"

---

[1]The trajectory matrix is formed by successively lagging the time series.

Table 16.5: Predictions of the number of North Atlantic hurricanes. Forecasts for the 5 years are made using the ARMA procedure (Algorithm 1) and the iterative method (Algorithm 2). Error bounds are based on the ARMA models.

Year	Algorithm 1	Algorithm 2	95% Error Bound
1997	4.4	4.7	± 3.1
1998	2.8	4.9	± 3.3
1999	3.4	5.3	± 3.3
2000	6.7	7.4	± 3.4
2001	6.4	5.5	± 3.5

policy on hindcasts is similar to the cross-validation strategy outlined in Chapter 14. The above approaches represent potential strategies for building multi-years models of hurricane activity. Caution is urged, however. There is significant vacillation in the variance of the dominant components. These variations are associated with fluctuations in the intensity of the quasi-regular oscillations, which limits the chance of significant forecast success with time-series models of hurricane activity. More work is needed in this area including extensive model testing and validation.

16.3 An Experimental Extended Range Forecast

An experimental forecast is made using the above algorithms. The forecast was issued in June of 1997, before the start of the 1997 hurricane season. The forecast predicts the frequency of hurricanes over the North Atlantic basin in each year for five years starting with 1997 (Table 16.5). Forecast values from both alogorithms are listed as a function of the future year. Error bounds are computed from the ARMA modeling approach. In general, the forecasts call for near or below average activity through 1999 (the average number of North Atlantic hurricanes over the period 1886–1996 is five) followed by a return to above normal hurricane activity for the year 2000. Both algorithms indicate below average activity for 1997. The forecasts for 1998 and 1999 are more uncertain as the two algorithms suggest different levels of activity. Both approaches agree in predicting above average activity during 2000 and 2001. As expected, the error grows with increasing lead time.

Assuming North Atlantic hurricanes follow a Poisson process, the forecast probability distributions for 1998 and 2000 are compared to climatology in Figure 16.1 using an average of the two forecast algorithms for the Poisson parameter. Neglecting the errors associated with forecasting the Poisson parameter, the graph indicates the prediction of above-normal activity for 2000 is more uncertain in terms of the exact number of hurricanes. These models are experimental and updates are likely. For instance, it is possible to extend the methodology to include other independent predictors like the QBO or possibly solar activity. Historical records of SST extend back to the second

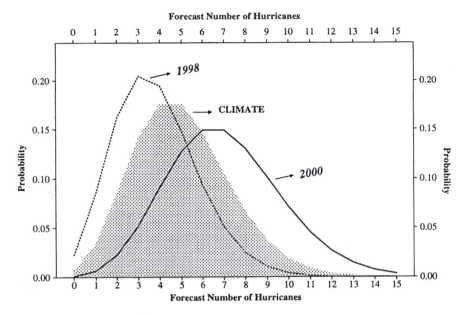

Fig. 16.1: Forecast probability distributions for the number of North Atlantic hurricanes. The distributions are based on the assumption that the seasonal number of North Atlantic hurricanes follows a Poisson random process. The Poisson intensity is the average forecast from the two algorithms for the years 1998 and 2000. A climatological distribution based on an average frequency over the period 1886–1996 is also presented.

half of the 19th century. Provisional records of winds in the stratosphere are possible by examining there relationship with sea-level pressures. Space-time modeling of North Atlantic hurricane activity represents an area of future research. In particular, the utility of methods such as multichannel SSA and neural networks needs to be explored. Improvements in multi-season statistical forecasts are likely forthcoming.

16.4   Prospects For Decadal Outlooks

Time-series models can be used as a guide to predicting hurricane activity a few seasons in advance. The potential for longer lead outlooks depends on a better understanding of the climate system. Of course a better understanding of climate would also likely improve the skill of the shorter lead models. The suggestion has been made that the key to understanding decadal scale variability in North Atlantic hurricane activity depends on knowledge of low-frequency variations in the major ocean circulation systems (Gray et al. 1997). Since warm SSTs are a necessary ingredient for tropical cyclone formation and development the suggestion appears reasonable. On average a significant net northward transport of heat occurs in the ocean surface layer over the Atlantic basin. The North Atlantic gulf stream current provides some of this transport. Estimates put the percentage of the Atlantic northward heat transport at between 15 and 30% of the total hemispheric transport at a latitude of 40°N. Connected to this northward heat transfer is a net water loss at the ocean surface to evaporation. Evaporation leaves behind relatively salty surface waters at high latitudes. High salinity combined with a cold ocean surface creates dense water that sinks toward the ocean floor. Subsidence of the cold water helps to maintain the net northward transport of surface waters thereby keeping a constant supply of salty waters at high latitudes. The cold water subsidence is called *deep water formation*. The complete cycle is called the *Atlantic conveyor belt*.[2]

Interruption of the average flow occurs rather abruptly if the surface waters become less dense through an influx of fresh water. A reduction of the deep water formation will, in turn, slow the northward heat transport of the ocean surface on decadal time scales. This is suggested in Bjerknes (1964) and supported by others including Broecker (1991) and Weaver et al. (1994). Fresh water influxes (or "freshenings") occur through substantial melting of the North Atlantic ice pack (Aagaard and Carmack 1989) and through increases in precipitation (Walsh and Chapman 1990). The discharge of fresh water from melting icebergs are recorded in *Heinrich layers* (or Heinrich events) observed in ice cores taken from Greenland. The Heinrich layers mark both the transition from glacial to interglacial periods and the transitions from interglacial to glacial periods (Broecker 1994).

As mentioned in Chapter 10, a pronounced freshening event appears to have occurred during the late 1960s, and which has persisted for the past several decades (Lazier 1980, Gray et al. 1997). This so-called *Great Salinity Anomaly (GSA)* has been linked qualitatively to cooling of the Northern Hemisphere SSTs and a warming of the SSTs in much of the Southern Hemisphere (Street-Perrott and Perrott 1990). It is also coincident with an increase in baroclinically-enhanced hurricane activity over the North Atlantic basin (see Chapter 6) and to a reduction in the occurrence of intense hurricanes over the basin (see Chapter 7). Figure 16.2 shows some of the atmospheric and oceanic linkages to changes in the Atlantic conveyor belt circulation. Weaker conveyor circulation is associated with cold SSTs over the North Atlantic and more frequent warm ENSO events over the Pacific. In contrast, a stronger circulation is coincident with a warm Atlantic and fewer El Niños. It is possible that a similar salinity anomaly with a corresponding weaker conveyor belt circulation occurred during the first 15 years or so of the 20th century (Dickson et al. 1990, Kushnir 1994, Hurrell 1995).

It is conceivable that the detention of the Atlantic conveyor belt leads to an Atlantic SST dipole with cooler temperatures north of the equator and warmer temperatures to the south, which may indirectly lead to fewer tropical-only hurricanes. Looking at the numbers, the average annual frequency of tropical-only hurricanes during the two negative salinity epochs (1900–1915 and 1970–1994) is 2.0 (81/41) compared to 4.0 (277/70) during the "normal" years. In contrast baroclinic activity is somewhat enhanced during the salinity anomalies, perhaps as a compensation to the lack of tropical-only storms. The compensation in hurricane activity vis-á-vis tropical-only

---

[2]The conveyor belt represents normal circulation conditions over the region.

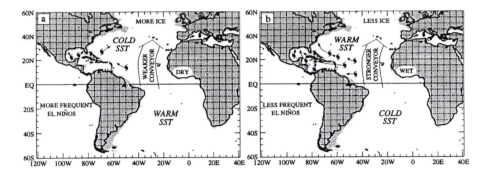

Fig. 16.2: Schematic of atmosphere and ocean linkages to the Atlantic conveyor belt circulation. "More ice" at high latitudes refers to an abundance of sea ice over the Davis Strait. Adapted from Gray et al. (1997).

and baroclinically-enhanced storms suggests a weak limit to the maximum potential frequency over the North Atlantic on time scales ranging from several years to decades. A better understanding of the linkage between North Atlantic basin hurricane activity and low frequency changes in Atlantic SSTs on multi-decadal time scales will allow extended-range tropical cyclone outlooks.

On still longer time scales, ocean anomalies are important for global climate since they play a fundamental role in the global thermohaline circulation. The dense North Atlantic surface water reaches the ocean floor before moving southward and exiting the basin as deep Southern Hemisphere circumpolar currents. Low-frequency changes in the global thermohaline circulation can seriously disrupt global climate patterns. In fact, paleoclimate data suggests that massive changes might have occurred in association with the retreat of the glaciers on the multi-century time scale (Broecker 1991). Under the assumption that SSTs are the limiting factor in hurricane generation, reconstructions of tropical SSTs from the Mesozoic era[3] through the Holocene era[4] indicate the potential for substantial variations in hurricane activity (Barron 1989). It is possible that during some glacial periods, SSTs were below the minimum threshold for hurricane formation. On the other hand, during the middle Cretaceous,[5] when SSTs were warm, it is possible that hurricanes were more frequent. New discoveries in paleotempestology (see Chapter 3) will help to piece together a better understanding of the linkages between climate and hurricanes.

---

[3] The Mesozoic era (65–225 million years before present) was characterized by the appearance of grasses, flowering plants, birds, and dinosaurs.

[4] The Holocene era followed the Mesozoic to the present time.

[5] The Cretaceous is the last period of the Mesozoic era.

17

People at Risk

A hurricane results in a disaster to the extent of its interaction with humanity. Consequently, the cause of the hurricane hazard is rooted in social conditions, actions, and policies. The number of residents living along the U.S. Gulf and Atlantic coasts is increasing. Population increases are noted in Mexico and over much of the Caribbean. Large numbers of people living in small areas create significant crowding. Population densities are particularly high in Puerto Rico, with about 430 people per square km. Saint Lucia in the Windward Islands has a population density of 258 persons per square km. Crowding is also significant in Haiti and Jamaica; both with about 238 people per square km. El Salvador is the most densely populated of the Central American countries. Population density stresses the environment. The need to build houses and clear farmland creates a landscape more prone to devastation from wind and floods. The problem is most acute in the less affluent nations.

In the United States the most dramatic rises in coastal population have occurred in the south from Texas to Florida northward to North Carolina. Increases in the number of weekend vacationers are noted all along the coastline, including New England (Sheets 1990). Many vacationers opt for the stretch of land along the immediate shoreline or the barrier islands. As we have noted, these areas are subject to inundation from the storm surge. Population increases also bring dramatic rises in property values to coastal communities of the United States. These changes, coming at a time of reduced frequency of major hurricane landfalls (see Chapter 7), lead to shifts in the societal vulnerability to hurricanes. Economic losses from hurricanes have risen dramatically during the past several decades. Deaths from hurricanes in the United States have dropped as substantially as economic losses have risen. The potential for large loss of life still exists, particularly for less developed nations.

The application of science and technology leads to improvements in the observations, analysis, and forecasting of hurricanes. Improved forecasts save lives and property. The skill level of the Tropical Prediction Center's NHC at forecasting 24 and 36 hr tropical cyclone intensity and position changes has improved over the last 30 years. Greatest improvements are noted in track forecasts. Twenty-four hour track errors are about 150 km. Improvements in recent years have averaged about 1% annually (Pielke et al. 1997). These improved forecasts allow for better preparedness and more precise evacuation statements. Hurricane hazard mitigation involves recognizing and adapting to the associated natural forces caused by hurricanes. Mitigation is defined as any sustained action taken to reduce and eliminate longterm risk to human life and property. Mitigation is often the most cost-effective, and environmentally sound, approach to reducing losses. The protection of life is the highest priority of the hurricane forecast and warning process. Preparedness is knowing the precise effects of storm surge in a community. Getting people out of harm's way requires adequate time and organizational effort. The time needed for evacuations determines the warning lead time. In places were evacuations take awhile, the potential for overwarning is greatest. Several of the more congested areas, like Galveston Island and the Florida Keys may require as much as 36 hr for complete evacuation.

An early attempt at hurricane-hazard mitigation was through a program of cloud seeding. The purpose of cloud seeding was to reduce a hurricane's intensity and make it less destructive. The first systematic attempt by the United States of this kind occurred in 1947 under the title *Project Cirrus*. The October hurricane of 1947 was seeded after it crossed Florida and was heading over the open waters of the North Atlantic. The initial program was followed by *Project Storm Fury* in 1962. Full-scale hurricane-modification experiments were conducted on hurricanes *Esther* in 1961, *Beulah* in 1963, *Debby* in 1969, and *Ginger* in 1971. Results from the experiments were mixed and controversial. It is difficult to detect small changes in hurricane intensity against a background of large natural variations. Modern efforts of hurricane-hazard mitigation focus on early warnings and adequate preparations rather than on storm modification. This chapter surveys the deadliest North Atlantic hurricanes in the historical record. We look at the deadliest storms over the basin and the deadliest storms to affect the United States. We also examine population and population changes along the hurricane-prone U.S. coastline. Projected population changes over the Caribbean are also a cause for concern. The rising tide of people in areas prone to hurricane strikes will increase the potential for disasters regardless of changes in climate.

17.1   Deadliest North Atlantic Hurricanes

Hurricane destruction along the coastline results from wind, rainfall, tornadoes, and storm surge. Of these, storm surge is responsible for 90% of the deaths. Most deaths are due to drowning. Table 17.1 lists the ten deadliest hurricanes in the historical record for the North Atlantic basin. The number of deaths is approximate and different reports are sometimes conflicting. Reports from the 1900 Galveston, TX hurricane range from 6,000 to 12,000 and reports for hurricane *Fifi* range between 8,000 and 10,000. The 1780 hurricane that devastated Martinique and other islands in the Caribbean resulted in an estimated 22,000 deaths. This storm tops the list of deadliest North Atlantic hurricanes. Estimates indicate approximately 9,000 deaths occurred in Martinique alone,

Table 17.1: Deadliest hurricanes of the North Atlantic. The year, begin and end (month/day) of each hurricane along with the location of most deaths are provided. The number of deaths is approximate. The table does not include hurricane *Mitch* in 1998, which resulted in at least 9,000 deaths in Central America.

Year	Begin	End	Deaths	Location Name
1780	10/10	10/16	22,000	Martinique & Caribbean
1900	9/8	.	10,000	Galveston, TX
1974	9/14	9/19	9,000	Honduras (Hurricane *Fifi*)
1930	9/1	9/6	8,000	Dominican Republic
1963	9/30	10/8	8,000	Haiti & Cuba (Hurricane *Flora*)
1776	9/6	.	6,000	Pointe-a-Pitre Bay, Guadeloupe
1775	9/9	9/12	4,000	Newfoundland Banks
1899	8/8	8/19	3,433	Puerto Rico & Carolinas
1928	9/12	9/17	3,411	Puerto Rico, FL, & Caribbean
1932	11/4	11/10	3,107	Cuba, Jamaica, & Cayman Isles

with another 4,000 to 5,000 in St. Eustatius and at least 4,000 in Barbados. Most of the deaths were island natives; however, the storm ravaged dozens of European settlements and killed thousands at sea. Barnes (1998) notes that the storm may have helped the Americans in their victory over England by curtailing the powerful British fleet. The island nations of the Caribbean have been particularly hard hit by loss of life from hurricanes. Tropical-storm-generated heavy rainfall is often the biggest threat to the nations of Central America. Strong winds dissipate on the coast, but orographically-enhanced precipitation can be disastrous. Destruction in Honduras from hurricane *Fifi* was exacerbated by the denuded hillsides that resulted in devastating mudslides under copious rainfall.[1] Of the ten deadliest, seven were felt in one part of the Caribbean or another. This fact is partially due to their location in a region commonly visited by intense tropical cyclones, and partially due to poorer social and economic conditions. Puerto Rico was particularly hard hit in 1899 and again in 1928. Seven of the ten deadliest hurricanes occurred in September and one each in August, October, and November.

The total number of deaths from all North Atlantic tropical cyclones over the period 1492–1996 is estimated at somewhere between 300 and 500 thousand (Rappaport and Fernández-Partagás 1995). As devastating as North Atlantic hurricanes can be, total deaths from Atlantic hurricanes represent only a small fraction of the deaths recorded from tropical cyclones in the Bay of Bengal region. The nation of Bangladesh along the north coast of the Bay of Bengal, has 119 million people and the highest average population density of any country in the world. Most of the country sits in the flood plain of the Brahmaputra, Ganges, and Meghina rivers, where flooding is common and

---

[1] Hurricane *Mitch* in late October 1998 devastated Honduras, Nicaragua, and other nations of Central America with massive flooding. Early estimates put the total number of dead at more than 9,000. In the final analysis, *Mitch* will go down as one of the worst natural disasters of all time in the Caribbean.

Table 17.2: Deadliest U.S. hurricanes. The rank, year, location, category, type, and number of deaths are provided for each hurricane. Note that the Florida Keys hurricane in 1919 and the New England hurricane in 1938 rank third in total number of deaths.

Rank	Year	Cat.	Type	Deaths	Location Name
1	1900	4	TO	10,000	Galveston (TX)
2	1928	4	TO	1,836	$FLse$
3	1919	4	TO	600	Keys (FL) & TX
3	1938	3	TO	600	New England
5	1935	5	BE	400	Keys (FL)
6	1957	4	TO	390	TX, LA
6	1944	3	TO	390	Northeast U.S.
8	1909	4	TO	350	Grand Isle (LA)
9	1915	4	TO	275	New Orleans (LA)
9	1915	4	TO	275	Galveston (TX)

often disastrous. Over the period 1963–1991 Bangladesh suffered seven major tropical cyclone disasters each resulting in at least 10,000 deaths (Zebrowski 1997). In particular, in late April and early May 1991 a single tropical cyclone killed 131,000 people, mostly due to catastrophic flooding. The storm exacted a significant setback to social and economic development in one of the world's poorest countries. When losses are expressed as a percentage of the Gross National Product (GNP), climate hazards such as floods, tropical cyclones, and droughts disproportionately impact poorer countries by a factor of 20 or more (Kates 1980).

The Galveston hurricane of September 8, 1900 represents the second deadliest hurricane of the North Atlantic basin in modern times. It represents the worst loss of life in the United States from any hurricane (see Table 17.2). In fact it ranks as the worst natural disaster in United States history. The Galveston hurricane was likely a category four on the Saffir/Simpson scale when it came onshore. Of the 37,789 residents of the island city of Galveston, an estimated 8,000 drowned in the hurricane storm surge. The second deadliest year was 1893 when 5 hurricanes made landfall in the United States. Of the two deadliest 1893 hurricanes, one struck Savannah on August 27, the other New Orleans on October 2.

Along with Texas, Florida seems particularly vulnerable to deadly hurricanes. Florida was affected by three of the five deadliest U.S. hurricanes. The worst occurred in September 1928 two years after the great Miami hurricane. After raking the Leeward Islands and Puerto Rico (*San Filipe II*) as a category five hurricane, it took aim on southeast Florida making landfall near West Palm Beach. As the hurricane tracked across southern Florida, 130 kt winds blew the water out of Lake Okeechobee. The community of Belle Glade to the south of the lake was particularly hard hit with destruction and loss of life. The other two deadly Florida hurricanes affected the Keys.

The great hurricane of 1919 passed through the Straits of Florida before devastating Key West. It continued westward tracking over the Dry Tortugas[2] before smashing into southern Texas. The storm intensified over the Gulf of Mexico. It ranks as one of the most intense U.S. hurricanes on record. Another deadly Florida hurricane hit Miami in September of 1926. More than 250 persons were drowned near Clewiston and more than 130 at Moore Haven. Statewide property damage from the hurricane was estimated at between $27 and $37 million dollars (Marth and Marth 1998). The worst hurricane disaster in the northeast occurred in September, 1938. The storm surge brought water to a depth of 3 m into downtown Providence, and more than 600 people were killed on Long Island and in other parts of New England. The disaster was caused by a strong tropical-only hurricane that swung northward from the western North Atlantic. The 1938 New England hurricane ranks in the top ten of the costliest hurricanes on record in the United States (Chapter 18). Like many of the most intense North Atlantic hurricanes, the deadliest storms tend to be tropical-only hurricanes. In fact, only the Labor Day storm of 1935—a category five storm at landfall over the Florida Keys—was not a tropical-only hurricane. Of the ten deadliest U.S. hurricanes, only one (*Audrey* of 1957) occurred during the second half of the 20th century.

Hurricane *Camille* in 1969, which struck Mississippi and Louisiana, is the only other category five hurricane to make landfall in the United States; hurricane *Camille*, with its 7.6 m (25 ft) storm surge, killed 256 people making it the eleven deadliest U.S. hurricane on record. All ten of the deadliest hurricanes were major hurricanes at landfall with all but the two northeastern hurricanes at category four or higher. A category three hurricane moving with considerable forward speed can have sustained winds at landfall of similar magnitude as a slower moving category four hurricane. This was certainly the case for the 1938 New England hurricane as it was traveling at approximately 15 m s^{-1} at landfall. Hurricane *Audrey* in 1957 and the northeast hurricane in 1944 also had significant forward speeds at landfall.

Local, regional, and state governments currently devote significant resources to emergency management related to the hurricane threat in the United States. Florida has the largest number of islands of any state sans Alaska. The number of islands of 10 acres or larger is estimated at 4,510 (Marth and Marth 1998). Residents and tourists on barrier islands in the path of an approaching hurricane are asked to evacuate. Hurricanes visiting the northeast, like the storm of 1944, pose a particular threat to people living on Long Island. The line of evacuation from the island is westward through the Bronx and Manhattan. When a hurricane approaches the island from the south evacuation routes through New York City's traffic bottleneck become clogged. Another bottleneck is the Florida Keys where the multi-bridge Overseas Highway connects the city of Key West from the southern peninsula. Evacuation notices are needed well in advance of the storm. Moreover, in some locations like northwestern Florida, the major escape route runs parallel to the coastline. This direction is not optimal for evacuations, particularly when the hurricane's direction is uncertain. A sudden and unforeseen shift

---

[2]The Dry Tortugas are small uninhabited coral islands 115 km west of Key West.

in the path of an intense hurricane raises the possibility that a storm surge will inundate a traffic-clogged highway threatening many lives. Historically, most hurricane deaths are from drowning. Vertical evacuations into high-rise buildings are a partial solution in areas where roads are inadequate for widespread exodus.

17.2   A Hurricane Problem

The United States and nations of the Caribbean have a hurricane problem. The Atlantic and Caribbean shorelines attract large numbers of people to live and vacation in areas prone to hurricane strikes. The numbers of residents are increasing and so is the crowding. As shown in Chapter 7, major hurricane strikes in the United States have been less frequent during the 1970s into the early 1990s, compared with the 1940s and 1950s. With large increases in coastal population, significant losses to life and property are serious threats. In the United States, approximately 45 million people live in coastal communities from Brownsville, Texas to Eastport, Maine ($\approx$16% of total U.S. population) with another nearly 3.8 million living in Puerto Rico. Moreover, the population of the hurricane-prone U.S. Virgin Islands is about 110,000, with most living on St. Croix and St. Thomas. Nations of the Greater Antilles are also quite crowded. Nearly 11 million people line in Cuba, over 8 million in the Dominican Republic, and 6.6 million in Haiti. In Central America, Honduras has a population of 5.8 million and Nicaragua a population of 4.4 million. The problem is compounded by the fact that the tourist season coincides with the time of greatest hurricane threats, particularly along the southeast coast of the United States. The additional people (larger by perhaps as many as 10 times) compounds problems associated with evacuation. Because infrastructure like new hurricane routes has not kept pace with population growth, it takes much longer to evacuate a threatened area then it did a decade or so ago (Sheets 1990). Increases in population have far exceeded improvements in forecasts of a hurricane's track and intensity changes. Moreover, a large majority of new residents to the U.S. coastline have never experienced the direct impact of a major hurricane. Social and economic factors play a role in the human vulnerability to hurricanes. In 1993, an estimated 17.8% of Floridians were living in official poverty according to the U.S. Bureau of the Census. This represents an increase from 15.7% in 1992 (Marth and Marth 1998).[3] A scenario featuring mass casualties from a hurricane continues to be a real threat along much of the U.S. coastline.

Residents in low lying coastal zones are at greatest risk to storm surge. However, the threat of hurricane fatalities and damage extends well beyond the immediate coastal counties. Flooding rains are the most significant concern for people living farther inland. For instance, hurricane *Camille* in 1969 killed 109 people in Virginia after making landfall in Louisiana. Most of the deaths were due to flooding. Floods caused by the remnants of hurricane *Agnes* in 1972 were also responsible for deaths and destruction throughout much of Pennsylvania after the storm made a rather unremarkable land-

---

[3]Poverty in 1993 was defined as an income of $14,763 for a family of four.

fall in northwestern Florida. Due to the mountainous terrain, tropical-cyclone induced flooding is often the biggest threat to cities and towns in Central America. Significant wind effects are sometimes felt a considerable distance inland, however. Hurricane *Hugo* in 1989 battered Charlotte, North Carolina (290 km inland) with wind gusts near 90 kt, knocking down trees and powerlines and creating significant social and economic problems. Hurricane-spawned tornadoes can produce locally heavy damage in areas far removed from the coast. All 11 of Louisiana's coastal counties as well as 29 inland counties were declared federal disaster areas following the visit of hurricane *Andrew* in 1992. Interestingly, the threat from lightning is relatively minor during hurricanes. Cloud-to-ground strokes, which are a ubiquitous during summer thunderstorms, are not as common in hurricanes. The lack of lightning is related to the relatively weak updrafts in the imbedded thunderstorms. Lightning is more likely to occur in the outer rainbands (feeder bands) than in the inner core.

Individual human response to a hurricane threat depends upon perception and personal evaluation as to what a warning means. Response is a function of an individual's assessment of their own experience and judgment relative to the warnings given (Simpson and Riehl 1980). Response also depends on the credibility of the warnings. The NHC routinely uses probability of landfall forecasts to warn coastal residents and businesses of hurricane threats. Unfortunately the Saffir/Simpson scale, which is a good indicator of a hurricane's destruction potential, does not indicate the potential for loss of life as precisely. A hurricane casualty potential scale would need to consider a multitude of environmental and social factors, including things as population, population density, landuse, evacuation plans, communication services, as well as many others.[4]

A 1983 survey of the people of Pinellas County, Florida (Clearwater and St. Petersburg) indicates that landfall probability statements have an affect on the public's response to evacuate (Baker 1984). People compare the probability that their own area will get hit to the probability that some other location will be affected. If people perceive their threat to be higher than others, then evacuation rates increase. Probability statements that allow people to compare their situation against their neighbors appears to provide a mechanism for response. Statements that do not allow for easy comparison of relative threats are less effective. However, the most important information influencing public response is *local* officials' statements, regardless of whether or not probabilities are used. People respond to specific instructions rather than to what they perceive as vague warnings. Interestingly, a large majority of people who evacuate unnecessarily in one hurricane will still leave in future threats (Baker 1991). Complacency in the face of repeated false alarms does not appear to be significant.

In the long run complacency increases with increasing time between significant events. Areas that have not experienced a major hurricane for some time may have a collective inertia that slows evacuations. Conversely in areas that were recently struck by a major storm, like south Florida, the problem may be that more people than necessary will evacuate next time creating massive traffic jams. Or, people's perceptions can

---

[4]A casualty potential scale could be used as part of the hurricane warning process.

Table 17.3: Hurricane safety tips. The tips are provided by FEMA for residents under a hurricane warning.

	Suggested Advice During a Hurricane Warning
1	Listen for weather updates and stay informed.
2	Keep portable radio and flashlight on hand—with fresh batteries.
3	Clear your yard of loose objects.
4	Move your boat to safe harbor and moor securely. Do not stay with boat.
5	Store drinking water in clean containers.
6	Shutter or board all windows and secure double-door entrances.
7	Plan your evacuation route, know where to go, and fill your car's gas tank.
8	If ordered to evacuate—obey immediately. Take a hurricane evacuation kit (see Table 17.4). Turn off water, gas, and electricity.
9	Do not enter evacuated areas until local officials have issued an all-clear.
10	Evacuate manufactured (mobile) homes for more substantial shelter.

be tainted by the "gambler's fallacy." The fallacy assumes that if a hurricane has occurred in one year it is less likely to occur the next year. This leads to a miscalculation. If hurricane landfalls are independent events (see Chapter 11), then the probability of a strike next year is the same as any year regardless of whether a strike has occurred this year. The occurrence of a hurricane this year does not change the odds of one occurring next year. The miscalculation is worse in regions where the occurrence of hurricanes can not be described as random independent events (e.g., North Carolina). In these cases, the appearance of a hurricane at one time may actually foreshadow the appearance of another one. The probability of getting hit in successive years may be higher than simple randomness dictates and considerably higher than the subjective probability of an individual who is mistaken by the gambler's fallacy.

The Federal Emergency Management Agency (FEMA) lists hurricane safety tips for residents in hurricane-prone areas (Table 17.3). The tips are a source of information for personal preparation in the event of a hurricane assault. The preparations provide a level of safety against loss of life and property. Along with preparation strategies it is advisable to have a readily-accessible hurricane evacuation kit. The kit should include the items listed in Table 17.4. A hurricane watch means hurricane conditions are a real possibility for an area. A hurricane warning means a hurricane is expected within 24 hr and immediate precautionary action is advised. In the case of an emergency, any questions about safety or evacuations should be directed at the local emergency management or civil defense office.

17.3   County Population Changes

The census population is a count of the number of persons residing in an area. The census is taken every 10 years (e.g., 1970, 1980, etc.). Between census years, popula-

Table 17.4: Hurricane evacuation kit. The suggested kit items are provided by FEMA.

Suggested Evacuation Supplies
• First aid kit
• Two-week supply of medicine
• Blankets or sleeping bag
• Extra clothing
• Lightweight folding chairs/cots
• Personal items including books, toys, and snacks
• Infant necessities
• Important papers (valid ID) and money

tion is estimated from demographic components of a change model that incorporates information on natural change (births and deaths) and net migration (net domestic migration and net movement from abroad) occurring in the area since the previous census count. Population and population density are critical components in the assessment of a region's vulnerability to a hurricane disaster. A change in population reflects a change in vulnerability. With more people comes more infrastructure and wealth. Population values presented here are actual census counts for years ending in zero and are model estimates in years between counts.

17.3.1 Growth Rates

Population change is specified in terms of growth. A useful concept in understanding population growth is that of "doubling time." A county with an increasing population will eventually double its number of residents after an amount of time. Consider for example a coastal county with 50,000 residents and an annual growth rate of 4%. Growth rate (G) is considered as

$$G = B - D + I, \qquad (17.1)$$

where B, D, and I are the birth, death, and migration rates in percent, respectively. If the birth rate is 6%, the death rate is 3% and the migration rate (number of people moving in from somewhere else) is 1%, then the growth rate is 4%. A sustained annual growth of 4% leads to a population exceeding 100,000 residents after only 18 years. The doubling time (D) can be estimated as $D = \frac{70}{G}$ where G is the annual growth rate in percent. Thus it takes less than two decades for a county to go from having relatively few people to one having a modest population, even if the growth rate is less than 5% annually. Doubling times on the order of decades are commensurate with time scales associated with significant fluctuations in hurricane activity. Population is expected to double in several Central American countries during the first half of the 21st century. Many new residents in coastal areas of the United States have never experienced hurricane-force winds, and many fewer have ever experienced major hurricane winds.

17.3.2 Texas and Louisiana

Figures 17.1 through Figure 17.7 show the estimated 1996 population in coastal counties and parishes from Texas to Maine and the percentage increase over the period 1990 through 1996 based on the 1990 census. Data are from the U.S. Bureau of the Census. Counties, parishes, and independent cities are the same as those used in Chapter 11 for estimating return-period statistics. Coastal county population is high from New Jersey to Massachusetts. In general, areas with the fewest number of people have the greatest percentage increases in numbers of new residents. An ominous exception is southeastern Florida where counties have substantial population and sizable increases in population.

Along the Texas coast population is concentrated more heavily in the north (Figure 17.1). Harris County, which includes the city of Houston, has a population exceeding 3.1 million residents. Cameron County (Brownsville) along the Mexican border and Nueces County (Corpus Christi) have populations approaching 315 thousand. Kenedy County, which covers an area nearly the size of Harris County, has less than 450 residents. Orange County (Orange) along the Louisiana border has more than 84 thousand residents. Chambers County, on the eastern shore of Galveston Bay has a population of 22,789. Percentage increases in population over 1990 values are shown on the adjacent map. Aransas (Rockport) and Cameron counties, have population increases exceeding 20%. Ten to twenty percent increases are noted in the populated counties surrounding Houston. Significant increases are also noted in San Patricio County (Sinton) on the north shore of Corpus Christi Bay. Population in Jefferson County (Beaumont and Port Arthur) is increasing. Counties with significant increases in population are often the ones most vulnerable to a hurricane catastrophe as many of the new residents are not aware of the damage hurricanes cause.

Population along coastal Louisiana is concentrated in the east toward New Orleans. Cameron Parish in the west has the fewest residents (less than 9 thousand). Vermilion Parish (Abbeville) has a population of just over 51 thousand. Iberia Parish (New Iberia) has a population of nearly 72 thousand. Farther to the east, Terrebonne Parish has 102 thousand residents and Plaquemines Parish has nearly 26 thousand residents. Both these counties have long coastlines that extend into the Gulf of Mexico. Orleans Parish (New Orleans) has a 1996 population of 476,625. St. Bernard Parish has nearly 67 thousand residents, and Jefferson Parish, which includes the southern New Orleans' suburb of Gretna has a population of 455 thousand. Population decreases are noted in St. Mary, Cameron, St. Bernard, and Orleans parishes. Though the city of New Orleans shows a decrease in population over the period 1990 through 1996, St. Tammany Parish (Covington) along the northern coast of Lake Pontchartrain shows an increase in the number of residents exceeding 20%. The 1996 population estimate for Saint Tammany Parish is 178,483. This area of southeastern Louisiana has seen significant increases in population over the past 5 years. The annual growth rate (1990–1994) for the city of Baton Rouge is 3.6%.

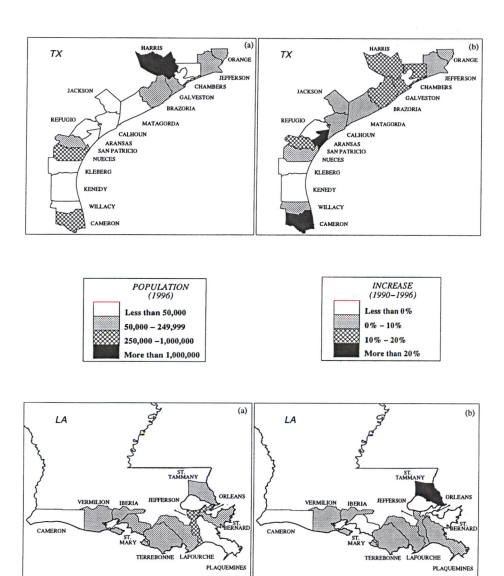

Fig. 17.1: (a) 1996 population estimates for coastal counties in Texas (TX) and parishes in (LA) based on the 1990 U.S. census and (b) the percentage increase over the period 1990–1996 for the same counties and parishes. Darker shading indicates more people or larger increase in population.

17.3.3   Mississippi, Alabama, and Florida

County population in coastal Mississippi and Alabama (Figure 17.2) ranges between 120 and 400 thousand. The exception is Hancock County (Bay Saint Louis) which has a population less than 39 thousand. The 1996 population in Harrison County (Gulfport) is 176,613 and in Jackson County (Pascagoula) is 128,267. Mobile County (Mobile) has a population of nearly 396 thousand residents. Baldwin County (Gulf Shores) has 123 thousand residents. All the counties in this area show increases in population over the period, with Hancock and Baldwin counties having increases exceeding 20%. The population in northwestern Florida is concentrated in the western Panhandle counties and along the counties of the central peninsula. Escambia County (Pensacola) has a population exceeding 277 thousand. Pasco County (New Port Richey) has more than 311 thousand residents. Counties along the Big Bend region from Gulf to Levy have populations less than 50 thousand. Dixie County (Cross City) has the fewest residents (12,352) of the coastal counties in Florida. The fastest-growing counties over this period include Santa Rosa and Walton in the Panhandle and Wakulla in the Big Bend region.[5] Interestingly, Florida's population in the seven westernmost counties grew only half to two-thirds as fast as the entire state in the 1970s and 1980s. But in the 1990s, population growth in this part of Florida surpassed the statewide rate. Santa Rosa and Walton are the fastest growing coastal counties in this part of Florida, with growth rates exceeding 20%.

   In southwestern Florida population ranges from highs of nearly 869 thousand in Pinellas (St. Petersburg) and nearly 898 thousand in Hillsborough (Tampa) counties in the north to a low of less than 81 thousand in Monroe County (Key West) in the south (Figure 17.3). A large percentage of these residents are susceptible to a storm surge. Sarasota County (Sarasota) has a population of nearly 300 thousand. Lee County (Fort Myers) has 380 thousand residents. The largest increases over the period from 1990 through 1996 occur in Charlotte and Collier counties, though population increases are noted in all counties of southwestern Florida. Counties along the east coast of Florida are crowded, particularly the southeast counties of Palm Beach, Broward, and Dade, where the number of residents per county is at or exceeds one million. Dade County (Miami) leads the list with over 2 million residents. Broward County (Fort Lauderdale) is second with nearly 1.5 million people. Flagler County (Bunnell) has the fewest residents (42,142). These counties also show large population increases (>10%) over the period 1990 through 1996. Both Dade and Broward have increases exceeding 20%. Approximately the same number of people living in Dade and Broward counties during 1996 lived in all coastal counties from Texas to Virginia in 1930 (Pielke and Pielke 1997a). Other rapidly growing counties along the Atlantic coast of Florida include St. Johns (St. Augustine) and Flagler in the north. Population in Flagler County increased by nearly 47% from 1990 to 1996.[6]

---

[5]The Big Bend region of northwestern Florida extends from Franklin to Dixie counties.
   [6]Orange County (Orlando) has a population of nearly 759 thousand (12% increase since 1990). Many of these new residents have never experienced a significant hurricane.

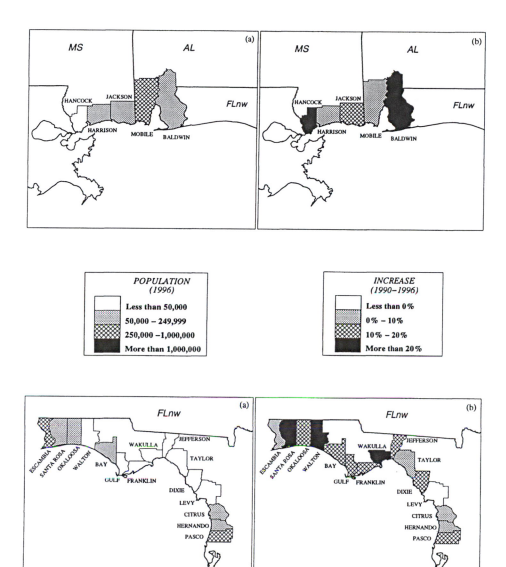

Fig. 17.2: (a) 1996 population estimates for coastal counties in Mississippi (MS), Alabama (AL), and northwest Florida ($FLnw$) based on the 1990 U.S. census and (b) the percentage increase over the period 1990–1996 for the same counties. Darker shading indicates more people or larger increase in population.

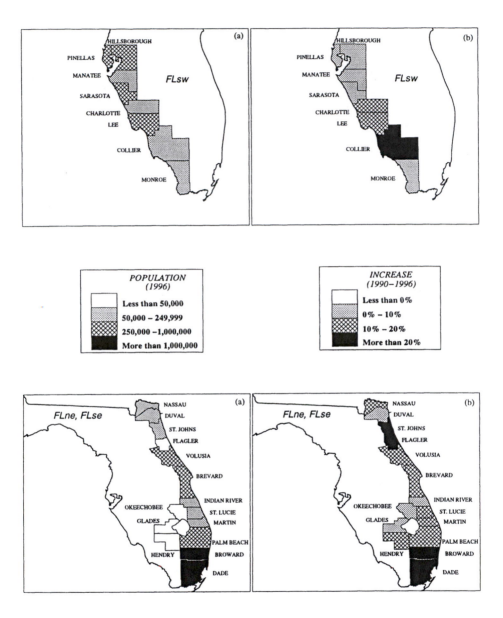

Fig. 17.3: (a) 1996 population estimates for coastal counties in southwest (*FLsw*) and east Florida (*FLse* and *FLne*) based on the 1990 U.S. census and (b) the percentage increase over the period 1990–1996 for the same counties. Darker shading indicates more people or larger increase in population.

17.3.4   Georgia to Maryland

In comparison to Florida, coastal counties from Georgia to North Carolina have relatively moderate populations (Figure 17.4). In Georgia, Chatham County (Savannah) has by far the greatest number of residents. The 1996 population estimate is 226,961; Glynn County (Brunswick) is second with nearly 66 thousand residents. Glynn County includes the resort area of St. Simons Island. Liberty County has 59 thousand residents and Bryan County has just over 22 thousand people. Mcintosh County has the fewest residents (9,592). Charleston County (Charleston) in South Carolina has a population of 277,721, which is the largest population of any county in the region. Horry County (Myrtle Beach) has a population of nearly 164 thousand. Colleton County has the fewest residents of the coastal counties in South Carolina. Beaufort County, which includes the resort area of Hilton Head Island, has a population of almost 103 thousand. Most coastal counties in North Carolina have less than 50 thousand residents. The exceptions include the counties south of the Outer Banks. Onslow County (Jacksonville), surrounding New River Inlet, and New Hanover County (Wilmington), including Wrightsville Beach, both have populations exceeding 143 thousand. Both Tyrrell and Hyde counties have fewer than 6 thousand residents. Dare County has a 1996 population estimate of 26,803. Largest increases in population are found in the counties of Georgia and southern North Carolina. In particular, Camden and Bryan counties in Georgia and Brunswick and Pender counties in North Carolina show increases exceeding 20% over the period 1990 through 1996. Dare County posted a population increase of nearly 18%. The relatively sparsely populated Camden County posted a 22% increase in the number of residents.

The counties and independent cities of Virginia and Maryland along the middle Atlantic coast are more sparsely populated (Figure 17.5). Most of the people are concentrated in the tidewater region along the south shore of the Chesapeake Bay. Independent cites of the region include Hampton, Newport News, Norfolk, Portsmouth, Chesapeake, Suffolk, and Virginia Beach. Virginia Beach has a population exceeding 430 thousand and annual growth rate (1990–1994) of 9.5%. This is short of the 18.8% growth rate for Chesapeake. In contrast, Norfolk had a negative annual growth rate of 7.0% over the same period. Chesapeake has more than 192 thousand residents and Suffolk has just shy of 59 thousand people. York County farther to the north has a population of 55 thousand. Mathews County has a population of less than 9 thousand. Accomack County has a population of 32 thousand and Northampton County has nearly 13 thousand residents. In Maryland, the counties of Baltimore and Anne Arundel (Annapolis) have the largest populations. The city of Baltimore has a population of 675,401. The District of Columbia has a population of 543,213. Most of the areas are showing population increases. York County in Virginia posted a 30% increase in population since 1990, as did Calvert County in Maryland. Population in the city of Baltimore dropped from 736 thousand in 1990. The number of residents living in Worcester County (Salisbury) increased by more than 17%. Annual growth rate in Washington, D.C. is a −6.6%.

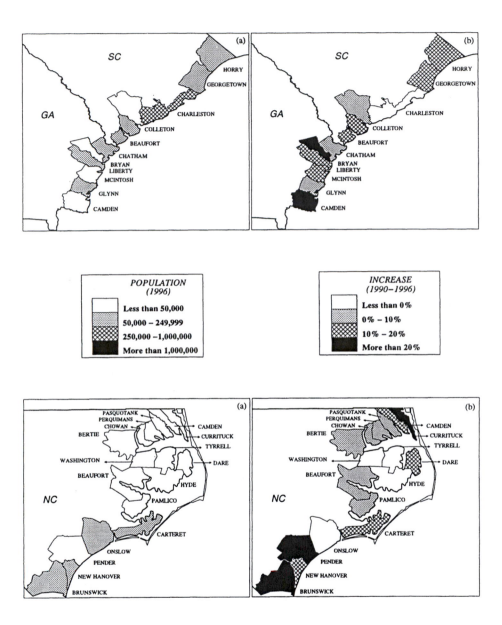

Fig. 17.4: (a) 1996 population estimates for coastal counties in Georgia (*GA*), South Carolina (*SC*), and (*NC*) based on the 1990 U.S. census and (b) the percentage increase over the period 1990–1996 for the same counties. Darker shading indicates more people or larger increase in population.

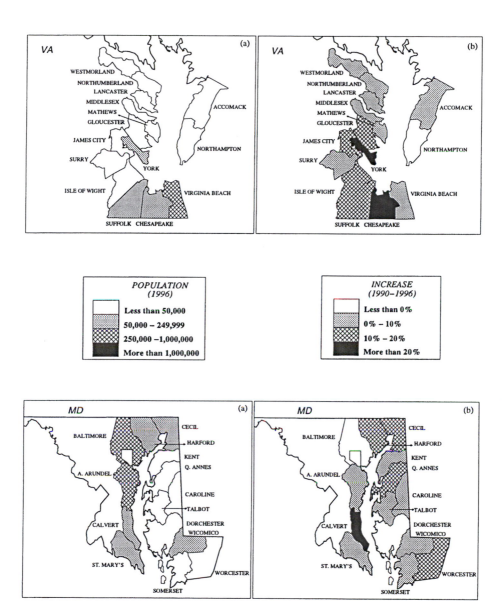

Fig. 17.5: (a) 1996 population estimates for coastal counties in Virginia (VA) and Maryland (MD) based on the 1990 U.S. census and (b) the percentage increase over the period 1990–1996 for the same counties. Darker shading indicates more people or larger increase in population.

17.3.5 Delaware to Maine

Coastal counties in the northeast are the most crowded. Figure 17.6 shows the population statistics for counties in Delaware, New Jersey, New York, and Connecticut.
The population in New Castle County (Wilmington), Delaware exceeds 471 thousand.
Kent and Sussex counties have populations in excess of 120 thousand. The number of
residents in the coastal counties of New Jersey range between a high of 846 thousand
in Bergen (Hackensack) to 67 thousand in Salem (Salem). Cape May County has over
98 thousand residents and Burlington (Mount Holly) has nearly 411 thousand people.
The 1996 population of Ocean County (Toms River) is 474,102 and the population of
Monmouth County (Freehold) is just over 591 thousand. The counties on Long Island
all exceed 1 million residents. Kings County (Brooklyn) tops the list with 2.27 million
people. Suffolk County, which has one of the highest hurricane strike probabilities of
anywhere in the northeast, has nearly 1.36 million residents. The city of New York
itself consists of 5 counties: Bronx, Kings, New York (Manhattan), Queens, and Richmond (Staten Island). Of these, only Staten Island has fewer than a million residents.
Fairfield County (Bridgeport), Connecticut has a population of almost 834 thousand.
This compares to New Haven's (New Haven) population of nearly 795 thousand and
New London's (Norwich) population of almost 251 thousand. Population has generally increased over the area. Largest percentage increases are noted in Kent and Sussex
counties in Delaware. Negative growth rates are noted in Brooklyn and the Bronx.
Decreases in population are noted in Hew Haven and New London counties of Con-
necticut.

Population in coastal New England is more moderate, but the trend is toward more
crowding (Figure 17.7). Providence County (Providence), Rhode Island has a population of nearly 578 thousand. Newport County (Newport) on the coast has a population
of 82,746. Approximately 34% of the county's population lives in the city of Newport.
More than 3.27 million people live in the state of Connecticut and over 990 thousand
more live in the state of Rhode Island. All counties in these two states are vulnerable to a hurricane strike. In Massachusetts, Essex County (Salem) leads the area with
687 thousand residents. Nearly 457 thousand people live in Plymouth County (Plymouth). Another 514 thousand live in Bristol County (Taunton). Nantucket Island has
the fewest number of people (less than 7,300). Dukes has a population of more than
13 thousand. Suffolk County (Boston), on the western shore of the Boston Bay, has
a population of 645 thousand. Farther to the north in New Hampshire, Rockingham
County (Portsmouth) has nearly 263 thousand residents. Cumberland County (Portland), Maine has just over 251 thousand people. Approximately 26% of these live in
the city of Portland. Lincoln County, Maine has less than 32 thousand people. Washington County on the northern coast has more than 36 thousand residents. Washington
is the largest of the coastal counties in Maine. With the exception of the heavily populated counties of Providence and Newport in Connecticut, all the coastal New England
counties show a trend toward more people. Dukes and Nantucket islands show substantial growth.

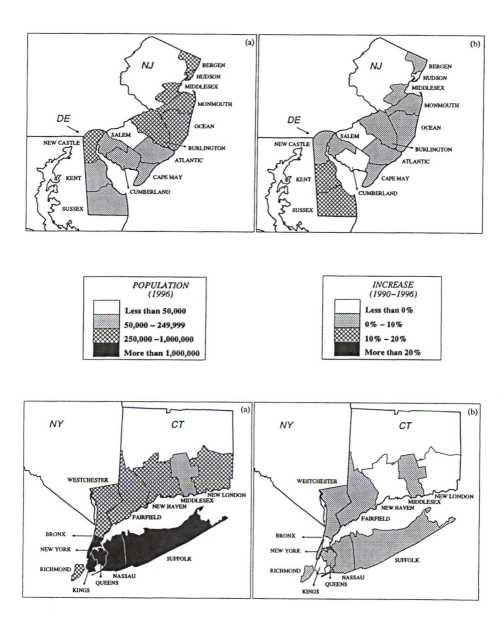

Fig. 17.6: (a) 1996 population estimates for coastal counties in Delaware (*DE*), New Jersey (*NJ*), New York (*NY*), and Connecticut (*CT*) based on the 1990 U.S. census and (b) the percentage increase over the period 1990–1996 for the same counties. Darker shading indicates more people or larger increase in population.

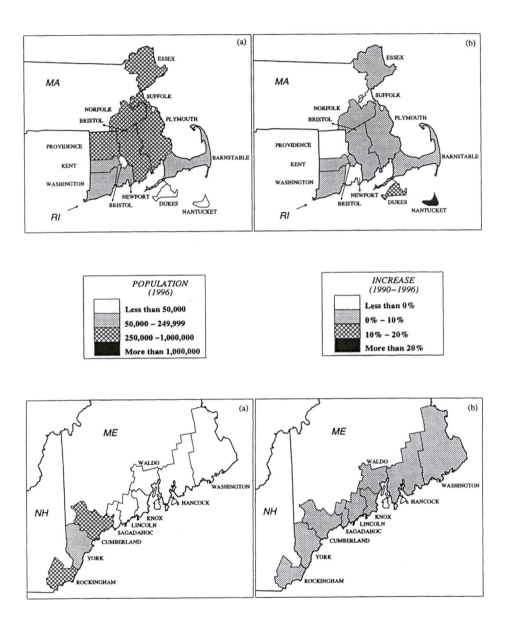

Fig. 17.7: (a) 1996 population estimates for coastal counties in Rhode Island (*RI*), Massachusetts (*MA*), New Hampshire (*NH*), and Maine (*ME*) based on the 1990 U.S. census and (b) the percentage increase over the period 1990–1996 for the same counties. Darker shading indicates more people or larger increase in population.

17.3.6 Population Density

Population changes as people move from one place to another. Other factors include birth and death rates. A major redistribution in population and population density happens as people move from the countryside to the cities and as city dwellers move to the suburbs. Even on the spatial dimension of counties, large gradients in population are found all along the hurricane-prone U.S. coastline. In particular, the Gulf coast from Texas to Florida shows wide variations in the number of residents per county. More broadly, increases in the number of U.S. coastal residents over the period 1950 through 1996 has led to record population densities for areas such as south Florida, and has resulted in many rural areas becoming substantially more congested. In the coastal counties from Texas to North Carolina, for instance, the number of counties with population in excess of 250,000 went from three to nine from 1950 to 1970 and then to eighteen by 1990. Moreover, the number of coastal counties with a population less than 50,000 residents went from 75 in 1950 to 54 in 1970 and to 38 by 1990 (Pielke and Pielke 1997a).

A convenient way to describe the degree of population clustering is through the *Lorenz curve* (see Gastwirth 1970 and Jones and Mars 1974). Given a set of n ordered numbers, the empirical Lorenz curve [$L(p)$] is defined at the points $p = i/n$, for $i = 0, 1, 2, \ldots, n$, where $L(0) = 0$ and

$$L(i/n) = S_i/S_n, \tag{17.2}$$

where

$$S_i = x_1 + x_2 + \cdots + x_i. \tag{17.3}$$

A uniformity in the distribution of some quantity appears as a diagonal line on a cumulative distribution graph. Lorenz curves are used here to compare uniformity of coastal county land area and coastal county population along the U.S. Gulf coast. The empirical Lorenz curve is defined at all points p. Here the set of n ordered numbers are the 1996 population estimates and land areas for the 56 coastal counties along the Gulf coast from Cameron County, Texas to Monroe County, Florida. The counties are outlined in Figure 17.8. Land areas appear fairly uniform across the counties. For land areas the S_i's are the cumulative percentage of land area and for population they are the cumulative percentage of the number of residents. The Lorenz curves for both land area and population are shown in Figure 17.9. The departure in the curve from a diagonal indicates a clustering in the spatial distribution. Clearly, land area per county is more uniformly distributed along the coastline than is population as the land area curve is closer to a diagonal. People prefer one county over another, creating a nonuniformity in population density across the region. On a smaller scale people tend to live in coastal cities rather that uniformly spread throughout the county. A cumulative distribution of population in zip codes within a county would reveal similar Lorenz curves for many of the counties. A highly nonuniform population density makes the problem of hurricane warning more critical. Small shifts in the track of an approaching hurricane could mean a significant increase in the number of residents under a direct threat.

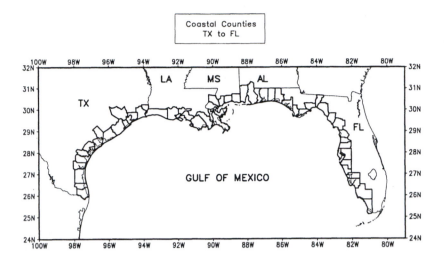

Fig. 17.8: Coastal counties from Texas (TX) to Florida (FL). Cumulative distributions of land area and population over these 56 counties are examined using Lorenz curves.

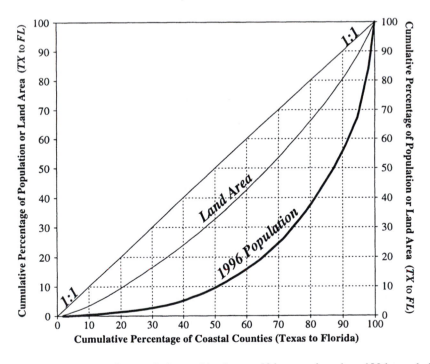

Fig. 17.9: Lorenz curves for population and land area. Values are based on 1996 population and land area estimates in the 56 Gulf coast counties from Texas (TX) to Florida (FL). The diagonal line indicates a curve for a strictly uniform distribution.

17.4   City Population Changes

County population statistics give an indication of the number of residents at risk to a hurricane strike. Yet as shown by the Lorenz curve, people within counties tend to cluster in cities where most of the infrastructure is located. This creates small areas with very dense concentrations of people. Hurricanes tend to generate fairly localized disasters, affecting only those people that have concentrated themselves in the area of hurricane-force winds. Destruction potential of hurricanes is more widespread when accompanied by excessive rainfall and flooding. Concrete streets, highways, buildings, and other impermeable surfaces focus the rainfall in localized regions of the city creating the potential for dangerous flash flooding. Evacuation times for warned communities is inversely proportional to population size and density. Crowded communities require more time for the evacuation process. Population densities are greatest in the major metropolitan areas.

Here population changes for cities along the Gulf and east coasts are examined. Figure 17.10 shows the population for selected cities in 1980, 1990, and 1994. The cities are all hurricane prone to one degree or another. The values for New York City are divided by 3 to fit on the scale. The 1994 estimate population of New York City is 7,333,253. Population growth was substantial from 1980 to 1990. Growth over the period 1990–1994 is 0.1%. Houston also has a substantial population. The 1994 estimate is 1,702,086. Population growth is a significant 4.4% since 1990. Houston is a growing metropolis that has not been hit by a hurricane since *Alicia* in 1983. Jacksonville has a large and increasing population. The 1994 population estimate is 665,070. This represents a 4.7% increase since 1990. Population growth in Jacksonville is juxtaposed with the fact that the area has not seen a major hurricane during the 20th century. Population densities are substantially different among these three populated cities. New York City has a density of 9159 persons per square km. This compares to 1216 and 338 persons per square km in Houston and Jacksonville, respectively.[7]

The southeast Florida cities of West Palm Beach, Fort Lauderdale, and Miami have a combined population exceeding 600 thousand residents. Miami's population was 373,024 in 1994 representing a 4% increase from 1990. Miami, with a population density of 4042 persons per square kilometer, is as crowded as many of the major cities in the northeast. In fact Miami's population density matches that of Newark, New Jersey. The large metropolitan areas of southeast Florida are only part of the picture. Many people in this area are clustered in smaller, but crowded communities like Hollywood, Pompano Beach, and Boca Raton. Another half million of Florida's residents live in the Tampa/St. Petersburg area along the eastern shore of the Gulf of Mexico. Most of Florida's coastal cities are susceptible to inundation from storm surges. The Tampa hurricane of late October 1921 produced a storm tide exceeding 3 meters and caused flooding throughout the waterfront neighborhoods at Palmetto Beach, Edgewater Park, and DeSoto Park (Barnes 1998). The hurricane of September

---

[7]New York is 27 times more crowded than Jacksonville.

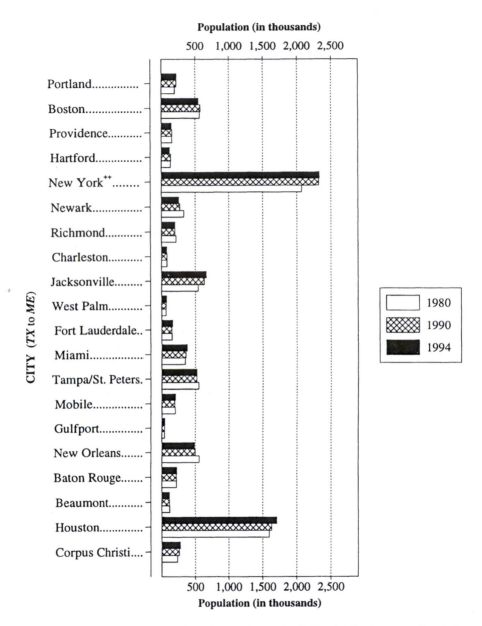

Fig. 17.10: Population in selected cities along and near the Gulf and Atlantic coasts. Population is given for the years 1980, 1990, and 1994 in thousands of residents. Values for New York City must be multiplied by three to get the actual population.

1926 sent a wall of water 4 meters above sea level through Miami's Biscayne Bay. The storm surge inundated parts of the city. Other densely populated areas prone to hurricane strikes include New Orleans and Boston. Most of New Orleans is below sea level. The city is kept dry by a extensive system of levees, canals, and pumps. New Orleans has been losing residents since at least 1980 and Boston since 1990. Boston is more than 4 times as crowded as New Orleans. In contrast, Baton Rouge had a growth rate of 3.6% over the period 1990–1994. Charleston and Richmond had negative growth rates over the period 1980–1994. Negative growth rates are also noted in Newark. Providence showed population increases during the 1980s, but the number of residents has dropped during the early 1990s. Portland on the other hand has posted two consecutive periods of positive growth exceeding 10%.

17.5   Projected Population Changes

Demographic changes cause the hurricane hazard to shift even in places where the hurricane threat remains the same. Coastal counties, especially the area within the counties along the immediate coastline, are becoming more crowded. The coastal counties of the northeast are the most densely populated of any in the United States, at approximately 750 persons per square mile. Estimates suggest this density could reach 830 persons per square mile by the year 2010. This represents a five-fold increase in density over noncoastal counties of the northeast. Largest population increases are projected for Suffolk County, New York and Fairfax County, Virginia (see Table 17.5). These coastal areas all have significant numbers of people. Hurricane evacuation plans are in place in many of the nation's coastal counties. Evacuation times increase with population density and officials estimate that in many locations the time required is close to 24 hr. For instance, much of Panama City's development is along the beach front with limited access to the mainland creating a situation in which complete evacuation even for this modest size city would take more than 22 hr. Experts estimate it would take about 15 hr to move residents of Pinellas County to inland safety.

Coastal population in the southeast United States is most concentrated in eastern Florida. In 1988 80% of the people living in the coastal southeast were located in eastern Florida. Some of the major cities of the southeast coastal region include Miami, Jacksonville, Savannah, and Charleston. Between 1988 and 2010, one-third of the coastal counties of this region are expected to have population increases of between 35 and 75%. Broward County, Florida is expected to lead the southeast coast in total number of new residents. Adjacent Dade and Palm Beach counties are also expected to see significant increases in population. Wait times for hurricane-force winds in these three southeastern Florida counties are less than five years (Chapter 11). The Orlando area of central Florida[8] is expected to continue its rapid growth. Orlando was rocked hard by the October hurricane of 1944 well before the area was transformed into a highly urbanized center.

---

[8]Orlando features the world famous Walt Disney World and many other tourist attractions.

Table 17.5: Projected population increases. Values for coastal counties are in thousands. County seat refers to the seat or courthouse. Data are from U.S. Department of Commerce (1990). Note that Seminole and Orange counties in Florida are inland from the immediate coastline.

Northeast County	County Seat	(1988–2010) Population Increase
Suffolk, NY	Riverhead	225
Fairfax, VA	Fairfax	210
Middlesex, MA	East Cambridge	144
Ocean, NJ	Toms River	124
Queens, NY	Jamaica	118

Southeast County	County Seat	(1988–2010) Population Increase
Broward, FL	Fort Lauderdale	436
Dade, FL	Miami	366
Palm Beach, FL	West Palm Beach	362
Seminole, FL	Sanford	145
Orange, FL	Orlando	131

Gulf County	County Seat	(1988–2010) Population Increase
Harris, TX	Houston	909
Pinellas, FL	Clearwater	192
Pasco, FL	New Port Richey	182
Lee, FL	Fort Myers	161
Hillsborough, FL	Tampa	158

Florida led the nation in new residents during the 1990s and the trend is expected to continue. Florida had 9 of the nation's 11 most rapidly growing metropolitan areas in 1990. Since 1955 most new residents to Florida have come from New York, Ohio, New Jersey, Pennsylvania, Illinois, Michigan, and Georgia (Marth and Marth 1998). It is estimated that by the year 2010, Florida's population will exceed 20,000,000 people. Many of the new residents to Florida live in hurricane-prone stretches of coastline. Approximately 80% of Florida's total population lives in 35 coastal counties, making Florida the leader in number of people at risk to a hurricane disaster. The problem is compounded by Florida's demographics: 18.6% of Florida's residents are 65 years of age or older, which is the highest percentage of any U.S. state. Wealthy retirees are increasingly opting for the newly built assisted-living communities throughout the state. As a comparison, the percentage of elderly residents in Texas is 10.2% and in North Carolina it is 12.5%. Florida ranks sixth in the nation with 92.9% of its residents living in metropolitan areas. This compares to 83.9% for Texas and only 66.6% for North Carolina. A significant 11% of Florida's residents live in mobile homes.

The Gulf coast accounts for 13% of the total coastal population (including California, Alaska, and Hawaii). Population in this region is expected to increase by 22% to

Table 17.6: Projected population changes. Population is a midyear value in thousands for countries of the North Atlantic affected by hurricanes. Data are from the Bureau of the Census, U.S. Department of Commerce.

Country	1997	2010	2050
Antigua and Barbuda	66	80	79
Bahamas	262	314	327
Barbados	258	284	276
Belize	225	356	489
Cape Verde	394	512	545
Colombia	37,418	49,266	55,798
Costa Rica	3,534	5,044	6,321
Cuba	10,999	11,699	10,565
Dominica	83	96	103
Dominican Republic	8,228	11,152	13,459
El Salvador	5,662	7,852	10,814
Grenada	96	141	210
Guatemala	11,558	18,131	20,034
Haiti	6,611	9,328	12,345
Honduras	5,751	9,042	12,528
Jamaica	2,616	3,213	3,712
Mexico	97,563	136,096	170,280
Nicaragua	4,386	6,973	9,714
Panama	2,693	3,619	4,418
Saint Kitts and Nevis	42	57	69
Saint Lucia	160	202	227
St. Vincent and the Grenadines	119	146	163
Trinidad and Tobago	1,273	1,409	1,447
Venezuela	22,396	30,876	37,773

almost 18 million residents by the year 2010. Some of the major cities of this region include Houston, New Orleans, Tampa, and St. Petersburg. Approximately one-third of all Gulf coast counties will increase in population by more than 30% over the next two decades. Harris County, Texas will lead the region in the number of new residents. The population of Harris County is expected to increase by nearly a million people over 1988 levels. Four counties along the western peninsula are also expected to see considerable increases in the number of new residents.

Population will increase across most of the hurricane-prone North Atlantic. Table 17.6 shows the projected population in nations affected by North Atlantic hurricanes. Values are given in thousands of people. Notable is the projected increase in population among the major islands of the West Indies. The Dominican Republic is expected to have more than 10 million residents by 2010 and more than 13 million by 2050. Haiti is expected to pass the 10 million mark before 2020. Significant increases are also anticipated in Central America. The population in Guatemala is expected to exceed 20 million by the middle of the 21st century. Percentage increases will be even larger in El Salvador, Honduras, and Nicaragua. Doubling time for the population in Honduras, based on the anticipated average annual growth rate of 4.4%, is approximately 16 years. An annual growth rate exceeding 4.5% is expected in Nicaragua. Nations of the Lesser Antilles are also expected to see increases in the number of residents during the next 15 to 40 years. Note that only a small percentage of the population of Mexico is prone to hurricane-force winds, although flooding from tropical storms originating over the eastern North Pacific and the North Atlantic can bring widespread destruction to areas of Mexico. Of the ten largest cities in the world, New York City and Mexico City have the potential to be significantly affected by a North Atlantic hurricane. For the period 1990–1995, the annual growth rates were 0.34% in New York City and 1.8% in Mexico City.

18

Property at Risk

Hurricanes kill and severely impact economies. Direct economic losses occur from property damage and from evacuation and preparation efforts. Population increases bring dramatic growth in property values, particularly in the United States. As a result, the economic losses in the United States caused by hurricanes have risen dramatically during the past several decades. In the early 1980s average annual losses approximated $1 billion. This amount is five times the average annual losses during the 1950s (Chen 1995). Despite the recent downturn in hurricane activity in the North Atlantic over the period 1991 through 1994 (see Chapter 12), frequent and damaging extra-tropical cyclone activity[1] combined with two intense U.S. hurricanes (*Andrew* of the North Atlantic and *Iniki*[2] of the eastern North Pacific) caused $40 billion in insured property losses (70% of which was covered by reinsurance companies). The financial losses seriously impacted the property insurance industry with a loss of nine firms to bankruptcy (Changnon et al. 1997). As a consequence, nearly $400 million of the losses went unpaid. The shakeup sent shivers through the insurance industry leading to a better appreciation of the key role climate hazards play in estimating potential losses from storms. Moreover, major hurricane disasters affect all Americans by diverting resources from other necessary public and private programs and by reducing the efficiency of the national economy. Property-loss potential across the West Indies is also on the rise. New resort developments featuring luxury hotels and tourist districts are being built along previously uninhabited beaches. The Caribbean coast of Mexico around Cancun is a spectacular example of new development in the path of future intense hurricanes. This chapter examines some of the economic impacts of U.S. hurricanes. We begin with a look at what causes the damage and factors related to the cost of preparation and evacuation efforts. We also examine the costliest hurricanes on record for the United States.

---

[1] A record snowfall from an extra-tropical storm during March of 1993, dubbed the "Storm of the Century," paralyzed the eastern one-third of the United States.

[2] Hurricane *Iniki* struck Hawaii in September of 1992.

Table 18.1: Saffir/Simpson scale and property damage. Wind speeds are in miles per hour (mph). Length or distance units are in feet (ft), yards (yd), and miles (mi). ASL stands for above sea level. Units are those routinely used for evacuation announcements.

Cat.	Wind Speed in mph	Natural and Manmade Effects from Wind and Sea
1	74–95	No damage to building structures. Minor damage to mobile homes, shrubbery, and trees. Some coastal road flooding and minor pier damage. Some roof, door, and window damage to buildings.
2	96–110	Considerable damage to vegetation, mobile homes, and piers. Coastal and low-lying escape routes flood 2-4 hour before arrival of eye wall. Small craft in open anchorage break moorings. Some structural damage to small residences and utility buildings with a minor amount of curtain-wall failures.
3	111–130	Mobile homes are destroyed. Flooding near the coast destroys smaller structures with larger structures damaged by floating debris. Terrain lower than 5 ft ASL may be flooded inland to 8 mi or more. More extensive curtain-wall failures with some complete roof structure failure on small residences.
4	131–155	Major erosion of beach. Major damage to lower floors of structures near shore. Terrain lower than 10 ft ASL may be flooded, requiring evacuation of residential areas inland to 6 mi. Complete roof failure on many homes and industrial buildings.
5	> 155	Some complete building failures with small utility buildings blown over or away. Major damage to lower floors of all structures located less than 15 ft ASL and within 500 yd of the shoreline. Evacuation of areas on low ground within 5 to 10 mi of shore may be required.

18.1   Hurricane Damage

All hurricanes have the potential to do damage with the threat of substantial damage rising with increasing intensity. However, the way in which storm surge, wind, and other factors conspire at one location determines the actual destructive potential. To aid in comparing hurricanes and to make the anticipated hazards of approaching hurricanes clear to emergency operation centers, the U.S. National Oceanic and Atmospheric Administration (NOAA) makes use of the Saffir/Simpson scale. The Saffir/Simpson scale assigns hurricanes to five categories (see Chapter 2) based on their destructive potential. The five categories and associated criteria for emergency management are reviewed in Table 18.1. The scale is used to provide an estimate of the potential property damage and flooding expected along the coast when a hurricane makes landfall. The scale is widely used by the media and emergency management operations.

The impacts of a hurricane are divided into three groups. The first-order (or direct) impacts are those most closely related to the hurricane (Pielke and Landsea 1998). Property loss from wind stress and highway washout from storm surge are examples of direct hurricane impacts. Direct impacts can themselves cause problems leading to what are known as secondary impacts. Disease from lack of potable water and injuries from cleanup operations are examples of secondary impacts. Still higher order impacts are those that happen after the storm has long passed. Elevated insurance rates is an example of a higher order impact. Costs incurred by hurricanes are typically only tallied for direct impacts. Secondary and higher order effects are more difficult to quantify, and are typically offset by government and industry grants to help the local economy during and after recovery. This has the effect of spreading the costs over a larger region than those areas directly impacted by the hurricane.

18.1.1   Storm Surge

Storm surge is responsible for much of the hurricane damage along the immediate coastal shoreline. Storm surge is the rise of sea water caused by wind and pressure forces of the approaching hurricane.[3] These forces produce ocean currents and swells that radiate from the storm center. Over the ocean these currents produce only a small storm surge because the converging and piling water is compensated by ocean currents at depth, which transport the water away (Rappaport and Fernández-Partagás 1995). However, as the hurricane approaches the coast it first encounters the continental shelf, which inhibits the deep compensating currents so the water rises. In many situations, the maximum storm surge heights measured relative to the average sea level are recorded at the head of bays. Typically the storm surge rises, peaks, and returns to normal within 6 to 12 hours. In a rapidly moving hurricane the storm surge cycle can be considerably faster, on the order of several minutes to a few hours. Waves on top of the high water of the storm surge can cause extensive damage as they crash into fixed shoreline structures.[4] Debris carried by the waves adds to the their battering effect.

Landuse changes can make property more vulnerable to the influence of storm surges. The development of ocean fronts can mean the loss of coral reefs as boat navigation is put at a priority. When beach access is increased, it is often at the expense of forests, mangroves, and sand dunes. These actions create a situation in which areas farther inland from the immediate coast are susceptible to damage from storm surges. Low-lying coastal cropland may take years to recover if inundated with salty ocean water (saltwater intrusion). Florida's Division of Emergency Management estimates that nearly 25% of the population of Florida lives in areas that would be seriously impacted by storm surge from a category three or stronger North Atlantic hurricane. Areas of southwestern Florida from Pinellas County southward are at highest risk as some 90% of the residents live in land that would be submerged by high storm tides (Marth and Marth 1998).

---

[3] See Chapter 2.
[4] The waves may reach to a height of 10 m.

18.1.2 Floods

Although hurricane-associated rainfall may be beneficial to agriculture, as in the case where a drought is broken, there is always the potential for serious flooding. As a general rule-of-thumb take 100 divided by the storm's forward motion in knots to get an estimate of the hurricane's potential maximum rainfall amount in inches. A hurricane moving inland at 20 kt has the potential to produce maximum rainfall amounts in the range of 5 inches (12.7 cm). A storm moving with only half that speed has the potential to produce twice as much rainfall. As the hurricane makes landfall, heaviest rainfall typically occurs to the east of the center of circulation. The quantities of pre-existing atmospheric and ground moisture in the area to the east of the storm center will influence the potential for flooding.

Heavy rainfall accompanying a hurricane presents an immediate stress to small stream and river beds where rapid rises in stream levels normally occur. A precise determination of a region's flood potential as a hurricane approaches depends on many factors including the area's physiographical, hydrological, and geological features. These factors combine in numerous ways with the meteorological and climatological features to make flooding potential quite contingent. However, there are some areas in the United States where river basins are more susceptible to flooding from hurricanes than others. For instance along the Texas Gulf coast, where most of the rainfall associated with the remnants of hurricanes spreads in the direction of the river-basin, the flood potential is greater than along the southeast Atlantic coast, where hurricane rainfall typically spreads across the river basins (Schoner 1968). Storm rainfall moving in the direction of the river basin spends more time over the watershed, thereby increasing the potential for catastrophic flooding. The problem for Texas is intensified by the fact that storm motion is often slower at these latitudes than at latitudes farther north. The "Hill Country Flood" in early August, 1978 from *Amelia* represents a spectacular example of inundation from tropical cyclone rainfall in Texas (see Caracena and Fritsch 1981).

Areas to the right of the hurricane as it makes landfall with respect to an observer standing at sea and looking toward shore often experience the heaviest rainfall, particularly for Gulf coast storms, though there is large variability to this rule depending on local factors. Rainfall amounts generally decrease with distance from the coastline. The decrease in rainfall is often quite abrupt, particularly along the U.S. Atlantic coastline. The exception to this rule of thumb is when the wind circulation of the hurricane reaches the Appalachian Mountains. As the air rides up over the mountains additional lift can produce a secondary maximum in rainfall. This orographically enhanced rainfall is usually quite intense (see Chapter 9). The naming of tropical storms by the NHC allows for less confusion when communicating the potential for destruction. Names of minor hurricanes are reused on a six-year cycle, but to make clear the potential destruction of a future storm, the NHC retires the name of a storm that takes a heavy toll in lives or property. Table 18.2 is a list of the retired hurricane names over the period 1953–1996. Both 1955 and 1995 had a record four hurricane names retired.

Table 18.2: Retired hurricane names. The list is for the period 1953–1996. The year in which the hurricane occurred is also given.

Year	Name(s)	Year	Name(s)
1954	Carol, Hazel, and Edna	1974	Carmen
1955	Janet, Connie, Diane, and Ione	1975	Eloise
1957	Audrey	1977	Anita
1960	Donna	1979	David and Frederic
1961	Carla	1980	Allen
1963	Flora	1983	Alicia
1964	Cleo, Dora, and Hilda	1985	Elena and Gloria
1965	Betsy	1988	Gilbert and Joan
1967	Beaulah	1989	Hugo
1969	Camille	1992	Andrew
1970	Celia	1995	Luis, Marilyn, Opal, and Roxanne
1972	Agnes	1996	Cesar, Fran, and Hortense

Three storm names were retired in 1996. Hurricane *Cesar* caused over 50 deaths and considerable destruction along its path through the southern Caribbean Sea and Central America. Flooding rains from hurricane *Hortense* over Puerto Rico and the Dominican Republic resulted in at least 21 fatalities. Property loss over North Carolina was significant from hurricane *Fran*.

18.2 Hurricane Warnings

Coastal counties are subdivided into zones according to their susceptibility to hurricane damage. Three zones most impacted by hurricane landfalls, in order of severity are barrier islands and beaches, the mainland surge areas exposed to over-wash flooding from the storm surge, and areas along the coast that are far enough inland to avoid the storm surge. Hurricanes cause significant damage to offshore installations such as boats, fisheries, and oil platforms (Burton et al. 1993). Mainland surge areas can be crowded with residents and tourists. The storm surge has historically caused the largest loss of life and extreme property damage. Moreover, as shown in Chapter 17, population densities in the United States are increasing most rapidly in coastal communities. Yet deaths from hurricanes in the United States show a remarkable decline during the 20th century (see Figure 18.1). The decline is largely a result of better hurricane watches and warnings issued by the NHC. Improved communication by emergency broadcast centers and the local and national media, particularly the Weather Channel, contribute to widespread awareness of the threat to safety hurricanes pose. An estimated 700,000 people fled the direct assault of hurricane *Andrew* in 1992. The NHC issues warnings when people are in danger from an approaching hurricane. Forecasters at NHC have several important prediction models available to provide guidance when making their

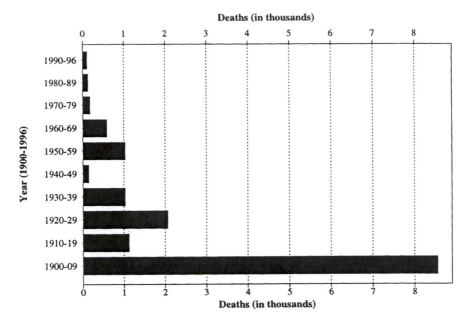

Fig. 18.1: Deaths from North Atlantic hurricanes in the United States. Values are in 10-year intervals, except for 1990–1996. A significant reduction in the number of fatalities associated with hurricanes has occurred over the 20th century. Improved hurricane warnings save lives and reduce property losses.

forecasts of future hurricane track and intensity changes. Table 18.3 lists the suite of prediction models along with a brief description. The listed scientific reference provides details on the workings of the model. Models are classified into three groups that include statistical, dynamical, and a combination of statistical and dynamical. The hybrid models incorporate numerically forecast data into a statistical prediction framework (Meisner 1995). Most models are subjected to refinement and new models are continually being tested and added.

Hurricanes that prompt preparations and evacuations can have substantial economic impact regardless of whether damage actually occurs. This is related to the problem of overwarning when a hurricane threatens. The issuance of hurricane warnings along the coast is usually accomplished 18 to 24 hours prior to the expected arrival of a hurricane. This time interval is an optimized trade-off between the desire to provide maximum warning lead time and the ability to keep the size of the warning area within reasonable limits (Neumann 1975). The most pressing concern is the lives of residents living along the coast. Lead times are determined to a large extent by the time required to get these people out of harm's way. Let the lateral extent of hurricane force winds of a hurricane be denoted l. Depending on storm size, l may vary from less than 100 km for a small hurricane to over 300 km for a large hurricane. The lateral extent (w) of the warning area must necessarily be greater than the hurricane size due to uncertainty

Table 18.3: NHC hurricane prediction models. References provide details on the model equations. GFDL is the Geophysical Fluid Dynamics Laboratory at Princeton University. NCEP is the National Centers for Environmental Prediction. Models are either statistical (S), dynamical (D), or a combination of statistical and dynamical (S-D). Information is from Meisner (1995).

Model Name	Type	Reference	Brief Description
CLIPER	S	Neumann 1972	Combination of climatology and persistence
SHIFOR	S	Jarvinen & Neumann 1979	Statistical intensity forecast model
SLOSH	D	Jarvinen and Neumann 1985	Sea, lake, and overland surge height model
AVN	D	Kanamitsu 1989	NCEP aviation medium range forecast model
NHC90	S-D	Neumann & McAdie 1991	Statistical-dynamical track prediction
BAM	D	Marks 1992	Beta and advection baroclinic dynamical track model
VICBAR	D	DeMaria et al. 1992	Nested barotropic track model
SHIPS	S	DeMaria & Kaplan 1994b	Statistical intensity prediction scheme
GHM	D	Tuleya et al. 1995	GFDL multiply-nested moveable mesh baroclinic track model

in track forecasts. The ratio w/l varies inversely with the size of the hurricane with values generally in the range of 2 to 4 (Sugg 1967, AMS 1986), with 3 as an average ratio.

For a typically hurricane with dimensions of 160 km ($w = 480$ km), the hurricane warning creates an over-warn distance covering 320 km. Assuming a uniform population density within a coastal county, the outcome of the large overwarn distance is that 67% of those who evacuate do so unnecessarily. The problem is compounded for nonuniform population densities.[5] The greater the number of people who are over-warned, the larger the credibility problem during subsequent warnings. However, in reality over warning may not have a significant affect on a person's decision to evacuate (Chapter 17). It is important for residents to be aware of this fact and to realize that it takes *on average* three hurricane warnings before hurricane-force winds are experienced in their own backyard. Besides the possible psychological effect on residents, over warning is expensive to local governments. If we assume a cost per coastal kilometer of evacuation at $100,000, then on average the price tag for over warning is $32 million. The 1985 hurricane season featured 6 U.S. hurricanes and 2 U.S. tropical storms. Every state from Texas to Maine that year had one coastal region under a hurricane warning at one time or another. The goal of the NHC is to minimize over warning

---

[5]See Chapter 17.

Table 18.4: Costliest U.S. hurricanes. Data are for the period 1900–1996. Dollar amounts are adjusted to 1990 values, except for hurricane *Andrew*. Category (Cat.) indicates the Saffir/Simpson category. Damage location indicates where most of the destruction took place. Note that the two costliest hurricanes occurred a mere 4 years apart.

Rank	Year	Hurricane	Cat.	Damage Location	Damage (billions of $)
1	1992	*Andrew*	4	$FLse, LA$	30.50
2	1989	*Hugo*	4	SC	8.49
3	1972	*Agnes*	1	Northeastern U.S.	7.50
4	1965	*Betsy*	3	FL, LA	7.43
5	1969	*Camille*	5	LA, MS	6.10
6	1955	*Diane*	1	Northeastern U.S.	4.83
7	1979	*Frederic*	3	AL, MS	4.33
8	1938	New England	3	Northeastern U.S.	4.14
9	1996	*Fran*	3	NC	3.20
10	1995	*Opal*	3	$FLnw$	3.07
11	1983	*Alicia*	3	TXn	2.39
12	1954	*Carol*	3	Northeastern U.S.	2.37
13	1961	*Carla*	4	TX	2.22
14	1985	*Juan*	1	LA	2.11
15	1960	*Donna*	4	Eastern U.S., FL	2.10
16	1970	*Celia*	3	TXs	1.83
17	1985	*Elena*	3	$MS, AL, FLnw$	1.76
18	1991	*Bob*	2	Northeastern U.S., NC	1.75
19	1954	*Hazel*	4	SC, NC	1.67
20	1926	Unnamed	4	$FLse$	1.52

through improvements in forecast models and other technologies. As Sugg (1967) indicates, people living in coastal counties along the Gulf of Mexico must expect a minimum overwarn area of roughly 371 km (200 mi). The length of coastline from Brownsville to Key West is approximately 2409 km. Thus, the average number of people per 371 km of the coast for the 56 counties is about 1,680,000. A 10% increase in the size of an average minimum warning area would therefore affect approximately 168,000 additional people. As a result, it is necessary to estimate the average protection costs per average size warning area.

18.3   Costliest U.S. Hurricanes

The historical record of the costliest North Atlantic hurricanes is of considerable interest to the insurance industry. A portion of the hurricane wind damage is covered by private insurance contracts. Table 18.4 lists the twenty costliest North Atlantic hurricanes for the United States over the period 1900 through 1996, with costs adjusted to

1990 dollars.[6] Note that the two hurricanes *Andrew* and *Hugo*, which occurred only 4 years apart, rank at the top of the list. In fact three of the top ten costliest hurricanes occurred in the 1990s, with eight of the top ten occurring since 1965. Only two of the top twenty occurred before 1954. Although the northeast has relatively few hurricane strikes (see Chapter 11), it is well represented in the list of costliest storms. One third of the costliest top 15 hurricanes produced damage north of North Carolina. Only the 1938 hurricanes shows up on the list of deadliest and costliest U.S. hurricanes. Tracks, including dates and intensities, for the ten costliest hurricanes are given in Figures 18.2 through 18.6. Eight of the ten costliest storms originated as depressions south of 20°N latitude. Only hurricanes *Alicia* and *Carol* developed north near 25°N. Hurricane *Agnes* originated near the Yucatan Peninsula in the northwestern Caribbean Sea and hurricane *Alicia* originated in the central Gulf of Mexico. Hurricanes *Andrew* and *Betsy* are classified as baroclinically-enhanced hurricanes and hurricane *Agnes* is considered a baroclinically-initiated storm. Interestingly, the costliest storms are typically early-season hurricanes. Nine of the ten costliest North Atlantic hurricanes occurred before the end of August. This is 2 weeks before the peak of the North Atlantic hurricane season (see Chapter 4) and is in marked contrast to the fact that only one of the deadliest North Atlantic hurricanes occurred before September. Four of the ten costliest storms affected the state of Florida.

Dade County took the brunt of the damage from hurricane *Andrew* in 1992. Inland penetration of the storm surge was minimal due to the local topography. The rapid storm motion combined with a relatively compact size contributed to light rainfall amounts. In fact, most of the damage to south Florida from *Andrew* was caused by the wind (Powell and Houston 1996). Two-thirds of the damage from hurricane *Hugo* in 1989 took place in Charleston, Berkeley, and Dorchester counties of South Carolina. The coastal island resorts, including the Isle of Palms, Sullivan's Island, and Folly Island sustained particularly heavy damage from the storm surge and winds during *Hugo*'s rampage. Damage costs from hurricane *Agnes*, which made landfall over northwestern Florida in 1972, were due largely to severe flooding which accompanied copious rainfall over Pennsylvania and West Virginia. Hurricane *Betsy* in September, 1965 crossed the southern tip of Florida as a category three storm. Sustained winds in excess of 100 kt were reported throughout the northern Keys. Hurricane *Camille* in August, 1969 made landfall near Biloxi causing catastrophic damage. Winds were clocked in excess of 150 kt with gusts near 175 kt. Heavy rains fell in Tennessee, Kentucky, West Virginia, and Virginia. Flash floods and mudslides were widespread. The James River in Virginia reached record levels causing significant damage from Lynchburg to Richmond. Hurricane *Diane* in August, 1955 was a category one hurricane at landfall along the North Carolina coast. It remained at tropical storm intensity as it recurved through Virginia, Maryland, and New Jersey dropping tremendous amounts of rain. Hurricane *Frederic* intensified over the southeast Gulf of Mexico before slamming the Alabama and Mississippi coastline as a category three storm.

---

[6]The table is an update of the list in Hebert et al. (1996) by Brian Maher and Jack Beven.

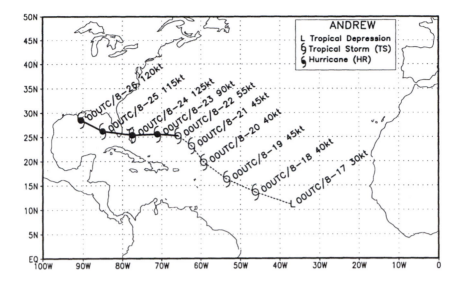

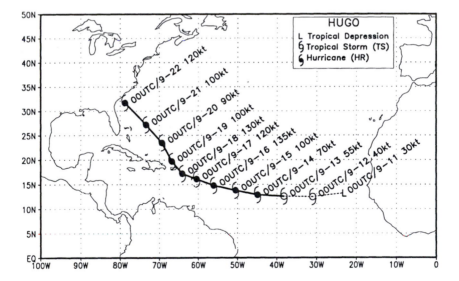

Fig. 18.2: Tracks of hurricanes *Andrew* in 1992 and *Hugo* in 1989. The track is indicated from depression stage (L) to hurricane landfall using selected dates (Time/Month-Day) and wind speeds (kt). Note that hurricane *Andrew* made landfall in Florida (*FL*) and in Louisiana (*LA*). Both storms began as easterly waves over the eastern North Atlantic.

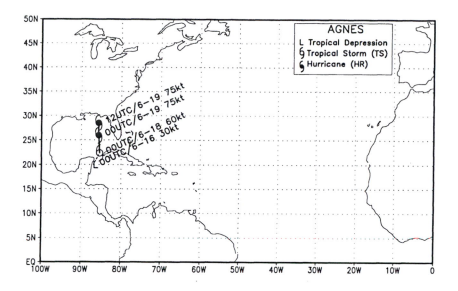

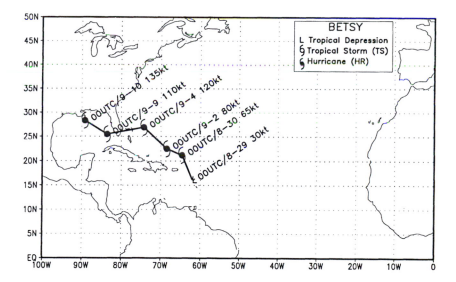

Fig. 18.3: Tracks of hurricanes *Agnes* in 1972 and *Betsy* in 1965. The track is indicated from depression stage (L) to hurricane landfall using selected dates (Time/Month-Day) and wind speeds (kt). Note that hurricane *Agnes* formed off the eastern coast of Yucatan, Mexico in the western Caribbean Sea. Early on *Betsy* appeared to be headed for the open waters of the northern North Atlantic.

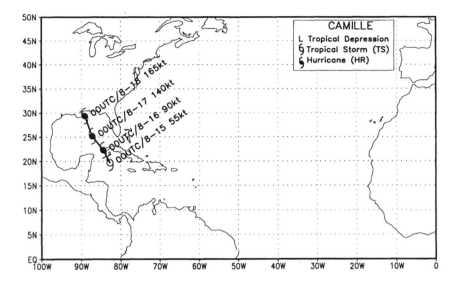

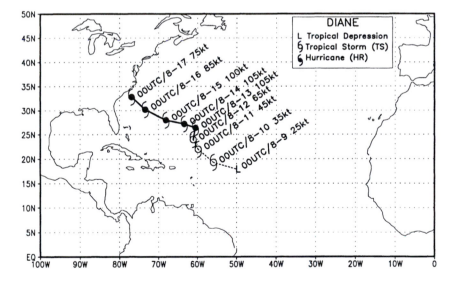

Fig. 18.4: Tracks of hurricanes *Camille* in 1969 and *Diane* in 1955. The track is indicated from depression stage (L) to hurricane landfall using selected dates (Time/Month-Day) and wind speeds (kt). Note that the depression stage for hurricane *Camille* is not shown. *Diane* appeared to be recurving out to sea before it made a left turn toward the United States.

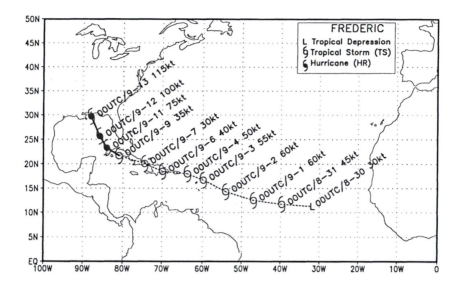

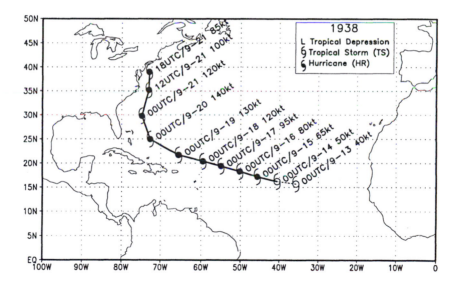

Fig. 18.5: Tracks of hurricane *Frederic* in 1979 and the New England hurricane in 1938. The track is indicated from depression stage (L) to hurricane landfall using selected dates (Time/Month-Day) and wind speeds (kt). Both storms originated over the eastern North Atlantic. Hurricane *Frederic* caused massive damage to the coastline of Gulf Shores. Escambia and Santa Rosa counties in northwestern Florida were impacted by gale-force winds and high surge. These same counties of Florida received a blow from hurricane *Opal* in October, 1995. The 1938 hurricane jogged to the north after it had recurved and was heading northeastward.

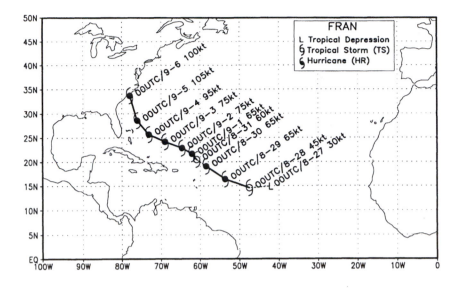

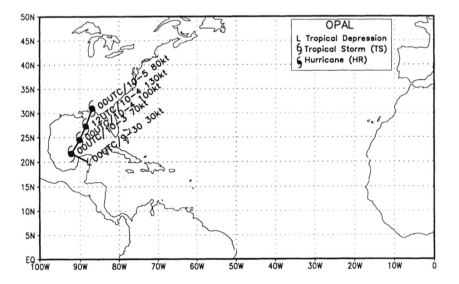

Fig. 18.6: Tracks of hurricanes *Fran* in 1995 and *Opal* in 1996. The track is indicated from depression stage (L) to hurricane landfall using selected dates (Time/Month-Day) and wind speeds (kt). *Fran* originated over the central North Atlantic from a tropical easterly wave and *Opal* originated off the eastern coast of Yucatan over the western Caribbean Sea.

18.4   Damage Costs

Damage costs from hurricanes are separated into two types: damage from wind and damage from water. Wind damage occurs as roofs are blown off, windows are broken, and foundations are compromised. Wind-driven flying debris causes considerable damage to structures. Water damage occurs with flooding from rainfall and storm surge. Post-hurricane surveys of the area are used to determine the extent of both damage types, though it is often not a clear-cut separation. Insurance against the cost of repairing from water damage in the United States is covered by the National Flood Insurance Program (see Chapter 19). Insurance against wind damage costs is done on a volunteer basis through private insurance underwriters. Hurricanes that sweep up the east coast and reach Long Island and New England are of greatest concern to reinsurance underwriters due to the enormous property values that exist there.

As indicated in Chapter 8, the frequency of U.S. hurricanes is rather steady over the period 1851–1996. The United States gets hit on average by one or two hurricanes a year. There is no significant trend to the average over the years. This fact, combined with the lack of important changes in hurricane intensity over the years, makes it unlikely that climate changes are the main cause of the increases in hurricane damage costs. Comparing hurricanes by damage cost can be done in a variety of ways. However, since both population and building costs change over time, it is difficult to make direct comparisons of hurricanes occurring during different decades. One method is to adjust the damage costs to a common base year. In this way hurricane costs are revised to a common dollar amount. Even when an adjustment for inflation is made, changes to population and infrastructure will make similar storms occurring during different decades reflect widely different damage costs.

The problem is compounded by the data records. Data collected on the costs of hurricanes is not done in a systematic way, and the type of data collected in one storm is often not consistent with data collected from a different storm. Additionally, pressure from the news media and television for precise damage numbers tends to result in an overestimation of damage costs (Hewings and Mahidhara 1996). In any case, the meaning of a particular loss estimate is difficult to ascertain. For instance, the replacement cost of a house destroyed in hurricane-force winds may or may not include upgrades to make the house safer against the next storm. The replacement cost to pre-hurricane levels and not the upgrade cost reflects the true damage cost of a particular storm. Exact replacement costs are not readily available. Nevertheless, this "with and without" principle is one of the clearest and most important economic guides in assessing natural hazard impacts (Howe et al. 1991). It is also the principle that is typically violated.[7]

Notwithstanding the limitations, it is interesting to compare the potential loss of an earlier hurricane on infrastructure as it exists in a later era. Moreover, potential future losses are often the primary concern to decisionmakers. To make a compari-

---

[7]Double counting losses is a common practice.

Table 18.5: Potential costliest U.S. hurricanes. Data are for the period 1900–1996. Dollar amounts are adjusted by inflation, personal property increases, and coastal county population changes according to Pielke and Landsea (1998). Values are in billions of dollars. Category (Cat.) indicates the Saffir/Simpson category. Notice that hurricane *Andrew* in 1992 is not at the top of this list, and hurricane *Hugo* in 1989 is not in the top ten.

Rank	Year	Hurricane	Cat.	Damage Location	Potential Costs (billions of $)
1	1926	Unnamed	4	$FLse$	72.30
2	1992	*Andrew*	4	$FLse, LA$	33.09
3	1900	Unnamed	4	TX	26.62
4	1915	Unnamed	4	TX	22.60
5	1944	Unnamed	3	$FLsw$	16.86
6	1938	New England	3	Northeastern U.S.	16.63
7	1928	Unnamed	4	$FLse$	13.80
8	1965	*Betsy*	3	FL, LA	12.43
9	1960	*Donna*	4	Eastern U.S., FL	12.05
10	1969	*Camille*	5	LA, MS	11.00
11	1972	*Agnes*	1	Northeastern U.S.	10.70
12	1955	*Diane*	1	Northeastern U.S.	10.23
13	1989	*Hugo*	4	SC	9.38
14	1954	*Carol*	3	Northeastern U.S.	9.07
15	1947	Unnamed	4	$FLse, LA, AL$	8.31
16	1961	*Carla*	4	TX	7.07
17	1954	*Hazel*	4	SC, NC	7.04
18	1944	Unnamed	3	Northeastern U.S.	6.54
19	1945	Unnamed	3	$FLse$	6.31
20	1979	*Frederic*	3	AL, MS	6.29

son it is necessary to account for changes in population. An example, presented by Burroughs (1997) compares the cost of hurricane *King* that struck southeast Florida in 1950 with a hypothetical identical storm in 1990. *King* was a powerful mid October storm that blasted the city of Miami. In the period from 1950 to 1990 building costs in Miami increased by a factor of 5.7 and population increased by a factor of 6. Overall change factor is 34.2 (5.7 × 6). Using the changes to Miami as a representation of the changes to the entire region of southeastern Florida affected by the hurricane, the potential damage cost of a similar hurricane in 1990 is obtained by multiplying this factor by the original $28 million in damage estimated for *King*. The amount of damage would be closer to $958 million if the same storm made landfall in 1990.

A comprehensive analysis of this kind is presented in Pielke and Landsea (1998). The potential impact of past hurricanes are examined by normalizing (adjusting) the costs for inflation, population, and wealth of the affected communities. Table 18.5 lists the top twenty U.S. hurricanes over the period (1900–1996) by potential costs. The

1926 southeast Florida hurricane (the Great Miami hurricane), which is listed as the 20th costliest U.S. hurricane ranks as the potential costliest of all times. Although Miami was a rapidly growing city in the early 1920s, today the metropolitan area ranks 11th largest in the nation.[8] The potential cost of a similar hurricane striking the city in 1996 is estimated at $72.3 billion dollars. An analogue to the Galveston hurricane of 1900 also looms as a storm with potential to cause catastrophic damage to modern development and infrastructure. The 1944 hurricane that raked the entire Florida Peninsula ranks as the fifth potential costliest. The storm originated over the western Caribbean Sea as a classic late-season hurricane and moved northward with the center passing within 20 km of Havana. Its large size, great intensity (category three), and direct path from the southwest through the northeast part of the state caused widespread damage. A similar storm today would create major economic and social hardships for the entire state. Some consider the 1944-type storm as the worst-case scenario for Floridians as evacuees from southern parts of the state would be evacuating in the direction of storm motion. Hurricane *Donna* in September, 1960 is noteworthy. It struck the Florida Keys as an intense category four hurricane, then recurved across the state and moved up along the eastern seaboard. Hurricane-force winds were felt in Florida, the middle Atlantic states, and New England. Damage was heavy in New York and New England. However, its inclusion in the top ten list of potential costliest hurricanes results from the damage a future storm of this kind would do in Florida.

The average Saffir/Simpson category for the 20 potential costliest hurricanes is 3.4 compared with an average category of 3.0 for the actual costliest. The difference might be related to the greater difficulty in representing the potential damage of flooding. Six of the top 10 potential costliest hurricanes affected Florida. This compares with only 3 of the top 10 actual costliest hurricanes. A return to a climate conducive to more intense hurricanes in Florida, as was the case during the 1940s, will have considerable social and economic impacts. In contrast to Florida, the actual and potential costliest storms affecting the northeastern United States tend to coincide. All 3 of the top 10 costliest hurricanes appear in the top 12 for potential costliest, although in slightly different order. The reason for this is that, although a powerful hurricane hitting New England would cause tremendous economic losses, the losses would be proportional to the kinds of losses historically experienced in the area, relative to the hurricane experience level of Florida. In the future, as the difference in the economic factors (wealth and population) between Florida and the northeast lessen, the distribution of the actual costliest hurricanes will reflect the differences in probability of occurrence with respect to the two regions. The past is evidently not as good of an indicator of the potential future losses in Florida as it is in New England.

The list of potential costliest hurricanes reveals that unlike actual costs, which show a dramatic increase through the last several decades, the potential costs were somewhat lower during the 1970s and 1980s compared to the period from the 1940s through the middle 1960s. Because of the tremendous increases in coastal population and wealth,

---

[8]Based on 1994 population estimates.

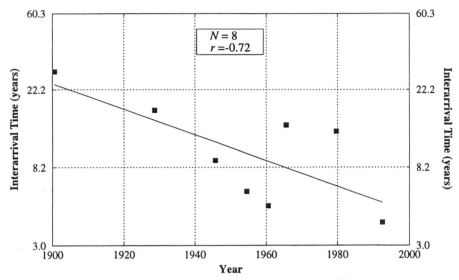

Fig. 18.7: Linear regression of interarrival times (log) within clusters of four events versus time of the event over the period 1900–1996. Events are defined as a normalized losses exceeding $2.2 billion. Pearson's correlation between the regression line and the observations is −0.72.

the potential costs are more closely linked to climate than are the actual costs. As noted, Florida hurricanes are on the top twenty potential cost list in nine different spots. Florida was particularly hard hit in the 1940s by intense hurricanes when population was considerably less.[9] The United States is becoming economically more vulnerable to North Atlantic hurricanes. Much of this is occurring in Florida. The rising vulnerability results from extensive changes in society and not from a significant rise in hurricane activity. Future changes in hurricane frequencies will change the equation.

It is not obvious from Table 18.5 whether or not significant damage events are increasing in frequency. A trend in normalized damage costs can occur from increases in storm frequencies or from relative economic changes. A coastal area that was earlier undeveloped will have a differentially higher change in vulnerability compared to an area that was previously developed. As mentioned in Chapter 11, if the distribution of events is described as Poisson, then the interarrival times are exponentially distributed. The dates of the 31 costliest hurricanes (>$2.2 billion in total costs) normalized by inflation, personal property increases, and coastal county population changes over the period 1900–1996 are used to test for a trend in frequency of events. The date on which the hurricane first reached hurricane or major hurricane intensity is used. The first 30 events over the period are from Pielke and Landsea (1998). Hurricane *Fran* of 1996 is included as the 31st event over the 97-year period. The mean annual number of costly hurricanes is 0.32 with a variance of 0.28. A significance test suggests the events are

---

[9]The significant changes that have occurred throughout the state makes Florida much more vulnerable to hurricanes than it was only a half century ago.

distributed as Poisson at an α level of 0.1. Figure 18.7 shows the interarrival times on a logarithm scale as a function of year. The negative trend line with a linear correlation of -0.72 has a p-value of 0.036. The analysis is important in showing an increasing frequency of catastrophic loss events. Interarrival times decrease from a value of more than 20 yr at the beginning of the century to a value of about 5 yr by the middle 1990s. An increase in interarrival times is noted over the period from the early 1970s through the 1980s with a decrease afterwards. The frequencies of the strongest hurricanes over the North Atlantic (>120 kt) during the period 1943 through 1996 show a similar fluctuation in interarrival times. This indicates that loss potential is related to the frequency of the most intense hurricanes. This is increasingly the case as economic development extends over a greater percentage of the coastline. The frequency of extreme events will likely continue to increase against the background of climate fluctuations until coastal development saturates. As a consequence, the U.S. economic vulnerability to hurricanes appears to be growing to match the climate variations in extreme events.

18.5   Property Values

Damage costs are a component of societal vulnerability to hurricanes. Costs include damages to public infrastructure, private enterprise, and personal property. These costs are not uniformly distributed along the nation's coastline. For most people the house is the most expensive personal property at risk to hurricane damage. The cost of repairing or rebuilding a damaged or destroyed house depends on the price of the house. Median housing prices provide an estimate of the average cost of living in an area. Housing prices also serve as way to compare the potential damage a hurricane would do across different locales. The total damage costs from two identical hurricanes hitting cities with different housing price distributions will be different. Table 18.6 lists the median price of existing single-family houses in 26 coastal or near-coastal cities at threat to North Atlantic hurricanes.

Cities in Table 18.6 are metropolitan statistical areas (MSAs) as defined by the U.S. Office of Management and Budget. The median value is an estimate of the central tendency in the price distribution over the metropolitan area. Housing is most expensive along the middle Atlantic and northeast coastlines. In fact, the top four median housing prices are north of Cape Hatteras. Boston ranks highest above New York and Washington. A recurrence of the 1938 New England hurricane on today's highly urbanized northeastern corridor would be catastrophic in terms of damage costs alone. In Florida, median single-family homes range in price from $126,600 in West Palm Beach to $73,000 in Daytona Beach. The southeastern coast of Florida represents a region where, increasingly, people of greater wealth are choosing to place themselves in harm's way. Median housing prices are lowest in cities along the northern Gulf coast from Corpus Christi to Pensacola. This region is occasionally visited by some of the more powerful storms of the North Atlantic. No consideration is given to a measure of the spread in the distribution. The largest one-year median price increase occurs in

Table 18.6: U.S. housing prices. Values are median prices of existing single-family houses for 1996 in 26 coastal or near-coastal U.S. cities. The list is ordered from most expensive to least expensive. Increases are expressed as a percentage of the 1995 value. Data are provided by the National Association of Realtors.

	City	State	Price (U.S. dollar)	% Increase
1	Boston	MA	189,300	5.8
2	New York	NY	174,500	2.8
3	Washington	DC	160,700	2.7
4	Hartford	CT	139,200	4.3
5	West Palm Beach	FL	126,600	4.4
6	Providence	RI	118,100	2.3
7	Miami	FL	113,200	5.7
8	Baltimore	MD	113,000	1.5
9	Ft. Lauderdale	FL	112,300	6.0
10	Norfolk/Virginia Beach	VA	110,200	5.6
11	Tallahassee	FL	109,800	10.0
12	Atlantic City	NJ	108,000	0.9
13	Sarasota	FL	107,700	3.1
14	Bradenton	FL	95,100	4.5
15	Charleston	SC	94,900	0.6
16	Jacksonville	FL	88,400	6.4
17	Baton Rouge	LA	87,000	2.8
17	New Orleans	LA	87,000	11.5
19	Houston	TX	84,700	6.9
20	Pensacola	FL	84,500	6.3
21	Mobile	AL	83,200	10.8
22	Tampa	FL	81,300	3.8
23	Corpus Christi	TX	79,600	2.6
24	Ft. Myers	FL	78,700	1.3
25	Biloxi/Gulfport	MS	73,800	1.0
26	Daytona Beach	FL	73,300	5.3

New Orleans. Large increases are noted in Mobile, in Houston, and in several cities of Florida. In particular, the southeast Florida cities of Miami and Fort Lauderdale combine high housing costs with large increases. Boston recorded a one-year 5.8% increase in median housing price.

Most property is either not insured or not insured adequately against hurricane damage. The value of insured property at risk to wind damage for coastal counties from Texas to Maine is estimated at $3.4 trillion for 1996. Figure 18.8 shows the insured property values of coastal counties by state for 1988 and 1993. The increase as a percentage of the 1988 value over the 5-year period is also shown. A doubling of values represents a 100% increase. Inflation accounts for just under 20% of the total increase in values over the period (Pielke and Pielke 1997a). Florida has the largest amount of property insurance at nearly $872 billion, followed by New York and

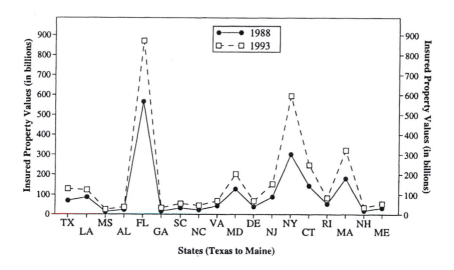

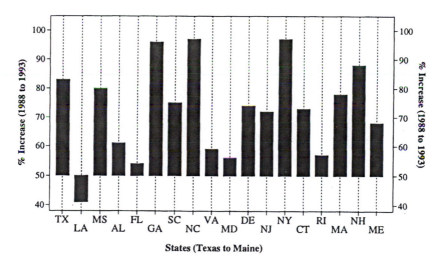

Fig. 18.8: Insured property values by state. Values are in billions of dollars for the years 1988 and 1993 (top). Percentage increases in property values over the period 1988–1993 (bottom). Insured property values are largest in Florida (FL) and New York (NY). Percentage increases are greatest for Georgia (GA), North Carolina (NC) and New York (NY). Data are from Pielke and Pielke (1997a).

Massachusetts. The highest insurance costs in Florida are in Broward, Dade, Monroe, and Palm Beach counties. These counties have large populations and the largest annual probabilities for the occurrence of a major hurricane (see Chapter 11). Property values in these areas are increasing above the national average. With the exception of Florida, insured property values are greatest along the northeast coast, where property values are highest.

Insured property values have increased in every state. The average increase is 68%, with all but Louisiana showing an increase of better than 50% over the 5-year period. Three states—Georgia, North Carolina, and New York—show increases exceeding 90%. Large variations in changes to property values are found for counties within the same state (Pielke and Pielke 1997a). Over the 5-year period, 24 coastal counties experienced more than a doubling of insured property values. Development in Florida and other parts of the coastal United States is at a level where the cost of rebuilding after a disaster might exceed the ability to pay for it. This poses a serious threat to local, regional, and state economies. A barrage of closely spaced major hurricanes or one super-hurricane could completely bankrupt a broad spectrum of insurance providers, which would send shockwaves through the United States and global economies.

19

Catastrophe Insurance

In his fascinating book entitled *Against the Gods: The Remarkable Story of Risk*, Peter L. Bernstein notes that whatever data we analyze, it will only be an approximation to reality. In the real world, except for games of chance, we never have all the information we need to achieve certainty about all possible future scenarios. Worst case futures include the potential for massive death and destruction. Unfortunately, hurricane disasters affect nations disproportionately. The poorest countries worry about loss of life, while the richest countries worry about economic costs. Improvements in the understanding of climate science will offer little in the way of benefits to areas with limited resources. Arguably a stronger global economy, bolstered by sound financial institutions creates an environment in which rich nations generously respond to natural disasters elsewhere.[1]

Rationality and quantitative analyses are important in making decisions under conditions of uncertainty. As understanding of climate increases, the ability to make skillful forecasts of hurricane activity several years in advance should improve (see Chapter 16). Hurricane climate forecasts will also become more specific in detailing landfall probabilities. Model skill will rise above skill based solely on climate averages. Forecasts couched in terms of probabilities are useful to decisionmakers. Individual companies may gain a competitive advantage over companies that are technically incapable of incorporating the forecasts or that choose to ignore them (Roth 1997). This is where a closer collaboration between climate scientists and the insurance industry can be productive. On the one hand, scientists are in a position to educate the companies about the limitations of climate data and climate prediction models and to aid in turning the data and forecasts into useful information. On the other hand, the insurance companies can inform climate scientists on what information is most relevant to their particular needs. In this chapter we examine how catastrophe insurance works and how companies might take advantage of hurricane information to improve decisions.

The modern commercial concept of property insurance was well established at the end of the 18th century. By the 1770s a young insurance industry had sprung up in

---

[1]The United States gave an $80 million aid package to help victims of hurricane *Mitch* in 1998.

429

the United States. In fact, as early as 1751 Benjamin Franklin was selling insurance to protect property from fire through a company called First America. However, most large policies at this time were still written overseas by Lloyd's of London. At the time of the New England hurricane in 1938, wind storm insurance coverage was not common either as a stand-alone policy or as a regular part of a fire insurance contract. The property damage inflicted by this devastating northeast storm created a significant public awareness of the potential impact of wind storms on property. By 1950, at a time when many returning veterans from World War II were buying houses, wind storm coverage was a routine part of the normal fire insurance policies on mortgaged property.[2] Today insurance coverage[3] for losses resulting from natural disasters in the United States such as earthquakes, tornadoes, hurricanes, and floods, run less than 20% of the actual loss due to limited participation in voluntary insurance programs. The U.S. federal government, through disaster aid packages, pays for a large portion of the remaining losses. The reported costs of hurricanes depends both on the extent of damage and on who pays. Here we are concerned with the private catastrophe insurance losses. Losses that are covered by individuals or by governments are not included. Dollar amounts given are expressed in terms of insured losses and do not represent the true losses.

19.1   Insured Losses

Of all natural disasters to occur in the United States, hurricanes account for approximately two-thirds of the insured property losses. Wind is responsible for much of the structural damage caused by hurricanes. The projected increases in wealth in coastal counties along the Gulf and southeast account for the large increases in expected natural hazard losses between 1970 and 2000 (Petak and Atkisson 1982). At least half of all hurricane damage from wind is covered by voluntary insurance programs. Table 19.1 lists the top catastrophe losses in terms of property insurance in the United States over the period 1950 through 1994. As seen, hurricanes are by far the costliest category in terms of aggregate losses to the insurance industry. Hurricane losses represent 62% of all catastrophic insurance losses. The number of extreme hurricane events is more than double the number of other catastrophes over the last half century. Earthquakes are second with 12% of all losses. Most flood events are not covered by private insurance companies. Though the number of insurance loss claims caused by tornadoes and other weather events is much larger than for hurricanes, the individual severity of each event is considerably less (Conning&Company 1994). The loss numbers for hurricanes may not represent future losses as the heavily developed coastal cities like Houston, Miami, and New York escaped direct blows by major hurricanes during this period. After hurricane *Andrew* of 1992, State Farm Fire & Casualty paid out $3.5 billion in damages. Allstate, not far behind, paid out $2.1 billion.

---

[2]C. C. Hewitt, Jr. (personal communication 1996).
[3]Coverage refers to the type of loss for which the insurance company will pay.

Table 19.1: U.S. insured catastrophe losses. Losses are by type over the period 1950–1994. Values are in billions of dollars. The number of ranked events, the losses, and the percentage of losses are given. Data are from Conning& Company (1994).

Catastrophe Type	Number	Catastrophe Loss (billions $)	Percentage Loss (billions $)
Hurricanes	13	32.95	62%
Tornadoes	5	4.00	8%
Other Weather	5	5.18	10%
Earthquakes	2	6.41	12%
Floods	2	1.40	2%
Manmade Destruction	3	3.38	6%
Total	30	53.32	100%

Hurricane damage due to flooding is part of the U.S. National Flood Insurance Program (NFIP). NFIP is administered by the U.S. Federal Emergency Management Agency (FEMA) and requires that communities pass a flood-plain management ordinance to participate in the program. In particular, it requires new and remodeled buildings be protected from damage by the 100-year flood event. In effect, the flood insurance program offers a mandate of mitigation in exchange for insurance coverage. Flood risk in commercial policies is generally covered by private insurers (Nutter 1994). Besides administering NFIP, FEMA works to reduce risks, strengthen support systems, and help people and communities prepare for and cope with disasters regardless of their cause. This is accomplished through the National Mitigation Strategy. The strategy is designed to increase public awareness of the risk to natural hazards in order to insure a public demand for safer communities in which to live and work. It aims to help reduce the risk to loss of life, injuries, economic costs, and the destruction of natural and cultural resources that result from natural hazards.

19.2   Components of Catastrophe Insurance

The property catastrophe insurance industry consists of numerous primary and reinsurance companies. Each company has a unique set of exposures and underwriting systems and a particular level of financial integrity. Since the ability of individual companies to measure and monitor risk varies, the effect of a hurricane catastrophe can be quite diverse across the industry. Faced with the potential of a catastrophic loss, each company tries to manage a diverse portfolio of exposures and investments so as to earn an adequate return. In short, insurance is rather straightforward as a means of spreading hazard risk. Companies charge annual premiums at a level based on known risks. Insurance companies use the premiums paid by people and corporations who have not sustained losses from a catastrophic event to pay off people who have. The same holds true of casinos. The pay-outs are taken from the coffer that is constantly

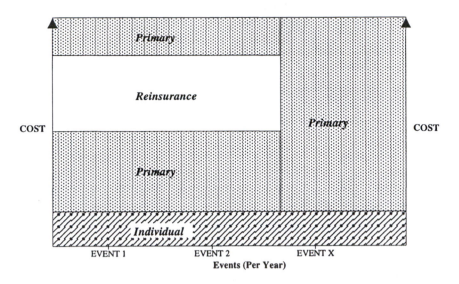

Fig. 19.1: Distribution of insured costs related a hurricane catastrophe on an individual home-owner. The distribution depends on the number of events over a prescribed period and the amount and level of reinsurance. Event X refers to any event after the second in the period.

being replenished by others. To the catastrophe insurance industry, hurricanes represent a peril because of their potential source of loss. This is in contrast to risk which is simply the uncertainty (or chance) associated with this loss (e.g., when or how large of loss?). Concerning the climate, the known risks are calculated by averaging out the claims made over as many decades as the data allow for various catastrophes. This works well when the records are representatively long and the climate responsible for the events is stationary. If suddenly hurricanes get more intense or are more likely to make landfall, then claims multiply and companies are prone to bankruptcy. Following hurricanes *Andrew* and *Iniki* in 1992, nine primary insurance companies went out of business. State Farm paid out $4 billion in claims, which was more than 7 times the previous record for payments by a single company. The possibility that the past is not a good indicator of the future is a serious concern to the insurance industry. As shown in Chapter 18 the concern is warranted in Florida regardless of climate change.

The value of an insurance policy is based on the promise to pay under the stipulations agreed upon. As such, it is necessary for insurance companies to be in sound financial shape. To ensure this, the insurance industry in the United States is regulated by state governments. Regulation is justified based on a desirability for adequate rates in an environment of stiff competition. The complexity of contracts and the need to share information on losses across the industry provides additional justification for government laws regulating insurance business conduct. Property catastrophe insurance against wind damage from a hurricane is structured in layers. At the bottom is the individual homeowner who pays out of pocket for minor damage repairs. Called

Table 19.2: Concentration of primary insurance volume. Values are in percentage for the five largest property underwriters in each state. The values represent percentage of total market share. The differences (Δ) in percentages between 1990 and 1993 are also provided. Data are from Conning & Company (1994).

Coastal State	1990	1993	Δ (1993 to 1990)
Texas	44.8%	42.6%	−2.2%
Louisiana	47.3%	43.3%	−4.0%
Florida	39.5%	40.1%	+0.7%
New York	32.1%	32.6%	+0.5%

a deductible, this cost is the amount of money the insured must pay under the policy before the insurance company (or carrier) is obligated to make a payment. Typically, the higher the deductible, the lower the insurance premium. Above the deductible, the damage from wind is paid by the homeowner's primary insurance coverage. The primary company has insurance against large losses from a reinsurance company, so a huge loss involving many houses is offset by a secondary level of insurance. The parties to a reinsurance contract are other insurers. Prior to hurricane *Andrew*'s visit to south Florida, most reinsurance companies did not set an upper limit (or cap) on their catastrophe coverage. This was justified based on the assumption of a maximum pay-out of no more than a few billion dollars as a worst-case disaster. Reinsurance companies now limit the loss amount on both the low and high ends (Figure 19.1). For a catastrophic loss the primary insurer is again responsible. Losses are also limited in time to a few events per annum. Thus the catastrophe contract is designed to reinsure the insurer against an accumulation of losses, which generally result from a single large event, such as a major hurricane strike to New York City.

The multiple-layer approach to catastrophe insurance offers some protection to the insurance industry. For example, the effect of market share from a large catastrophe on the major primary property insurance companies is illustrated in Table 19.2. It shows the estimated percentage of the premium market by the top five property insurers in each state 2 years before and 1 year after hurricane *Andrew*. Despite some companies suffering large losses, there is a good bit of protection in the market as a whole as registered by the small change in market share for the top companies. This cover is due to the reinsurance industry. Reinsurance effectively allows primary companies to overexpose by letting them write more policies than they can cover. This can be risky as the total reinsurance capacity is limited to $20 billion or so (Conning & Company 1994).[4] Differences are 4% or less in the four areas. Reduction in market share occurred in Texas and Louisiana. Texas was not affected by a significant hurricane in the period. In times of fewer hurricanes, competition from smaller companies tends to increase.

---

[4] A catastrophe on the order of $30 to $50 billion would significantly disrupt many of the industry's top insurers.

Table 19.3: U.S. costliest insured catastrophes. Loss values are in billions of dollars over the period 1950–1996. Both the original reported loss and the restated loss in term of 1994 dollars are given. Loss data are from Conning&Company (1994).

Year	Insured Catastrophe	Damage Location	Reported Loss	Restated Loss
1992	Hurricane *Andrew*	*FLse, LA*	$15.50	$16.07
1994	Earthquake *Northridge*	*CA*	9.00	9.00
1989	Hurricane *Hugo*	*SC*	4.20	4.85
1965	Hurricane *Betsy*	*FL, AL*	0.52	2.28
1993	Winter Storm	Eastern U.S.	1.75	1.81
1991	Fire	Oakland, *CA*	1.70	1.81
1992	Hurricane *Iniki*	*HI*	1.60	1.66
1979	Hurricane *Frederic*	*AL, MS*	0.75	1.44
1974	Tornadoes	Xenia, *OH*,	0.45	1.27
1983	Winter Storms	Eastern U.S.	0.88	1.27
Total			$36.35	$41.46

19.3   A New Awareness

The early 1990s were particularly active in terms of property catastrophes caused by weather events. This fact is surprising considering the dearth in North Atlantic basin hurricane activity over this time (see Chapter 12). Of the top ten costliest insured catastrophes in the United States over the period 1950 through 1996, half occurred in the short, 4-year span between 1991 and 1994 (see Table 19.3). The two North Atlantic hurricanes of consequence in this period were *Bob* in 1991 and *Andrew* in 1992. Hurricane *Bob* hit the heavily insured northeast states of Rhode Island, Massachusetts, New York, and Connecticut. The $16 billion in insured losses from hurricane *Andrew* was more than double the expected maximum loss experts had predicted back in 1986. Other catastrophes during the early 1990s included the Northridge earthquake, the Oakland fires, hurricane *Iniki*, and the east coast blizzard in 1993. Hurricane *Andrew* was the major contributor to the record insured losses over this time period (Changnon et al. 1997).

Actually, starting in the middle 1980s, the property insurance industry saw a significant increase in claims largely from more people of greater wealth living in exposed areas. Indeed, the capacity of some insurance markets could not keep up with the need for insurance coverage. Hurricane *Andrew* created a shortage of reinsurance capacity in the United States. The reinsurance industry responded by cutting back exposures in hurricane-prone areas. This forced primary insurers to pay higher prices for reinsurance. An influx of new capital into the reinsurance industry during 1993, largely from companies in Bermuda, helped in providing additional capacity. Presently the reinsurance capacity is still below what would be necessary under a super-catastrophe of $50 billion or more (Conning & Company 1994). A loss of this size would likely require significant help from the federal government. As we saw in Chapter 17, population along the southern Atlantic and Gulf coasts is increasing, and in many places the increases are substantial. Despite the risks, the desirability of living and working near the ocean continues to attract people to the coastline. Population increases along limited stretches of land has dramatically increased the population density living in areas most prone to a hurricane disaster. In particular, Florida has over 14 million residents (1996 estimate) with 80% of them living or working near the 2,174 kilometers of coastline. The net increase in the number of new residents to Florida over the period 1990–1996 was 11.3%.

Coastal counties are becoming more crowded. Population densities are rising fastest in the immediate coastal areas. Figure 19.2 shows the actual and projected population density by region in 10-year increments. The preference for coastal living is apparent by the large difference in densities between non-coastal and coastal locations. The projected population density for 2000 is nearly 400 people per square mile in coastal areas. This compares with a projection of 100 people per square mile in non-coastal areas. While density in the United States as a whole has risen by 38% over the period 1960–1990, population density in southeast coastal areas rose by as much as 129%. Population densities in other hurricane-prone coastal areas is also increasing considerably faster than non-coastal locations. Total coastal population density is expected to exceed 420 people per square mile by the year 2010.

With more people comes more property. During 1994, coastal regions accounted for more than 47% of all housing and over half the growth in construction during the previous two decades (Conning & Company 1994). Moreover, property values are rising, doubling and tripling the catastrophe exposure since the 1980s (Changnon et al. 1997). An estimate puts the total capacity of insurance (including reinsurance) in the United States at approximately $100 billion dollars. Nearly 70% of the $40 billion in U.S. property and crop losses during 1991–1994 were handled by reinsurance companies (Changnon et al. 1997). The trend toward more people and property along the U.S. coastline is expected to continue well into the first decade of the 21st century. This combined with limited insurance capacity raises questions concerning sustainable development.

Paradoxically, over the period from the middle 1960s to 1994, the frequency of North Atlantic hurricanes, particularly the most threatening tropical-only storms, was relatively low (see Chapter 5). This trend was largely unrecognized by the insurance industry (Changnon et al. 1997). In hindsight, the catastrophe insurance industry failed to relate insurance exposure to the real threat of hurricane perils. This failure, combined with leveraging,[5] caused widespread losses during the early 1990s. Insurance losses from catastrophes led to a new awareness in the insurance industry concerning hurricanes and climate. This was reflected in a major reassessment of risk, along with shifts

---

[5]Leveraging refers to the process where companies with limited capital are allowed to use reinsurance and manipulate assets to overexpose.

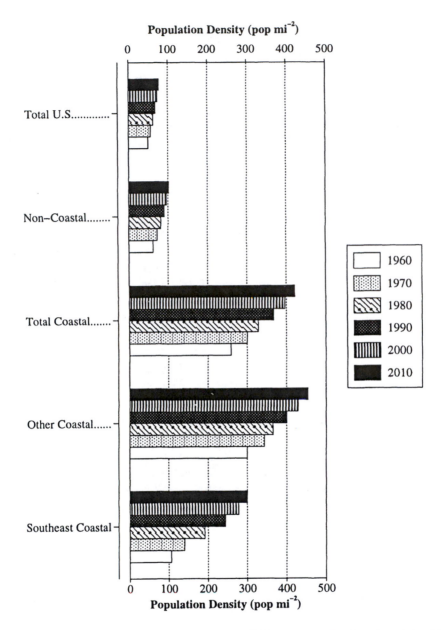

Fig. 19.2: Population densities in the United States. Values are actual and projected densities in areas over the period 1960–2010 in 10-year intervals. Southeast coastal refers to population densities in the hurricane-prone southeast Atlantic coast, other coastal refers to the other hurricane-prone coastal areas in the United States. Total coastal refers to the combination of the southeast Atlantic and other hurricane coastal areas. Non-coastal refers to the population densities in the remaining areas of the hurricane-prone states. Total U.S. population densities are also provided. Data are from Conning& Company (1994).

and restrictions in insurance availability in specific hurricane-prone locations.[6] Significant reaction in the reinsurance industry included higher rates (as much as 200% higher in some cases), stricter controls on underwriting, and a complete withdrawal from the catastrophe insurance business for some companies.

The re-evaluation of the hurricane threat had several repercussions including new state laws to prevent early withdrawals of insurance coverage in high-risk areas, and the emergence of additional funds to insurers so that extended coverage could be made. During the special legislative session in November of 1993 for example, the Florida Hurricane Catastrophe Fund (FHCF) was created to improve the availability and affordability of property catastrophe insurance to its coastal residents. The FHCF (Florida CAT fund) was established to provide reimbursement to insurance companies for a portion of their catastrophic hurricane losses. The CAT fund invests a small portion of each homeowner's monthly insurance premium into a pool as reinsurance to private insurers against a significant loss. It serves as an incentive to keep insurers in Florida and to keep rates affordable. It is envisioned that the CAT fund will eventually be able to repay damages caused by a hurricane (Barnes 1998). Clearly a better appreciation of the importance for understanding hurricane climate by catastrophe insurance and reinsurance companies was a consequence of the early 1990's catastrophes. This can be seen from the fact that insurance companies are now routinely using risk projection models to estimate potential losses over insured areas. These models vary widely in sophistication but most use some type of climate data to infer probabilities of peril. In Florida the law requires that companies use model projections of hurricane losses to insure against rates that are too excessive or not adequate enough.

19.4   Diversification

Pay-outs from insurance are typically done on a per event basis. For the case of wind damage the event is usually defined as a 72-hour period of damage from a single storm. Reinsurance companies will limit coverage to the first two events in a year, with the third event covered exclusively by the primary insurer. As a result, reinsurance companies are concerned about the aggregation of risk within a single event or over a few events within a short span of time. Costs incurred from a single major hurricane strike will be capped, but several moderate hurricane disasters in the same year (like *Carol*, *Edna*, and *Hazel* in 1954) could stress a company's ability to pay. Consequently companies are likely to diversify.

The mathematics of diversification help to explain its appeal. While the income generated on a diversified insurance portfolio[7] will be equal to the average income on all individual policies, its variance will be less than the average variance of the individual policies. According to Bernstein (1996) this means that diversification is a

---

[6]The person, usually elected, with responsibility for licensing and regulating the activities of insurance companies authorized to do business within a state is called the insurance commissioner. The employee of the insurance commissioner responsible for reviewing policies, procedures, and financial stability of insurance companies within the state is called the insurance examiner (Myers 1995).

[7]Portfolio means a collection of paper assets.

Table 19.4: Contingency table of hurricane incidence. Values are the number of years with no hurricanes and number of years with at least one hurricane landfall for Galveston County, Texas and Dare County, North Carolina over the period 1900–1996. Galveston and Dare were both hit in the same year only once over the 97-year period.

	Galveston County (TX)	
Dare County (NC)	No Hurricane	At Least One Hurricane
No Hurricane	64 years	16 years
At Least One Hurricane	16 years	1 year

type of free lunch at which you can connect an ensemble of risky policies with high expected returns into a relatively low-risk portfolio, as long as you minimize the correlation among the returns of the individual policies. This amounts to minimizing the correlation between perils. It should be kept in mind however, that diversification may justify higher exposure to risk but it does not guarantee against loss, only against losing everything all at once (Bernstein 1996).

Since an insurance company is concerned with aggregate risk, the main question is, over given time and space intervals, what is the largest possible loss that could occur? Consequently a strategy of diversification is necessary to insure that a single event or several events in a single year do not amount to a significant loss. Therefore a company insuring against wind damage would like to know if there is negative correlation among different areas affected by hurricanes. Consider for instance, the frequently hit counties of Galveston, Texas; Dade, Florida; and Dare, North Carolina. Each of these counties has a return period for hurricane-force winds (from hurricanes) of less than 8 years (see Chapter 11). But what is the annual correlation between hurricanes affecting Galveston County and Dare County? Since in most years, neither county gets hit, the Pearson product-moment correlation coefficient is not necessarily the best measure of this relationship.

Consider the contingency table (Table 19.4) for years with no hits and years with at least one hit for Galveston and Dare counties. From this we see that most years are without hits for both counties. Of the 17 years with at least one hurricane in Galveston County, only one of these years also had a hurricane in Dare County. A quantitative measure of this association is given by the correlation coefficient of a bivariate normal population from a two-way contingency table. Maximum likelihood estimates for the correlation coefficient are calculated from the bivariate normal density.[8] The bivariate normal correlation coefficient between hurricanes strikes in Galveston County, Texas and Dare County, North Carolina is −0.369. This indicates that years with hurricane strikes in Galveston tend to be years without strikes in Dare and years with strikes in Dare tend to be years without strikes in Galveston. The exclusiveness of hurricane

[8]The likelihood is specified from the multinomial distribution of the contingency table (Martinson and Hamdan 1972).

Table 19.5: Landfalling hurricanes in 50-year intervals. Frequencies are given for Puerto Rico, Jamaica, Bermuda, and Florida over the period 1501–1996, except for Bermuda where the record begins in 1609.

Years	Puerto Rico	Jamaica	Bermuda	Florida	Total
1501–1550	7	0	.	1	8
1551–1600	2	1	.	3	6
1601–1650	3	0	8	0	11
1651–1700	2	4	2	1	9
1701–1750	9	8	6	2	25
1751–1800	10	12	5	9	36
1801–1850	13	11	9	20	53
1851–1900	6	5	10	34	55
1901–1950	6	5	11	38	60
1951–1996	1	2	6	19	28
1501–1996	59	48	57	127	291

years for these two hurricane-prone counties suggest the potential for some diversification. The correlation between Galveston and Dade counties is +0.17, while the correlation between Dade and Dare counties is −0.16. Thus, there appears to be some potential for north-south diversification.[9]

Over the long term, the need for diversification may be even more important. Consider the correlation between typhoon activity in the western North Pacific and hurricane activity in the North Atlantic. Is the frequency of typhoons in Japan correlated with the frequency of hurricanes in Florida? Questions like this can be answered with long historical and proxy records. Kam-biu Liu of Louisiana State University has reconstructed a 1000-year record of typhoon activity in the Guangdong Province of China from historical documents. Records from Puerto Rico allow similar estimates of Caribbean hurricane activity back to the time of Columbus (see Chapter 9). Puerto Rico depends heavily on U.S. tourism and the Guangdong Province has one of the fastest growing economies in the world because of its investment relationships through Hong Kong. Tropical cyclone activity in both regions shows significant variations. Aggregating activity in 50–year intervals beginning with 1501 suggests a negative relationship between activity over southern China and the Caribbean. In fact, the most active 50-year period in southern China before the 20th century occurred from 1651 to 1700, when activity in Puerto Rico is estimated to have been at its lowest. Conversely, in the late 1700s and early 1800s, activity in the eastern Caribbean was high, but was low in southern China. A negative relationship in natural disaster potential on the multi-decadal time scale can be used for strategic social and economic planning.

The opportunity for diversification on the decadal time scale appears somewhat

---

[9]On a larger scale, there are similar relationships in hurricane activity between different coastal states (see Chapter 11).

more limited if restricted to the North Atlantic basin. Table 19.5 shows the number of hurricanes in 50-year intervals making direct hits on Puerto Rico, Jamaica, Bermuda, and Florida over the period 1501–1996. The numbers indicate the frequency of land-falling hurricanes rather than the frequency of landfalls. Overall, it appears that from the late 18th into the first half of the 20th century, North Atlantic hurricanes were more frequent compared with periods both earlier and later. Positive covariability is strongest for Florida and Bermuda ($r^2 = 0.58$) and also for Puerto Rico and Jamaica ($r^2 = 0.64$). The frequencies indicate the potential for some diversification across latitudes. That is, when hurricanes are more frequent at low latitudes (Puerto Rico and Jamaica) they tend to be slightly less common at higher latitudes (Florida and Bermuda). In fact the correlation is negative between Jamaica and Bermuda. This is consistent with our finding that tropical-only hurricanes and baroclinically-enhanced hurricane activity are inversely correlated on the annual time scale (see Chapter 10). Notice that multi-decadal variations in Puerto Rican hurricane activity is smaller than the variations in Florida activity. From the perspective of insurance coverage, this makes Florida more risky than Puerto Rico, as risk is directly related to variance.

19.5   Risk and Return

The working hypothesis for a private insurance company has to be that the measure-ment and management of risk can be utilized for a competitive advantage. Specifically, improvements in measuring underlying exposures and in modeling aggregate losses al-lows the development of superior insurance portfolios. These advantages may include such things as lower overall risks or more profits. Improvements in the measurement of financial risk allows for better use of capital, which translates into superior returns on investments. There are three basic forms of risks to an insurance company. They include liability risk, asset risk, and business risk. Both asset risk, which involves the risk that the actual value of assets will be less than projected (market value decreases, investments do not perform as expected, etc.) and business risk, which is largely the economics of all business (price competition, government regulations, etc.) are not logically connected to weather and climate. Liability risk (or obligation risk) is the risk that the cost of settling after disaster strikes will be greater than anticipated. Uncer-tainty and liabilities includes a random component to cover the inherent random nature of claim events and what is termed *parameter risk*, arising from the inability to know the frequency and magnitude of events such as hurricanes that give rise to claims. It is here were knowledge of the past, present, and future behavior of hurricane climate is important.

The framework that serves as a logical guide for the measurement and manage-ment of risk is the concept known as *efficient frontier*. The efficient frontier model was developed by Harry M. Markowitz[10] and it is the basis of modern investment portfo-lio theory. In its raw form, and as it relates to the insurance industry, a company is

---

[10]Markowitz received a Nobel Prize in Economic Science in 1990 for this work.

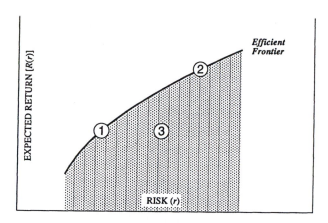

Fig. 19.3: Efficient frontier model of portfolio selection. The points define portfolio selections based on various levels of risk and expected return. The line represents the efficient frontier.

presented with several alternative options for selling insurance. For each option, the company knows the *expected return* and the *risk* associated with that return. The risk is commonly measured by the standard deviation of the return. Indeed, the standard deviation (as a metric of variance) is the essential parameter in most techniques for quantifying risk. Recall the argument that Florida is more risky than Puerto Rico with respect to insuring hurricane catastrophes due to greater variability of storm frequencies. The goal for a risk-averse company is to obtain a desired balance between risk and expected return. A company that does not consider risk is called risk neutral, in which case a company would invest in whatever asset promised the highest average expected return regardless of the volatility in actual returns (Miller and Downton 1993). The risk-averse company, on the other hand, is concerned about both expected return and the risk associated with that return. Higher risk investments will be chosen if the expected return is also higher by an amount sufficient to compensate the company for bearing the additional risk. Generally, the actions of risk-averse companies to maximize expected utility drives them to investment in a diversified portfolio. Besides expected return and risk, the company is aware of the correlation of returns with all other options. The problem is to choose an insurance portfolio by agreeing on a mix of options. The efficient frontier hypothesis states that not all option mixes are optimal. For any given mix, alternative mixes can be found that achieve a higher return at the same level of risk or the same return at a lower risk.

The frontier presents itself as a boundary consisting of a set of possible points representing option mixes that are efficient in the sense that one cannot improve upon them. Consider Figure 19.3 where efficient expected return $R(r)$ is plotted as a function of risk, r. The function $R(r)$ defines the efficient frontier beyond which no better portfolio is possible. Points one and two are both optimal in the sense of efficient mixes but they represent different ratios of risk to return. Point three is suboptimal in that for the accepted level of risk a higher return is possible. Points above the line are suboptimal in the sense that they represent unrealistic expectations for given risks.

Expected return (or performance) can be measured in a number of ways. A company must simply always want to choose a strategy that maximizes the expected value of whatever metric is chosen, all other constraints being equal. Variance is a statistical measure of how widely the returns on an asset swing about the average. The larger the standard deviation, the less the average return will signify about the likely outcome. Standard deviation as a measure of risk has its drawbacks since it focuses only on the dispersion of outcomes in a symmetrical way without special recognition of the more extreme events. The efficient frontier hypothesis allows for the measurement of risk to be any metric a company thinks is most relevant (probability of bankruptcy over the next 10 years, probability of a 20% reduction in revenues, etc.). Moreover, for both performance and risk measures, a combination of several different metrics are possible within this framework. An extension of the efficient frontier model allows for the analysis of the worst-case scenario for losses. The method calculates the probability of the maximum possible losses for every portfolio to obtain a value at risk for the company's overall exposure. The terms of this worst case are phrased as the worst loss that can be expected within, say, the next 5 years with a specified statistical confidence. This type of portfolio analysis reveals where risks are most concentrated.

An important observation from the standpoint of property catastrophe insurance for hurricanes is that, though expected return and risk metrics can reflect any chosen time scale (horizon), the efficient frontier is *not* fixed in time; it evolves. Suppose a company's portfolio is strictly wind-loss insurance; it is diversified in the sense that exposures are spread over the different tropical-cyclone basins, with particular attention to negative correlation in activity between basins. Now imagine that, in response to global climate changes, the world-wide frequency of hurricanes rises. This may translate to a shift in the efficient frontier toward higher risk and lower expected returns. Under such a circumstance, a portfolio that was on the frontier will become unacceptable in the sense that the expected return will be too high for the allowable level of risk. The choice is then to accept a lower level of return or increase the risk to obtain the same level of expected return. Of course, a portfolio that is suboptimal under the current climate conditions may become optimal under the new regime. Conversely, if the global frequency of hurricanes drops then it is likely that the efficient frontier will shift toward lower risks and greater expected returns, ignoring of course the likelihood of lower premiums due to increased competition. It is within this framework that knowledge of past, present, and future hurricane activity is useful to the insurance industry.

19.6 Coping Strategies

Despite the tremendous losses wrought on southern Florida by hurricane *Andrew*, the losses could have been much worse had the storm made a direct hit on downtown

Miami. Some estimates put the cost in excess of $50 billion under such a scenario. The capacity of reinsurers to develop and deliver necessary risk-distributing systems to occupants and investors in hurricane-prone areas can be undermined by the shear magnitude of the episodic losses that may occur during a major landfalling hurricane (Petak and Atkisson 1982). Because insurance companies routinely sell more insurance than they are able to cover at one time, the insurance industry might only be able to pay out a third or less of a $50 billion loss. Consequently, insurance companies are looking for improved managerial strategies and creative ways to finance possible future losses in order to stay in business.

19.6.1   Financial Risk Models

Modern insurance management is rooted largely in modeling strategies. A schematic of a portion of a reinsurance company's financial analysis based on the work of Stephen P. Lowe and James N. Stanard is presented here. The entire model includes three basic components including a liability scenario generator that produces distributions of aggregate underwriting results for the insurance portfolio, an asset generator that, when combined with the liability generator, gives a distribution of operating results for the combined insurance/investment portfolio, and a multi-period financial model that extends the distributions to a longer time horizon. The model is a set of linked programs and data sets allowing flexibility so that it can be used in different ways to help various users. The liability scenario generator is the only component sensitive to hurricane climate.

Since the core business is property catastrophe reinsurance, large emphasis is put on modeling the volatile claim's history (catastrophe or cat modeling). For each peril, in each part of the world, a set of catastrophic events is archived. The events vary according to size, location, intensity, and the likelihood of insured damage that might ensue. Probabilities based on the likelihood of particular combinations of event parameters occurring together are also assigned. In combination with the amount of insurance coverage for each event, the probabilities represent a severity distribution for a particular peril (e.g., wind damage). With each peril in each region, a frequency distribution is specified that reflects the probability of a given number of such events happening with a year. For instance, a frequency distribution for the number of U.S. landfalling hurricanes during an El Niño year can be made (see Chapter 10). Estimating the frequency and severity of hurricane-force winds at landfall is a critical component of the risk model. The HURISK is a public model, developed by Charles J. Neumann of Science Applications International Cooperation for the NHC, that determines probability of hurricane winds at given points along the coastline (Neumann 1987).

To estimate annual aggregate financial losses, the frequency and severity distributions of the catastrophe are combined. Current models use several tens of thousand scenarios of annual losses based on random, but stratified combinations (Latin hypercube approach). The losses in each scenario represent the losses to the primary insurance company. These losses are subsequently run through reinsurance contracts to

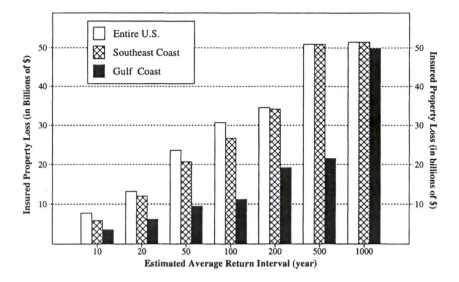

Fig. 19.4: Estimated insured property losses. Values are plotted as a function of return period in years and annual probabilities. Values are based on the catastrophe loss model of Applied Insurance Research (Clark 1997). The southeast coast is Florida (FL) to North Carolina (NC) and the Gulf coast is Texas (TX) to Alabama (AL). The entire United States includes the Gulf, southeast, and northeast coasts.

obtain corresponding losses to the reinsurance company. Worldwide databases allow the reinsurance company to get results for its entire portfolio for each scenario. Each of the simulated catastrophic events has an associated day of the year so that a pattern of losses throughout the year can be estimated. The model parameters are tested for sensitivity in a couple of ways to ensure robustness of the final results. For instance, the model can be run on different event archives from various data providers (vendors). Results and model parameters can be checked against other reinsurance companies. The underlying frequency and severity distributions can be altered based on current and future knowledge of events. For wind catastrophes this is particularly relevant given the ability to predict hurricane frequencies (see Chapters 13 through 16).

A proprietary hurricane catastrophe model was developed for a wide range of users at Applied Insurance Research (AIR). The model is used to estimate current loss potential as well as to estimate future loss projections (Clark 1997). Figure 19.4 is constructed from the model output which shows the hurricane loss potential for the entire United States together with loss potentials for the southeast and Gulf coasts separately. The loss estimates are given as probabilities of experiencing losses of various magnitudes from a single hurricane in any given year. The associated periods are also provided. The probability of the United States experiencing at least a $7.8 billion loss in any single year is estimated at 10%, while the annual probability of at least a $30.7 billion payout is 1%. The southeast coast dominates the hurricane loss projections for the country as a whole. The losses are estimated for insured property only and do not reflect the potential for total costs to be substantially higher. Losses from a substantial flood event associated with the hurricane do not show up in these estimates. The costs are not fixed, but continually change, usually in the direction of higher values as coastal properties and their values change (Clark 1997).

Risk Management Solutions (RMS) licenses financial risk models for the hurricane peril. Deterministic models are used to simulate historical or modified historical storms. Models can be used to permit user-defined storms based on a combination of attributes from past hurricanes. Stochastic models are based on specifying and combining probability distributions from a range of possible conditions. Output from the models are expressed as the probability of exceeding a loss in dollars for some specified amount. For RMS the financial modeling is one of three modules used for risk analysis, with the other two being hazard and damage modeling. Risk modeling is valuable for assessing insurance underwriting strategies and in managing and integrating capital. Individual risks can be evaluated along with how additional new risk will influence the overall structure of a portfolio.

19.6.2 Catastrophe Bonds

Selling bonds on the capital markets offers one possibility to large insurance companies looking to fund a possible catastrophic hurricane loss. An insurance business can sell a catastrophe bond to capable investors. Investors with large surplus capital might be willing to risk a substantial premium for a significant interest rate. As an illustration, a company sells a catastrophe bond worth $300 million that guarantees the interest rate, say 12%, but puts the premium at risk. The risk is whether or not a hurricane catastrophe of a specified magnitude, say less than $2 billion, will occur along the stretch of coastline heavily insured by the company or elsewhere. If no catastrophe occurs, then the investor keeps the premium plus the 12% interest. For a middle-range catastrophe, some reduction to the premium occurs and for the cataclysmic event the investor company forfeits all the premium, either immediately or to be paid back over a specified period.

In this way the insurance or reinsurance company is guaranteeing more liquidity to cover losses in the event of the rare, major catastrophe. This is done at the expense of much smaller annual losses paid out as interest on the bonds during most years. The key is to understand the likelihood of hurricane catastrophes (see Chapters 10 and 11). Although this kind of security, which views insurance risk as a tradable commodity subject to pricing is new, it has the potential to be a substantial market. In fact, in 1997 the first company to deal solely in the market of weather derivatives was formed. This New York-based Worldwide Weather Trading Company allows cash bets on the underlying weather in a manner that is analogous to other financial market derivatives, such as the price of foreign exchange. The Bermuda Commodities Exchange is scheduled to offer option contracts on catastrophe risks. Careful climatological analysis and forecasting play a key role in the use of catastrophe bonds.

19.6.3   Other Approaches

Insurers are formulating other approaches to handle the potential of future losses from hurricane catastrophes. Improvements in data collection and forecasting of hurricanes on all time scales is one area where substantial benefits may be realized, though traditionally they have not proven to be much help in avoiding large insured property losses. Such improvements combined with an open dialogue with weather and climate scientists offers the potential to reduce (but not eliminate) surprises from nature in the face of climate change. Scientists are keenly aware of the historical data sources as well as new data sources through advanced retrieval technologies from satellites and radar. Scientists are also involved with development and verification of hurricane forecasting technologies from track and intensity forecasts to future hurricane frequency distributions. This science along with future scientific advances will aid companies in setting appropriate rates.

Companies are also adopting catastrophic loss mitigation programs aimed at stricter building codes and statutes for development of individual and commercial properties along the coastline. Coastal buildings and infrastructure can be strengthened before a disaster strikes. Just as Californians have learned to build structures to withstand all but the most severe earthquakes, coastal residents will learn to build communities in ways to avert a hurricane catastrophe. New communities like Seaside in Florida that are built on the land side of the dunes will fair much better against the storm surge than communities built directly on the sand. For existing communities other mitigation strategies are possible. For instance, the Florida Department of Community Affairs (DCA) has created the Residential Construction Mitigation Program (RCMP). The program is designed to help homeowners retrofit their houses and protect their investments before disasters occur. Retrofitting generally takes the form of installing shatterproof glass, window shutters, hurricane clips,[11] and roof-shingle adhesives. If a roof can be repaired or strengthened to ensure it stays in place during a hurricane, then one can expect less damage to property both inside and outside the dwelling. Damage patterns observed after *Opal* hit northwestern Florida in 1995 are encouraging. The patterns suggest a strong relationship between destruction and buildings not built to code. Newer, stronger structures built on pilings were undamaged whereas adjacent older structures were totally destroyed (Barnes 1998).

The Natural Destruction Reduction Initiative (NDRI) is a partnership between the federal government and the private sector aimed at better building standards, better forecasts and warnings, and improved mechanisms for getting information out to people at risk. The partnership involves the National Institute of Standards and Technology (NIST), NOAA, and FEMA with the goal of hazard mitigation and the building of disaster-resistant communities. In the final analysis, the insurance industry is an efficient way to obtain the right balance for covering the risk of future hurricane damage (Burroughs 1997). To the degree that people make decisions to place themselves in

---

[11]Hurricane clips are used to connect the roof to the exterior and support walls of a house or building.

harm's way or fail to adopt mitigation strategies, they need to pay higher premiums for their insurance. At least in principle, if insurance premiums are adjusted to reflect the actual risk of covering a disaster, then costs may influence people's decision whether or not to locate in hurricane-prone areas. In the interim, however, a catastrophic event or events could overwhelm the insurance industry's ability to pay. In that case the federal government will be called upon to pick up the tab. In this regard it is noted that the hurricane peril—as a scientific problem—receives considerably less support in congress when compared to the support given to the National Earthquake Hazard Reduction Act (NEHRA) as a program in earthquake hazard mitigation.

20

Integrated Assessment

Economic losses in the United States caused by hurricanes have risen dramatically during the past several decades. In the early 1980s losses averaged approximately $1 billion annually, five times the losses in years during the 1950s. Concerning the U.S. hurricane problem, the challenge is to improve public and private decisionmaking with the help of scientific research (Pielke and Pielke 1997b). Knowledge of uncertainty and risk are important for understanding the economics associated with the hurricane peril. For example, the insurance industry (Chapter 19) depends on a careful assessment of various risks. It is essential to have an adequate record of the past and present behavior of hurricane climate as well as a proper interpretation of it. According to Changnon and Changnon (1990), there are four components of climate data important for the analysis of risks relevant to property insurance. The components include averages, variances, frequencies, and intensities of extreme events. A critical aspect of these components are long-term shifts. Beyond knowledge of the past and present climate, predictions of future climate changes are integral to risk analysis. Emphasis on the economics of hurricane activity affecting the United States is not meant to downplay the seriousness of the threat to human lives, both in the United States and elsewhere.

The ideas of risk and uncertainty have various meanings depending on context. Indeed, uncertainty is a term which by itself encompasses many notions. Here we are concerned with uncertainty as a component of decisionmaking. Under conditions of complete certainty, decisions produce certain consequences. However, under uncertainty, the consequences of a decision are not known precisely until a later time. The decisionmakers themselves are a broad and divergent group. Moreover, often their decisions will be driven by non-hurricane considerations. What are the relevant pieces of information concerning hurricanes that are important to decisionmakers? How does a company take advantage of this information to gain a competitive advantage over other companies when information exchange between companies is ubiquitous?

In this final chapter we examine a few aspects of this problem collectively known as *integrated assessment*. Integrated assessment is described as an interdisciplinary process which involves combining, interpreting, and communicating knowledge from

Table 20.1: Utility table for deciding on a Caribbean cruise during the hurricane season. The actions (a_i's) are a choice between cruise or don't cruise, while the states (s_i's) are a hurricane or no hurricane in the Caribbean Sea.

		s_1 Hurricane	s_2 No Hurricane
a_1	Cruise	1/4	1
a_2	Don't Cruise	1/2	1/2

different and often diverse fields. We begin by presenting some basic concepts relevant to the problem, including expected utility and the nature of probability. We emphasize the role of simulation technologies for addressing a host of questions about risk management and other opportunities with regard to the hurricane threat. As mentioned in Chapter 19, simulation models are now routinely used to assess probabilities of losses in the insurance industry. Similar technologies can be applied to assess questions of social vulnerability. The future is uncertain. This book has aimed at laying the groundwork for understanding the potential of reducing the uncertainty in terms of both science and society.

20.1 Expected Utility

Consequences are a result of a chosen act (action) and the realization of a particular state. Decision making under uncertainty involves coming up with a decision by choosing an action before the state is realized. As an example modified from McKenna (1986), suppose you need to decide today whether to book a Caribbean cruise. Ideally you would like to know whether or not a hurricane will occur in the Caribbean at the same time as your visit. To take full advantage of the opportunity you must decide before the arrival of the hurricane season. The choice of action is between "cruise" or "don't cruise" while the states are "hurricane" or "no hurricane." The degree of comfort you obtain depends on your decision and on nature. For instance, the greatest discomfort arises if you choose to go and the ship encounters a major hurricane causing sea sickness (or worse). Of course there is also discomfort in knowing that the weather was impeccable after deciding not to cruise. A choice to cruise met with a state of no hurricane affords maximum comfort. Arbitrary utilities can be assigned to each pair of action and states and arranged in a matrix as shown in Table 20.1. In this context, the decision amounts to choosing between a_1 and a_2 before either s_1 or s_2 is realized. If probabilities are assigned to the occurrence of a Caribbean hurricane (i.e., the states) then the analysis involves risk rather than uncertainties. This a hypothetical example of a decision-analytic model to handle a hurricane climate decision.

More formally, it is useful to characterize risk involved in choosing a course of action by a decisionmaker when the outcome is expressed as a range of possibilities weighted by associated probabilities. In its most transparent form this amounts to deciding between various alternatives. Here we compare a few toy portfolios. A toy portfolio with three expected returns is given by

$$L^1 = [l^{11}, l^{12}, l^{13}],$$ (20.1)

where

$$l^{11} = (0.2, 50), \quad l^{12} = (0.6, 25), \quad l^{13} = (0.2, -10).$$

Expected return l^{11} means there is a 20% chance of gaining $50, or an expected pay-out of $10. Summing the expected returns over the three choices determines that portfolio one has an overall expected value of $23. The choice is between portfolio one (L^1) or portfolio two (L^2) where

$$l^{21} = (0.3, 50), \quad l^{22} = (0.4, 25), \quad l^{23} = (0.3, -10),$$

so that the expected value of L^2 is $22. If maximizing the expected return is used to decide, then L^1 is the the choice. Other considerations come into play, however. For some investors L^2 is better simply because of the greater chance of the biggest return [l^{21} = (0.3, 50)]. Yet investors who are risk-averse will feel L^2 is less desirable because of increased chance of a loss [l^{23} = (0.3, −10)]. Thus, the various subjective choices are not completely covered by the concept of expected value.

A better concept is that of expected utility. Without going into the mathematical details (see McKenna 1986), we note that each expected pay-out is an ordered pair involving probability p_j and associated outcome y_j. For instance,

$$l^1 = (p_1, y_1) = (0.2, 50).$$ (20.2)

Consider a portfolio with four expected pay-outs (l^1, l^2, l^3, l^4), where the outcomes (y_j's) are ordered from largest gain (y_1) to largest loss (y_4). Now a different portfolio can be considered that includes only the best and worst outcomes (y_1 and y_4). Then there is an established principle that says for any y_j, there exists $0 < u_j < 1$, such that the outcome y_j is equally preferred ($\sim$) to a portfolio involving y_1 with probability u_j. This portfolio (L_j^*) is said to be a standard portfolio for y_j.

$$L_j^* = [l_j^{*1}, l_j^{*2}] \sim y_j,$$ (20.3)

where $l_j^{*1} = (u_j, y_1)$ and $l_j^{*2} = (1 - u_j, y_4)$. Here u_j is the probability of y_j when a company would be indifferent between choosing y_j and a portfolio involving only y_1 and y_4. Thus, u_j is a subjective evaluation made by individuals. Note that

$$L_1^* = [l_1^{*1}, l_1^{*2}] \sim y_1,$$ (20.4)

where $l_1^{*1} = (u_1, y_1)$ and $l_1^{*2} = (1 - u_1, y_4)$. For $u_1 = 1$, the expected pay-out

$$l_1^{*1} = (1, y_1) = y_1,$$

and the expected pay-out

$$l_1^{*2} = (0, y_4) = 0,$$

so that as anticipated,

$$L_1^* = [y_1, 0] \sim y_1.$$

Similarly for $u_4 = 0$,

$$l_1^{*1} = (0, y_1) = 0, \qquad l_1^{*2} = (1, y_4) = y_4,$$

so we can write

$$L_1^* = [0, y_4] \sim y_4.$$

One limitation of expected utility theory is the failure of invariance. Suppose portfolio L^1 is preferred to L^2 and L^2 to L^3, then invariance means that L^1 is preferred to L^3. This transitive relationship of preference is necessary under the theory of expected utility. Studies show, however, that people do not respond to identical choices in identical ways if presented the same options differently. For instance, the unhappiness people feel about a given loss is not compensated by the happiness felt about an equal amount of gain. As a result, the failure of invariance (or transitivity) is quite ubiquitous.

Other conceptions of rational choice are possible. An alternative theory is that of *bounded rationality*. The fact that real-world decisionmakers tend to want to simplify as much as possible is in contrast to the maximization hypothesis explicit in the concept of expected utility. Specifically, the decisionmaker strives to attain some satisfactory, though not likely optimal, level of achievement. This simpler strategy, called "satisficing" provides a cut-off point in the analysis. This may take the form of making an early decision then considering confirming evidence more strongly than contradictory evidence. Or it make take the form of following any course of action such that costs will never exceed a chosen threshold irrespective of an improvement in expected return. Satisficing amounts to taking the first satisfactory choice encountered out of a group of perceived options as opposed to setting an optimal choice in advance and searching until it is found (Wilson 1998). Moreover decisions are often made based on learned rules of thumb (heuristics) that have worked in the past. This can save a large amount of time and energy by reducing the calculations of probabilities and prediction to a series of judgments.

20.2   Probability

Probability is the language used to quantify risk and uncertainty. Overall our common-sense views of probability blend with the more formal and rigorous concepts used by experts, though there is a natural tendency to overestimate or underestimate the likelihood of extreme events. The rules for assigning and manipulating probabilities are

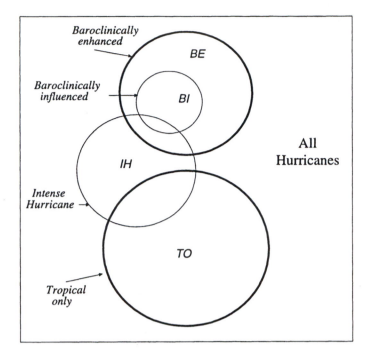

Fig. 20.1: Venn diagram illustrating the relationship between various categories of North Atlantic hurricanes used in this book. The diagram is used to depict the axioms of probability. TO stands for tropical-only, BE for baroclinically-enhanced, BI for baroclinically-initiated, and IH for intense hurricane. Note the categories of tropical-only and baroclinically-enhanced are mutually exclusive and exhaustive.

important in order to use the concepts in situations that may be complicated. Probabilities are real numbers in the closed interval between 0 and 1 that are mapped onto "events" or "states." If S represents the set of all hurricanes with intensities greater than 33 m s^{-1}, then the probability assigned to S, denoted as $P(S)$ is one. If two events are mutually exclusive (i.e., their intersection cannot occur—it has probability zero), then the probability of at least one of the two events occurring [i.e., $P(S_1 \cup S_2)$] is the sum of the probabilities of the two individual events. For example, the probability of a hurricane being either tropical-only or baroclinically-enhanced is equal to the probability of it being tropical-only plus the probability of it being baroclinically-enhanced (see Figure 20.1). Moreover, if the probability that the next hurricane is tropical-only is equal to p, then the probability that it is not tropical only is $1 - p$. These are axioms of probability.

Probabilities are interpreted in two rather distinct ways. One interpretation is based on the evaluation of past frequencies and the other is linked to forecasting the future. Neither of the two interpretations invalidates the mathematical axioms. The classical, or frequentist, view defines the probability of an event in a particular trial as the frequency with which it occurs in a long sequence of similar trials. For example, the probability of observing exactly one intense hurricane in any particular season is 27% since over the "long" sequence of 111 years there have been 30 years with one and only one intense hurricane. More precisely, the law of large numbers guarantees that as the number of years in the sequence goes to infinity, the probability that the ratio of intense hurricane years to total years will differ from the actual ratio by more than an arbitrary constant approaches zero. It is a theoretical concept where it is appropriate to think of some system generating intense hurricanes from year to year in a stochastic (statistical) manner. The assumption is that every year is exchangeable with every other year in the sequence.

The frequentist interpretation is sound and practical with a strong empirical foundation. The limitation is that oftentimes in climatology the notion of probability may seem justified, but the frequentist interpretation is useless. For instance, it is regularly not clear what the relevant sequence of trials of similar events is. If we include all years in the sequence for estimating the probability of a major hurricane, then we ignore the evidence that years of above normal rainfall in western Africa are associated with more intense hurricanes compared to the dry years. In other words, wet years are not identical to dry years for generating intense hurricanes. Note that similar arguments can be made for other variables, like wind shear, thus elevating each year to the status of a unique event, at the same time rendering the assumption of a sequence of identical events rather useless.

In contrast, from the personal viewpoint the statement that the probability is 20% that one major hurricane will occur next year represents a degree of belief. The statement of probability may arise from considering the output of an objective model, but is encompasses opinions about the reasonableness of the model and other factors. This subjective (or Bayesian) view is the amount of confidence an individual has that an event will occur given all the relevant information presently known to that person. It is useful to express this probability as a function of two arguments, $P(X|y)$, where X is the uncertain event, and y is the amount of information on which the probability is conditioned. In this way, probability is purely subjective in that no one need agree on its precise value. This represents a limitation since it allows different individuals to have different probabilities concerning a particular event that either will or will not occur. Yet it provides for more general applicability as well as being more aligned with intuition. The important thing to remember is that there is no such thing as *the* probability of occurrence. Different people or one person at different times may assign distinct probabilities to the same occurrence.

The subjective view does not imply probability assignments can be completely arbitrary. In order to be proper probabilities they must be consistent with the axioms of probability stated previously. Rather, the subjectivist view extends the use of probability to include a variety of situations.[1] More generally, the Bayesian viewpoint focuses on the frequent occasion when we have good intuitive judgment about the probability

---

[1] For example, in Chapter 14 we saw how Bayes theorem can be applied to make a probabilistic prediction.

of some event and want to understand how to alter these judgments as actual events unfold (Bernstein 1996). A thinking person will use the available data to establish a personal degree of belief. The probability of five category three hurricanes striking the coastline next year is extremely low from both objective and subjective perspectives. There will not necessarily be agreement on how many landfalling storms will occur next decade.

The idea of degree of belief is also quite useful in the construction of scientific hypotheses. My personal degree of belief in the hypothesis of negative feedback mechanisms operating to keep the maximum frequency of North Atlantic hurricanes in check on time scales of a year or so is based on my scientific investigations into hurricane activity and atmospheric scaling behavior. Understanding hurricane climate from these two areas of inquiry provides a unique background that will assign a probability to the hypothesis. My probability will no doubt differ from someone else's probability assignment based on their unique insights. Science advances by repeatedly analyzing and testing claims made by a hypothesis. Results from the tests are used to modify probability statements concerning the hypothesis. Only from the subjective viewpoint can probabilities be used in this important way.

20.3   Social Vulnerability to U.S. Hurricanes

Fluctuations in North Atlantic hurricane activity have had, and will continue to have, social and economic consequences in the United States. Vulnerability to hurricanes will certainly grow as population and crowding continues to intensify along the coastlines. Fortunately, casualties from U.S. hurricanes has dropped dramatically throughout the 20th century. Hurricane *Audrey* in 1957 was the last U.S. hurricane to result in more than 300 deaths. The frequency of U.S. hurricanes from the middle 1940s to the middle 1950s prompted congress to invest in improved hurricane warnings and preparedness. Government actions toward public safety and recovery from hurricanes are part of a national hurricane policy. The policy came about progressively through a series of event-driven crises management efforts from the Galveston hurricane of 1900 to hurricane *Camille* of 1969 (Simpson 1998). In order to address the important questions related to what the future holds, multidisciplinary research is needed to establish linkages between the natural and social sciences. According to Pielke (1996), the recommendations made in the *1995 Assessment of the Intergovernmental Panel on Climate Change* (IPCC) point to a need to better understand the impact of hurricanes by addressing the interrelation of physical and societal dimensions of these impacts. Such multidisciplinary knowledge of hurricane impacts is crucial for politicians who need to make policy decisions over various space and time scales.

According to Pielke and Pielke (1997a), one year after hurricane *Andrew* devastated parts of southern Florida, Robert C. Sheets explained to congress how much worse the situation might have been had *Andrew* hit a mere 30 km to the north. Under that scenario, the cost to Florida would have been two to three times more than it was. In fact, as shown in Chapter 7, the U.S. Atlantic and Gulf coasts as a whole have fared well over the past several decades compared to the decades of the 1940s and 1950s. Increases in population and property along the nation's coastlines coming at a time when major hurricanes have been less frequent than the long-term average, suggests that the United States is vulnerable to a near-future hurricane disaster. Societal vulnerability to hurricanes is a function of exposure and frequency (Pielke 1995, 1997). It makes sense that if people and property are not exposed to the threat of hurricanes or if no hurricanes occur then societal vulnerability is nil. Exposure is defined by the number of people in harm's way and by the cost of properties which could be destroyed. It is a function of three factors as indicated in (Pielke and Pielke 1997a). They include population at risk, property at risk, and preparedness. Chapter 9 gives population statistics for Puerto Rico, Jamaica, and Bermuda, while Chapter 17 examines the population statistics for U.S. coastal counties from Brownsville, Texas to Eastport, Maine. Chapter 18 considers insured property values for coastal states. Preparedness refers to the ability to reduce vulnerability. It is a general concept that incorporates factors such as planning, mitigation, response, restoration, as well as others. Mitigation has various components including technical, practical and political, all of which are subject to local conditions and resources making preparedness variable from one community to the next (Pielke and Pielke 1997a). With population and property values on the rise in most communities, preparedness is the principal means for reducing societal vulnerability to hurricanes. This requires investment in emergency management operations as a key element to community preparedness. This book has stressed various aspects of North Atlantic hurricane frequencies (Chapters 4, 5, and 6), including the occurrence of devastating landfalling storms (Chapters 8 and 9). Emphasis has been on past, present, and future activity. Forecasts of future activity will get better, becoming more specific and allowing for longer lead times.

A responsibility of the U.S. Global Change Research Program (USGCRP) is to quantify the potential consequences of climate change and climate variability on social and natural systems. The quantification is performed in terms of scientific uncertainties. Integrated assessment studies investigate individual components of a larger system (e.g., changes in hurricane landfall probabilities) and show how changes in the individual components interact and affect other parts of the system (e.g., landuse, insurance rates, etc.) Assessments are carried out by integrating vertically from the climate system through socioeconomic impacts and by integrating horizontally to illuminate and identify climate change impacts both locally and by resource. The impact of climate change on regional landfall frequencies will influence human health, ecosystems, biodiversity, tourism and recreation, as well as other social and economic systems.

20.4   Value of Seasonal Predictions

Knowledge of potential future scenarios is crucial. The essence of risk management lies in maximizing the areas where knowledge or control of future outcomes is possible, while minimizing the areas where no information exists or where control is unlikely. The traditional view of seasonal climate predictions is that although they contain some information, they are not skillful enough to be of economic benefit. Yet the relationship between forecast skill and forecast utility is not direct (Krzysztofowicz 1992, Roebber and Bosart 1996). Forecasts with only marginal skill can have important economic value under certain situations. For instance, Katz et al. (1987) demonstrate that under some circumstances, relatively small increases in forecast quality can be of significant economic value to wheat farmers in the northwestern U.S. Great Plains faced with a decision to plant a crop or let the land lie fallow (Livezey 1990). Moreover, the energy and insurance industries are able to take advantage of seasonal climate forecasts.

20.4.1   Potential Benefits

Perhaps the best response to the concern that both weather and economic forecasts are wrought with so much uncertainty as to render them useless is found in the fundamental nature of decisionmaking in a free society. Since those who make decisions are generally the ones accountable for the choices at election time, they are constrained to take into account all available options. This includes how the various options might be influenced by future conditions. For the hurricane hazard, these future conditions are related to climate change. The success or failure of policy or business decisions ultimately involves making a reasonable assessment of the future, which in itself is a form of prediction. Moreover, in a democratic forum the choice of ignoring available forecasts, even ones wrought with uncertainty, is often far more difficult to justify (Burroughs 1997).

According to Robert E. Livezey of the U.S. Climate Prediction Center, there are two necessary conditions that must be met by scientists in order for a broad range of users including government officials and business leaders to take advantage of skillful, long-range climate predictions. The first involves the creation of a perception in the user community that forecasts are skillful enough to be of some value. This requires the developers of models to provide accurate analysis of skill variability including a description of the variability and a demonstration of reproducibility in independent forecast situations. The latter of which is often accomplished through cross-validation techniques when statistical models are involved (see Chapter 14). The second condition involves matching of reliable forecasts with users who can take advantage of them and the development of mechanisms to facilitate coordination between model developers and model users.

One such mechanism has recently emerged to facilitate exchange between hurricane climate scientists and the catastrophe reinsurance industry. The Risk Prediction Initiative (RPI) founded by Anthony Knap, Anthony Michaels, and Mark Johnson in 1994 is a nonprofit organization of the Bermuda Biological Station for Research (BBSR). RPI organizes meetings between reinsurance company personnel and climate scientists for the purpose of keeping the insurance industry informed about new findings in climate research. In exchange RPI sponsors scientists to study various aspects of the hurricane problem. Focus is on issues most relevant to the insurance industry, but the exchange provides scientists an opportunity to better advance climate science. For instance, reliable hurricane return-period estimates along the U.S. coastline are of significant interest to the industry to warrant funding the development of proxy hurricane records (see Chapter 3). Results from this work will naturally lead to new scientific hypotheses concerning mechanisms responsible for millennial scale fluctuations in hurricane activity.

Important decisions related to the impacts of hurricanes are grounded in the notion of future likelihood (Pielke 1996). Changes to the statutes of the South Florida Building Code, for example, have recently been made based on projected expected hurricane incidence (Englehardt and Peng 1996). The projections are extrapolations from historical data. An improved understanding of past hurricane climate variability and improvements in forecasting future activity coupled with integrated assessment models and simulations will lead to a better understanding of hurricane impacts beyond what can be obtain through extrapolation. The judicious and rigorous development and utilization of forecast models of hurricane activity is certainly an important component. Yet better decisions and a reduction in societal vulnerability will not necessarily follow directly from improvements in long-lead forecasts of hurricane activity (Pielke 1996). For instance, multi-year forecasts are valuable to the reinsurance industry only to the extent that the companies are *confident* in the underlying science and have the *knowledge* to be able to incorporate the forecasts into their risk analysis models (see Chapter 19). Reduced vulnerability will result only when the new information is used in an adequate and appropriate way.

The discussion thus far has concentrated on what are called prescriptive models of the value of hurricane forecasts. These are based on what the user *ought* to decide in a given situation. A rational analysis leads to an optimal decision. Given the free cost of forecast information, prescriptive models cannot produce a negative value of information. In an optimal world, useless information is disregarded. Information that does not change any decision has zero value. Thus, prescriptive models represent the potential value of forecast information, but may not give reliable estimates of the actual value to a user (or to a group of users) given decision practices already in place (Stewart 1997). For instance, there is little guarantee that decision makers will behave optimally in practice (Wilks 1997). On the other hand, descriptive models can estimate the actual value of information but require knowledge of a user's decisionmaking model. A limitation of the descriptive approach is that it may overlook the potential of additional value resulting from redesigning current decision procedures.

20.4.2 Descriptive Approach

Of course, perfect predictability will doom the insurance industry since risk is eliminated. On one level, assessing the value of improved hurricane forecasts can be broken down into two steps. The first involves estimating the consequences of large differences in year-to-year or multi-year hurricane frequencies and variances on net profits using

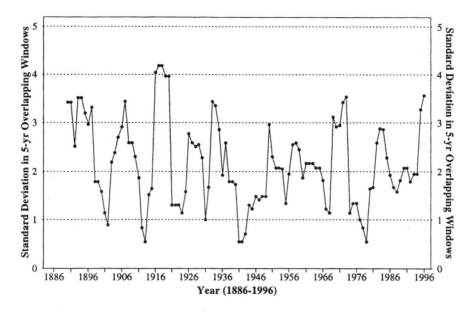

Fig. 20.2: Variability in the frequency of North Atlantic hurricanes over the period 1886–1996. Values are running 5-year standard deviations (*s.d.*). The standard deviation in a 5-year interval is plotted on the last year in the interval. Note the large swings in variance from year to year. There is some indication of a correspondence between high variance and warm ENSO events, particularly since 1960.

economic simulation models. Large variations in the interannual variability of hurricane frequency do indeed exist as shown in Figure 20.2. A 5-year running standard deviation of the number of North Atlantic hurricanes over the period 1886–1996 shows periods of large year-to-year variance interspersed with periods of less variance. The middle 1990s is a period of large volatility in hurricane activity. The variability rivals the variability seen during the early 1970s, but is less than the variability during the late 1910s due to the very active 1916 season among a string of below-average years. Similar results are obtained with 10-year intervals. A 7-year trimmed standard deviation that eliminates the most and least active year in each interval indicates that the 1990s are unparalleled with respect to interannual hurricane volatility. Not surprisingly, there appears to be some correspondence between high interannual hurricane variability and warm ENSO events. The dominant period in the interannual variability is 12 to 14 years. Although the precise frequency is related to the 5-year window length, the oscillation time corresponds to the periodicity in the movement of sea-surface temperature anomalies over the North Atlantic (Sutton and Allen 1997) and to the pattern of North Atlantic sea-level pressure variability (Deser and Blackmon 1993). The large interannual variability in hurricane activity during the 1990s may be a signal of important changes in hurricane climate.

The second step toward assessing utility of seasonal forecasts involves the integration of the differences and variability in profit margins into a decision-making model. The model can be used directly to assess the overall value of hurricane forecasts to a particular company and to project future profitability. The situation is complicated however, by the fact that people's behavior is influenced by their perceptions. Thus economic forecasts arising from predictions of future hurricane activity will feedback into the economic analysis as changes are made in response to the prognostications. In this context, the question of whether policy and business decisions will be rational or amount to an overreaction is relevant (Burroughs 1997).

20.5   Chaos and Complexity

The value of improved hurricane forecasts to individual users can be understood from both the prescriptive and descriptive approaches. However, the value to a group of users, not all sharing the same interests, is not as straightforward. Help in this regard may come from modern ideas in chaos and complexity theories. The dynamic efficient frontier model outlined in Chapter 19 allows for a characterization of free market behavior based on complexity theory (Prigogine 1997). Complexity theory is rooted in the notion of far-from-equilibrium conditions, where a system is capable of displaying emergent behavior. A healthy and robust insurance market, characterized by many individual interacting companies, is better conceptualized as in a state of perpetual change rather than one of stability due to equilibrium constraints. The latter view is the basis of much of the classical scientific and economic analyses. It should be noted that scientists understand complexity as the behavior of systems containing many interacting agents existing on the interface of regularity and chaos. This is also known as the *edge of chaos*.

All chaotic systems are essentially regular (or periodic). What makes them appear random is a continual transition from one periodic orbit to another. Under the influence of nonlinear interactions, only a relatively few agents are needed to produce chaotic behavior. The development of hurricanes and patterns of hurricane frequency may be the result of underlying chaotic behavior. The self-similar scaling relationships identified in Chapter 8 for landfalling hurricanes suggest this may be so. This would severely limit the utility of classical statistics. In fact, to paraphrase Zebrowski (1997), chaos throws a wrench into the very definition of hurricane climate. If the atmosphere and oceans are inherently chaotic, then averages and norms have no physical basis. Climate scientists, for example, might say it is rather meaningless to talk about global climate conditions half way between an ice age climate and an interglacial climate. The "average" climate state may be inherently unstable. Fortunately, there are a number of alternatives to classical statistical analysis, many of which have recently emerged from chaos and complexity theories.

Complexity theory and the idea of deterministic chaos are being used in many diverse fields including the study of weather and climate (Elsner and Tsonis 1992) and the

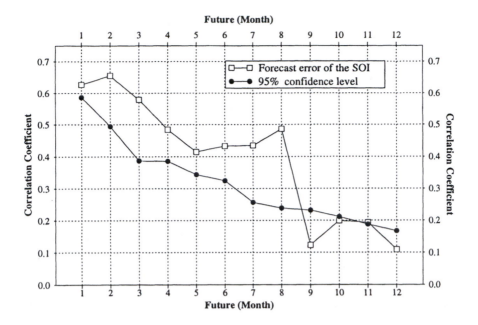

Fig. 20.3: Prediction error from a model of the Southern Oscillation Index (SOI). Values are correlation coefficients between actual and predicted SLP anomalies for "future" months in a hindcast experiment. The 95% confidence level on the correlation is computed using the method of surrogate data (Theiler et al. 1992). Adapted from Elsner and Tsonis (1993). The decay in prediction skill with time is suggestive of chaos in the ENSO.

analysis of financial markets (Brock 1986). This revolution has two thrusts. One represents a fundamental shift in thinking away from the ideas of stability and order toward the concepts of instability, adaptability, and change. The other is the development and adoption of new techniques for analysis and prediction. Based on a theoretical model of the atmosphere and ocean interactions, it is suggested that ENSO might be chaotic (Vallis 1986). Evidence coming from new techniques in data analysis provides support for the hypothesis (see Figure 20.3). Decay of the prediction skill using a nonlinear time-series model suggests underlying chaotic dynamics in the ENSO. Tools such as these provide new insights into the climate of hurricanes.

We speculate that if the efficient frontier is modeled as a state of a chaotic dynamical system approaching but never in equilibrium, then concepts and tools from complexity theory can be used to make predictions about where the frontier might be at some future time relative to a particular preference state of risk and expected return. New data-analytic tools can be used to document and predict future volatility of any state of the system allowing for better decisions concerning the direction and turbulence of the insurance market. Under this paradigm, the necessary requirements of risk management according to Chorafas (1994) are rapid access to current information and algorithms for manipulating large data streams in a meaningful way. This should lead to a consistent level of performance despite the occasional inevitability of set backs. One particularly promising set of algorithms for use in risk management are based on simulation technologies.

20.6 Simulation

Computer simulation of complex systems is an emerging field of study. A formal and rigorous approach to integrated assessment can be based on explanatory simulation. Explanatory simulation is achieved through implementation of artificial worlds. The creation of artificial worlds is accomplished through a sequence of stages as outlined by Melanie Mitchell at the Sante Fe Institute (SFI).

• Simplify the real-world problem as much as possible, keeping only what appears essential to answering the questions being asked.

• Write a program that simulates the individual agents of the system, their individual rules for action and interaction, along with whatever random elements appear appropriate.

• Run the program many times with different random numbers and collect the data and statistics from the various runs.

• Try to understand how the simple rules used by the individual agents give rise to the observed global behavior of the system.

• Change the parameters in the system to identify sources of behavior and to learn the effects of different parameter settings (like an improved level of forecast skill in a long-lead model of landfalling hurricanes).

• Simplify the simulation further, if possible, or add new elements that appear necessary.

One simulation approach is called *swarm* for its analogy with the way insect colonies result in collective behavior in the absence of a central authority. The swarm software system, developed by Chris Langton and colleagues at SFI, is a multiple-agent software platform for the simulation of complex adaptive systems. As formulated, swarm is a standardized and flexible set of software to achieve the stages of an artificial world simulation with minimal hassles from purely engineering and programming matters. This is possible because swarm offers a wide spectrum of tools and a kernel to drive the simulation.

The modeling formalism used in swarm is that of a collection of independent but interacting agents. There are no assumptions about what the agents are or what kind of model is being designed. Moreover, there are no specific requirements concerning particular environments nor are their constraints in the types of interacting mosaics that evolve. As noted in Minar et al. (1996), swarm simulations perform in a variety of different fields such as chemistry, economics, ecology, and political science. The atom (or primitive) of a swarm simulation is the agent. An agent is a player in the system that can produce events, which in turn influence other agents or itself. The events happen only at discrete time steps. In ecology, the prototypical agents are the set of predators, the set of prey, and the food supplies; whereas in an economic simulation, relevant to the insurance industry agents can be companies, stockbrokers, shareholders, and banking institutions (Casti 1997).

The swarm is a collection of agents that display emergent behavior unpredictable by examining local interaction rules. The swarm represents the entire model. Yet it is possible to build a meta-model where the collective behavior of the swarm becomes a single agent interacting with other swarm agents and primitive agents. The capacity to build meta-models is powerful, flexible, and analogous to real-life evolution and competition. Since swarms can be created and destroyed as the simulation proceeds, they can be used to model systems where multiple levels of description emerge in the absence of a central control. The purpose of swarm, or swarm-like simulations for decisionmakers is to provide an analytic infrastructure for understanding possible futures. As economic conditions change, financial opportunities expand, and forecasts of climate perils improve, the management of risk should reflect the evolving and often vague market conditions. In the context of investment opportunities, simulations could be used for predicting the details of large market swings on individual portfolios. Simulation technologies represent the wave of the future for many tasks associated with integrated assessment. It is likely that these technologies will be applied with success to the hurricane problem.

References

Aagaard, K., and E. C. Carmack, 1989: The role of sea ice and fresh water in the Arctic circulation. *J. Geophys. Res.*, 94, 14485–14498.

AMS, 1986: Is the United States headed for hurricane disaster? *Bull. Amer. Meteor. Soc.*, 67, 537–538.

Angell, J. K., and J. Korshover, 1985: Surface temperature changes following six major volcanic episodes between 1780 and 1980. *J. Climate Appl. Meteor.*, 24, 937–951.

Arkin, P. A., 1982: The relationship between interannual variability in the 200-mb tropical wind field and the Southern Oscillation. *Mon. Wea. Rev.*, 110, 1393–1404.

Avila, L. A., 1991: Atlantic tropical systems of 1990. *Mon. Wea. Rev.*, 119, 2027–2033.

Avila, L. A., and G. B. Clark, 1989: Atlantic tropical systems of 1988. *Mon. Wea. Rev.*, 117, 2260–2265.

Avila, L. A., and R. J. Pasch, 1992: Atlantic tropical systems of 1991. *Mon. Wea. Rev.*, 120, 2688–2696.

——, 1995: Atlantic tropical systems of 1993. *Mon. Wea. Rev.*, 123, 887–896.

Baker, E. J., 1984: Public response to hurricane probability forecasts. NOAA Tech. Memo. NWS FCST 29.

——, 1991: Hurricane evacuation behavior. *Int. J. Mass Emergencies and Disasters*, 9, 287–310.

Ballenzweig, E. M., 1959: The relation of long period circulation anomalies to tropical storm formation and motion. *J. Meteor.*, 16, 121–139.

Barnes, J., 1998: *Florida's Hurricane History*. The University of North Carolina Press.

Barnes, S. L., 1964: A technique for maximizing details in numerical weather map analysis. *J. Appl. Meteor.*, 3, 396–409.

Barnston, A. G., and H. M. van den Dool, 1993: A degeneracy in cross-validated skill in regression-based forecasts. *J. Climate*, 6, 963–977.

Barron, E. J., 1989: Severe storms during Earth history. *Geological Society of America Bulletin*, 101, 601–612.

Basher, R. E., and X. Zheng, 1995: Tropical cyclones in the southwest Pacific: Spatial patterns and relationships to Southern Oscillation and sea surface temperature. *J. Climate*, 8, 1249–1260.

Bengtsson, L., M. Botzet, and M. Esch, 1995: Hurricane-type vortices in a general circulation model. *Tellus*, 47A, 175–196.

—, 1996: Will greenhouse gas-induced warming over the next 50 years lead to higher frequency and greater intensity of hurricanes? *Tellus*, 48A, 57–73.

Berger, J. D., 1985: *Statistical Decision Theory and Bayesian Analysis*. Springer-Verlag.

Bernstein, P. L., 1996: *Against the Gods: The Remarkable Story of Risk*. John Wiley and Sons.

Bjerknes, J., 1964: Atlantic air-sea interaction. *Advances in Geophysics*, Academic Press, 1–82.

Blackman, R. B., and J. W. Tukey, 1958: *The Measurement of Power Spectra*, Dover Publications.

Blodget, L., 1857: *Climatology of the United States, and of the Temperate Latitudes of the North American Continent*. Lippincott.

Bosart, L. F., and B. E. Schwartz, 1979: Autumnal rainfall climatology of the Bahamas. *Mon. Wea. Rev.*, 107, 1663–1674.

Bosart, L. F., and J. A. Bartlo, 1991: Tropical storm formation in a baroclinic environment. *Mon. Wea. Rev.*, 119, 1979–2013.

Bowie, E. H., 1922: Formation and movement of West Indian hurricanes. *Mon. Wea. Rev.*, 50, 173–179.

Box, G. E. P., and G. M. Jenkins, 1976: *Time Series Analysis: Forecasting and Control*. Holden-Day, Inc.

Brennan, J. F., 1935: Relation of May-June weather conditions in Jamaica to the Caribbean tropical disturbances of the following season. *Mon. Wea. Rev.*, 63, 13–14.

Brock, W. A., 1986: Distinguishing random and deterministic systems. *J. Economic Theory*, 40, 168–195.

Broecker, W. S., 1991: The great ocean conveyor. *Oceanography*, 4, 79–89.

—, 1994: Massive iceberg discharges as triggers for global climate change. *Nature*, 372, 421–424.

Broomhead, D. S., and G. P. King, 1986: Extracting qualitative dynamics from experimental data. *Physica D*, 20, 217–236.

Brockwell P. J., and R. A. Davis, 1991: *Time Series Theory and Methods*. 2nd. Ed., Springer-Verlag.

Bunting, A. H., M. D. Dennett, J. Elston, and J. R. Milford, 1975: Seasonal rainfall forecasting in West Africa. *Nature*, 253, 622–623.

Burroughs, W. J., 1992: *Weather Cycles Real or Imaginary?* Cambridge University Press.

—, 1997: *Does the Weather Really Matter? The Social Implications of Climate Change*. Cambridge University Press.

Burton, I., R. W. Katz, and G. F. White, 1993: *The Environment As Hazard*. 2nd Ed., The Guilford Press.

Caracena, F., and J. M. Fritsch, 1981: Focusing mechanisms in the Texas flash floods of 1978. NOAA Tech. Memo. ERL OWRM–12.

Carlson, T. N., 1971: An apparent relationship between the sea-surface temperature of the tropical Atlantic and the development of African disturbances into tropical storms. *Mon. Wea. Rev.*, 99, 309–310.

Carr, L. E., III, M. A. Boothe, and R. L. Elsberry, 1997: Observational evidence for alternative modes of track-altering binary tropical cyclone scenerios. *Mon. Wea. Rev.*, 125, 2094–2111.

Carson, D. A., 1992: *The Hadley Centre Transient Climate Change Experiment.* U.K. Meteorological Office.

Carter, M. M., and J. B. Elsner, 1996: Convective rainfall regions of Puerto Rico. *Int. J. Climatol.*, 16, 1033–1043.

——, 1997: A climatology of tropical cyclone rainfall over Puerto Rico. Preprints, *22nd Conf. on Hurricanes and Tropical Meteorology*, Ft. Collins, CO, Amer. Meteor. Soc., 596–597.

Carton, J. A., X. Cao, B. S. Giese, and A. M. daSilva, 1996: Decadal and interannual SST variability in the tropical Atlantic Ocean. *J. Phys. Oceanogr.*, 26, 1165–1175.

Casti, J. L., 1997: *Would-Be Worlds: How Simulation Is Changing The Frontiers of Science.* John Wiley and Sons.

Caviedes, C. N., 1991: Five hundred years of hurricanes in the Caribbean: Their relationship with global climate variabilities. *GeoJournal*, 23.4, 301–310.

Chang, P., L. Ji, and H. Li, 1997: A decadal climate variation in the tropical Atlantic Ocean from thermodynamic air-sea interactions. *Nature*, 385, 516–518.

Changnon, S. A., and J. M. Changnon, 1990: Use of climatological data in weather insurance. *J. Climate*, 3, 568–576.

Changnon, S. A., D. Changnon, E. R. Fosse, D. C. Hoganson, R. J. Roth Sr., and J. M. Totsch, 1997: Effects of recent weather extremes on the insurance industry: Major implications for the atmospheric sciences. *Bull. Amer. Meteor. Soc.*, 78, 425–435.

Chapel, L. T., 1934: The significance of air movements across the equator in relation to development and early movement of tropical cyclones. *Mon. Wea. Rev.*, 62, 433–438.

Chen, L., 1995: Tropical cyclone heavy rainfall and damaging winds. *Global Perspectives on Tropical Cyclones.* WMO Tech. Rep. TCP-38.

Chorafas, D. N., 1994: *Chaos Theory in Financial Markets: Applying Fractals, Fuzzy Logic, Genetic Algorithms, Swarm Simulation & The Monte Carlo Method to Manage Market Chaos & Volatility.* Probus Publishing Company.

Clark, K. M., 1997: Current and potential impact of hurricane variability on the insurance industry. *Hurricanes: Climate and Socioeconomic Impacts*, H. F. Diaz and R. S. Pulwarty, Eds., Springer-Verlag.

Cohen, T. J., and E. I. Sweetser, 1975: The "spectra" of the solar cycle and of data for Atlantic tropical cyclones. *Nature*, 256, 295–296.

Colebrook, J. M., 1978: Continuous plankton records: Zooplankton and environment, North East Atlantic and North Sea, 1948–1975. *Oceanologica Acta*, 1, 9–23.

Colón, J. A., 1953: A study of hurricane tracks for forecasting purposes. *Mon. Wea. Rev.*, 81, 53–66.

——, 1956: On the formation of Hurricane Alice, 1955. *Mon. Wea. Rev.*, 84, 1–14.

Conning & Company, 1994: *Lighting candless in the wind: Industry response to the catastrophe problem*. Conning & Company.

Cook, E. R., D. M. Meko, and C. W. Stockton, 1997: A new assessment of possible solar and lunar forcing of the bidecadal drought rhythm in the western United States. *J. Climate*, 10, 1343–1356.

Cry, G. W., 1965: Tropical cyclones of the North Atlantic Ocean. U.S. Weather Bureau Tech. Paper No. 55, 148 pp.

Cry, G. W., and W. H. Haggard, 1962: North Atlantic tropical cyclone activity, 1901-1960. *Mon. Wea. Rev.*, 90, 341–349.

Cubasch, U., R. Voss, G. C. Hegerl, J. Waszkewitz, and T. J. Crowley, 1997: Simulation of the influence of solar radiation variations on the global climate with an ocean-atmosphere general circulation model. *Climate Dynamics*, 13, 757–767.

da Silva, A. M., C. C. Young, and S. Levitus, 1994: *Atlas of surface marine data 1994*. Volume 1: Algorithms and procedures. National Oceanic and Atmospheric Administration.

Davis, C. A., and M. L. Weisman, 1994: Balanced dynamics of mesoscale vortices produced in simulated convective systems. *J. Atmos. Sci.*, 51, 2005–2030.

Davis, R. A., Jr., S. C. Knowles, and M. J. Bland, 1989: Role of hurricanes in the Holocene stratigraphy of estuaries: Examples from the Gulf coast of Florida. *J. Sedimentary Petrology*, 59, 1052–1061.

DeMaria, M., and J. Kaplan, 1994a: Sea surface temperature and the maximum intensity of Atlantic tropical cyclones. *J. Climate*, 7, 1324–1334.

——, 1994b: A statistical hurricane intensity prediction scheme (SHIPS) for the Atlantic basin. *Wea. Forecasting*, 9, 209–220.

DeMaria, M., S. D. Aberson, K. V. Ooyama, and S. J. Lord, 1992: A nested spectral model for hurricane track forecasting. *Mon. Wea. Rev.*, 120, 1628–1643.

Dennett, D. R, 1995: *Darwin's Dangerous Idea: Evolution and the Meanings of Life*. Touchstone.

Deser, C. and M. L. Blackmon, 1993: Surface climate variations over the North Atlantic ocean during winter: 1900-1989. *J. Climate*, 6, 1743–1753.

Devore, J. L., 1991: *Probability and Statistics for Engineering and the Sciences*. Brooks/Cole.

Dickson, R., E. M. Gmitrowicz, and A. J. Watson, 1990: Deep water renewal in the North Atlantic. *Nature*, 344, 848–850.

Druyan, L. M., and P. Lonergan, 1997: Implications of GCM-projected climate change for tropical cyclone climatology. Preprints, *10th Conf. on Applied Climatology*, Reno, NV, Amer. Meteor. Soc., 62–66.

Dunn, G. E., 1940: Cyclogenesis in the tropical Atlantic. *Bull. Amer. Meteor. Soc.*, 21, 215–229.

——, 1956: Areas of hurricane development. *Mon. Wea. Rev.*, 84, 47–51.

——, 1963: The hurricane season of 1962. *Mon. Wea. Rev.*, 91, 199–207.

——, 1964: The hurricane season of 1963. *Mon. Wea. Rev.*, 92, 128–138.

Dunn, G. E., W. R. Davis, and P. L. Moore, 1956: Hurricane season of 1956, *Mon. Wea. Rev.*, 84, 436–443.

Dunn, G. E., and B. I. Miller, 1964: *Atlantic Hurricanes.* 2nd ed., Louisiana State University Press.

Dvorak, V. F., 1973: A technique for the analysis and forecasting of tropical cyclone intensities from satellite pictures. NOAA Tech. Memo. NESS 45, 19 pp.

——, 1975: Tropical cyclone intensity analysis and forecasting from satellite imagery. *Mon. Wea. Rev.*, 103, 420–430.

Easterbrook, G., 1995: *A Moment On Earth: The Coming Age of Environmental Optimism.* Penguin Books.

Elsner, J. B., 1996: A probabilistic model of the number of intense Atlantic hurricanes for 1996. *Experimental Long-Lead Forecast Bulletin*, 5/1, 29–31.

Elsner, J. B., and A. A. Tsonis, 1992: Nonlinear prediction, chaos and noise. *Bull. Amer. Meteor. Soc.*, 73, 49–60.

——, 1993: Nonlinear dynamics established in the ENSO. *Geophys. Res. Lett.*, 20, 213–216.

——, 1996: *Singular Spectrum Analysis: A New Tool in Time Series Analysis.* Plenum.

Elsner, J. B., and C. P. Schmertmann, 1993: Improving extended-range seasonal predictions of intense Atlantic hurricane activity. *Wea. Forecasting*, 8, 345–351.

——, 1994: Assessing forecast skill through cross validation. *Wea. Forecasting*, 9, 619–624.

——, 1995: Multiple least-squares regression and Poisson model forecasts of Atlantic tropical storm activity for 1996. *Exp. Long-Lead Fcst. Bull.*, 4, (4) 19–20.

Elsner, J. B., G. S. Lehmiller, and T. B. Kimberlain, 1996a: Objective classifications of Atlantic hurricanes. *J. Climatol.*, 9, 2880–2889.

——, 1996b: Early August forecasts of Atlantic tropical storm activity for the balance of the 1996 season, using Poisson models. *Experimental Long-lead Forecast Bulletin*, 5 (3), 26–28.

Elsner, J. B., X. Niu, and A. A. Tsonis, 1998: Multi-year prediction model of North Atlantic hurricane activity. *Meteorol. Atmos. Phys.*, 68, 43–51.

Elsner, J. B., K.-B. Liu, and B. L. Kocher, 1999a: Trends in hurricane landfall probabilities in the U.S. Preprints, *23nd Conf. on Hurricanes and Tropical Meteorology*, Dallas, TX, Amer. Meteor. Soc.

Elsner, J. B., A. B. Kara, and M. A. Owens, 1999b: Fluctuations in North Atlantic hurricanes. *J. Climate*, 12, 427–437.

Emanuel, K. A., 1991: The theory of hurricanes. *Annual Review of Fluid Mechanics*, 23, 179–196.

Emanuel, K. A., 1997: Climate variations and hurricane activity: Some theoretical issues. *Hurricanes: Climate and Socioeconomic Impacts*, H. F. Diaz and R. S. Pulwarty, Eds., Springer-Verlag.

Emanuel, K. A., K. Speer, R. Rotunno, R. Srivastava, and M. Molina, 1995: Hypercanes: A possible link in global extinction scenarios. *J. Geophys. Res.*, 100, 13755–13765.

Enfield, D. B., and D. A. Mayer, 1997: Tropical Atlantic sea sea surface temperature variability and its relationship to El Niño-Southern Oscillation. *J. Geophys. Res.*, 102, 929–945.

Englehardt, J., and C. Peng, 1996: A Bayesian benefit-risk model applied to the south Florida building code. *Risk Analysis*, 16 (1), 81–91.

Epstein, E. S., 1985: *Statistical Inference and Prediction in Climatology: A Bayesian Approach*. Amer. Meteor. Soc.

Evans, J. L., 1993: Sensitivity of tropical cyclone intensity to sea surface temperature. *J. Climate*, 6, 1133–1140.

Evans, J. L., and K. S. McKinley, 1997: Maximum cyclone intensity and recurvature. Preprints, *22nd Conf. on Hurricanes and Tropical Meteorology*, Fort Collins, CO, Amer. Meteor. Soc., 356–357.

Eyre, L. A., 1989: Hurricane Gilbert: Caribbean record-breaker. *Weather*, 44, 160–164.

Ferguson, E. W., 1973: Comments on "The unnamed Atlantic tropical storms of 1970." *Mon. Wea. Rev.*, 101, 378–379.

Fernández-Partágas, J., and H. F. Diaz, 1995a: A Reconstruction of Historical Tropical Cyclone Frequency in the Atlantic from Documentary and Other Historical Sources: Part I: 1851–1870. Climate Diagnostics Center, Environmental Research Laboratories, NOAA, April 1995.

——, 1995b: A Reconstruction of Historical Tropical Cyclone Frequency in the Atlantic from Documentary and Other Historical Sources: Part II: 1871–1880. Climate Diagnostics Center, Environmental Research Laboratories, NOAA, June 1995.

——, 1996a: A Reconstruction of Historical Tropical Cyclone Frequency in the Atlantic from Documentary and Other Historical Sources: Part III: 1881–1890. Climate Diagnostics Center, Environmental Research Laboratories, NOAA, March 1996.

——, 1996b: A Reconstruction of Historical Tropical Cyclone Frequency in the Atlantic from Documentary and Other Historical Sources: Part I: 1891–1900. Climate Diagnostics Center, Environmental Research Laboratories, NOAA, November 1996.

——, 1996c: Atlantic hurricanes in the second half of the nineteenth century. *Bull. Amer. Meteor. Soc.*, 77, 2899–2906.

Fitzpatrick, P. J., J. A. Knaff, C. W. Landsea, and S. V. Finley, 1995: Documentation of a systematic bias in the aviation model's forecast of the Atlantic upper-tropospheric trough: Implication for tropical cyclone forecasting. *Wea. Forecasting*, 10, 433–446.

Foster, D. R., and E. R. Boose, 1992: Patterns of forest damage resulting from catastrophic wind in central New England. *USA Journal of Ecology*, 80, 79–98.

Fraedrich, K., 1986: Estimating the dimensions of weather and climate attractors. *J. Atmos. Sci.*, 43, 419–432.

Frank, N. L., 1970a: Atlantic tropical systems of 1969. *Mon. Wea. Rev.*, 98, 307–314.

——, 1970b: On the Nature of Upper Tropospheric Cold Core Cyclones over the Tropical Atlantic. Ph.D. thesis, Dept. of Meteorology, Florida State University. [Available from the Florida State University, Department of Meteorology, Tallahassee, FL, 32306-4520.]

Frank, N. L., and G. B. Clark, 1977: Atlantic tropical systems of 1976. *Mon. Wea. Rev.*, 105, 676–683.

Fugiwhara, S., 1931: Short note on the behavior of two vortices. *Proc. Physico-Mathematical Society of Japan, Series. 3*, 13, 106–110.

Garriott, E. B., 1895: Tropical storms of the Gulf of Mexico and the Atlantic Ocean in September. *Mon. Wea. Rev.*, 23, 167–169.

——, 1906: The West Indian hurricanes of September 1906. *Mon. Wea. Rev.*, 34, 416–417.

Gastwirth, J. L., 1970: The estimation of the Lorenz curve and Gini index. *The Review of Economics and Statistics*, 52, 306–316.

Gedzelman, S. D., 1980: *The Science and Wonders of the Atmosphere*. John Wiley and Sons.

Gentry, R. C., 1983: Genesis of tornadoes associated with hurricanes. *Mon. Wea. Rev.*, 111, 1793–1805.

Goldenberg, S. B., and L. J. Shapiro, 1996: Physical mechanisms for the relationships between El Niño, West African rainfall, and North Atlantic major hurricanes. *J. Climatol.*, 9, 1169–1187.

Gray, W. M., 1968: Global view of the origins of tropical disturbances and storms. *Mon. Wea. Rev.*, 96, 669–700.

——, 1979: Hurricanes: Their formation, structure, and likely role in the tropical circulation. *Meteorology Over the Tropical Oceans*, D. B. Shaw, Ed., Royal Meteor. Soc., 155–218.

——, 1984a: Atlantic seasonal hurricane frequency: Part I: El Niño and 30 mb quasi-biennial oscillation influences. *Mon. Wea. Rev.*, 112, 1649–1668.

——, 1984b: Atlantic seasonal hurricane frequency: Part II: Forecasting its variability. *Mon. Wea. Rev.*, 112, 1669–1683.

——, 1990: Strong association between West African rainfall and U.S. landfall of intense hurricanes. *Science*, 249, 1251–1256.

——, 1992: Summary of 1992 Atlantic Tropical Cyclone Activity and Verification of Author's Forecast. Report of Department of Meteorology, Colorado State University, 18 pp. [Available from Department of Atmospheric Sciences, Colorado State University, Ft. Collins, CO, 80523.]

Gray, W. M., and C. W. Landsea, 1992: African rainfall as a precursor of hurricane-related destruction on the U.S. East Coast. *Bull. Amer. Meteor. Soc.*, 73, 1352–1364.

Gray, W. M., C. W. Landsea, P. W. Mielke Jr., and K. J. Berry, 1992: Predicting Atlantic seasonal hurricane activity 6-11 months in advance. *Wea. Forecasting*, 7, 440–455.

——, 1993: Predicting Atlantic basin seasonal tropical cyclone activity by 1 August. *Wea. Forecasting*, 8, 73–86.

——, 1994: Predicting Atlantic basin seasonal tropical cyclone activity by 1 June. *Wea. Forecasting*, 9, 103–115.

Gray, W. M., J. D. Sheaffer, and C. W. Landsea, 1997: Climate trends associated with multidecadal variability of Atlantic hurricane activity. *Hurricanes: Climate and Socioeconomic Impacts*, H. F. Diaz and R. S. Pulwarty, Eds., Springer-Verlag.

Haarsma, R. J., J. F. B. Mitchell, and C. A. Senior, 1992: Tropical disturbances in a GCM. *Climate Dynamics*, 8, 247–257.

Hagemeyer, B. C., 1997: Peninsular Florida tornado outbreaks. *Wea. Forecasting*, 12, 399–427.

Haggard, W. H., 1958: The birthplace of North Atlantic tropical storms. *Mon. Wea. Rev.*, 86, 397–404.

Haigh, J. D., 1996: The impact of solar variability on climate. *Science*, 272, 981–984.

Hamill, T. M., 1997: Reliability diagrams for multicategory probabilistic forecasts. *Wea. Forecasting*, 12, 736–741.

Hand, D. J., 1981: *Discrimination and Classification.* John Wiley and Sons.

Hastenrath, S., 1990: Tropical climate prediction: A progress report, 1985–1990. *Bull. Amer. Meteor. Soc.*, 71, 819–825.

Hebert, P. J, 1977: Intensification criteria for tropical depressions in the western North Atlantic. NOAA Tech. Memo. NWS NHC-3, 22 pp.

Hebert, P. J., and N. L. Frank, 1973: Atlantic season summary, 1972. *Mon. Wea. Rev.*, 102, 456–465.

Hebert, P. J., and K.O. Poteat, 1975: A satellite classification technique for subtropical cyclones. NOAA Tech. Memo. NWS SR-83, 25 pp.

Hebert, P. J., and G. Taylor, 1975: Hurricane experience levels of coastal county populations from Texas to Maine. U.S. Department of Commerce, NOAA, NWS, Community Preparedness Staff and Southern Regions.

——, 1979a: Everything you always wanted to know about hurricanes. *Weatherwise*, 32, 60–67.

——, 1979b: Everything you always wanted to know about hurricanes. *Weatherwise*, 32, 100–107.

Hebert, P. J., G. Taylor, and R. A. Case, 1990: The deadliest, costliest, and most intense United States hurricanes this century (and other frequently requested hurricane facts). NOAA Tech. Memo., NWS NHC 31, Miami, FL.

Hebert, P. J., J. D. Jarrell, and M. Mayfield, 1992: The deadliest, costliest, and most intense United States hurricanes this century (and other frequently requested hurricane facts). NOAA Tech. Memo., NWS NHC 31, Miami, FL.

——, 1993: The deadliest, costliest and most intense United States hurricanes of this century. NOAA Tech. Memo. NWS NHC-31.

——, 1996: The deadliest, costliest and most intense U.S. hurricanes of this century. NOAA Tech. Memo. NWS TPC-1.

Henry, A. J., 1929: The frequency of tropical cyclones (West Indian hurricanes) that closely approach or enter continental United States. *Mon. Wea. Rev.*, 57, 328–331.

Henry, J. A., K. M. Portier, and J. Coyne, 1994: *The Climate and Weather of Florida*, Pineapple Press, Inc..

Hess, J. C., and J. B. Elsner, 1994a: Extended-range seasonal hindcasts of easterly-wave origin Atlantic hurricane activity. *Geophys. Res. Lett.*, 21, 365–368.

——, 1994b: Historical developments leading to forecasts of annual tropical-cyclone activity. *Bull. Amer. Meteor. Soc.*, 75, 1611–1621.

Hess, J. C., J. B. Elsner, and N.E. LaSeur, 1995: Improving seasonal hurricane predictions for the Atlantic basin. *Wea. Forecasting*, 10, 425–432.

Hewings, G. J. D., and R. Mahidhara, 1996: Economic impacts: Lost income, ripple effects, and recovery. *The Great Flood of 1993: Causes, Impacts, and Responses*, S. A. Changnon, Ed., Westview Press.

Ho, F. P., J. C. Su, K. L. Hanevich, R. J. Smith, and F. P. Richards, 1987: Hurricane Climatology for the Atlantic and Gulf Coasts of the United States. NOAA Tech. Rep. NWS 38.

Holland, G. J., 1997: The maximum potential intensity of tropical cyclones. *J. Atmos. Sci.*, 54, 2519–2541.

Hope, J. R., and C. J. Neumann, 1971: Digitized Atlantic tropical cyclone tracks. NOAA Tech. Memo. NWS HYDRO SR–55.

Howe, C. W., H. C. Cochrane, J. E. Bunin, and R. W. Kling, 1991: *Natural Hazard Damage Handbook: A Guide to the Uniform Definition, Identification, and Measurement of Economic and Ecological Damages from Natural Hazard Events*. Project Report for NSF Grant CES 8717115.

Hoyt, D. V., and K. H. Schatten, 1997: *The Role of the Sun in Climate Change*. Oxford University Press.

Hughes, P., 1987: Hurricanes haunt our history. *Weatherwise*, 40, 134–140.

——, 1994: The great leap forward. *Weatherwise*, 47, 22–27.

Hurrell, J., 1995: Decadal trends in the North Atlantic oscillation and regional temperature and precipitation. *Science*, 269, 676–679.

Jarrell, J. D., P. J. Hebert, and M. Mayfield, 1992: Hurricane experience levels of coastal county populations from Texas to Maine. NOAA Tech. Memo. NWS NHC-46.

Jarvinen, B. R., and C. J. Neumann, 1979: Statistical forecasts of tropical cyclone intensity. NOAA Tech. Memo. NWS NHC–10.

——, 1985: An evaluation of the SLOSH storm surge model. *Bull. Amer. Meteor. Soc.*, 66, 1408–1411.

Jones, B. G., and J. H. Mars, 1974: Regional analysis for development planning in disaster areas. OWRT Project No. A-045-NY, Grant No. 14-31-0001-4032, Cornell University, NY. [Available from the Center for Urban Development Research, Cornell University, 726 University Avenue, Ithaca, New York 14853.]

Jordan, C. L. and T. -C. Ho, 1962: Variations in the annual frequency of tropical cyclones, 1886-1958. *Mon. Wea. Rev.*, 90, 157–164.

Kanamitsu, M., 1989: Description of the NMC global assimilation and forecast system. *Wea. Forecasting*, 4, 335–342.

Kaplan, J. and M. DeMaria, 1995: A simple empirical model for predicting the decay of tropical cyclone winds after landfall. *J. Appl. Meteor.*, 34, 2499–2512.

Kates, R. W., 1980: Climate and society: Lessons from recent events. *Weather*, 35, 17–25.

Katz, R. W., B. G. Brown, and A. H. Murphy, 1987: Decision-analytic assessment of the economic value of weather forecasts: The fallowing/planting problem. *J. Forecasting*, 6, 77–89.

Keim, B. D., and J. F. Cruise, 1998: A technique to measure trends in the frequency of discrete random events. *J. Climate*, 11, 848–855.

Kendall, M.G. and A. Stuart, 1977: *The Advanced Theory of Statistics*. Griffin, London, 4th ed.

Kimberlain, T. B., 1996: Baroclinically-Initiated Hurricanes of the North Atlantic Basin. M.S. Thesis, Dept. of Meterology, The Florida State University. [Available from the Florida State University, Department of Meteorology, Tallahassee, FL, 32306-4520.]

Kimberlain, T. B., and J. B. Elsner, 1998: The 1995 and 1996 North Atlantic hurricane seasons: A return of the tropical-only hurricane. *J. Climate*, 11, 2062–2069.

Knaff, J. A., 1997: Implications of summertime sea-level pressure anomalies in the tropical Atlantic region. *J. Climate*, 10, 789–804.

Kocin, P. J., and J. H. Keller, 1991: A 100-year climatology of tropical cyclones for the northeast United States. Preprints, *19th Conf. on Hurricanes and Tropical Meteorology*, Miami, FL, Amer. Meteor. Soc., 152–157.

Konrad, H. W., 1985: Fallout of the wars of the Chacs: The impact of hurricanes and implications for prehispanic Quintana Roo Maya processes. Status, structure and stratification: Current archaeological reconstructions. University of Calgary.

Kraft, R. H., 1961: The hurricane's central pressure and highest wind. *Mariner's Weather Log*, 5, 155.

Krzysztofowicz, R., 1992: Bayesian correlation score: A utilitarian measure of forecast skill. *Mon. Wea. Rev.*, 120, 208–219.

Kushnir, Y., 1994: Interdecadal variations in North Atlantic sea surface temperature and associated atmospheric conditions. *J. Climate*, 7, 141–157.

Landsea, C. W., 1993: A climatology of intense (or major) Atlantic hurricanes. *Mon. Wea. Rev.*, 121, 1703–1713.

Landsea, C. W., and W. M. Gray, 1992: The strong association between western Sahel monsoon rainfall and intense Atlantic hurricanes. *J. Climatol.*, 5, 435–453.

Landsea, C. W., W. M. Gray, P. W. Mielke, Jr., and K. J. Berry, 1992: Long-term variations of western Sahelian monsoon rainfall and intense U.S. landfalling hurricanes. *J. Climate*, 5, 1528–1534.

Landsea, C. W., N. Nicholls, W. M. Gray, and L. A. Avila, 1996: Downward trends in the frequency of intense Atlantic hurricanes during the past five decades. *Geophys. Res. Lett.*, 23, 1697–1700.

Landsea, C. W., G. A. Bell, W. M. Gray, and S. B. Goldenberg, 1998: The extremely active 1995 Atlantic hurricane season: Environmental conditions and verification of seasonal forecasts. *Mon. Wea. Rev.*, 126, 1174–1193.

Lapham, I.A., 1872: List of the great storms, hurricanes and tornadoes of the United States (1635–1870). *J. Franklin Institute*, 63, 210–216.

Lawrence, J. R., and S. D. Gedzelman, 1996: Low stable isotope ratios of tropical cyclone rains. *Geophys. Res. Lett.*, 23, 527–530.

Lawrence, M. B., B. M. Mayfield, L. A. Avila, R. J. Pasch, and E. N. Rappaport, 1998: Atlantic hurricane season of 1995. *Mon. Wea. Rev.*, 126, 1124–1151.

Lazier, J. N. R., 1980: Oceanographic conditions of weathership *Bravo*, 1964-1974. *Atmos.-Ocean*, 18, 227–238.

Lehmiller, G. S., T. B. Kimberlain, and J. B. Elsner, 1997: Seasonal prediction models for North Atlantic basin hurricane location. *Mon. Wea. Rev.*, 125, 1780–1791.

Lighthill, J., G. Holland, W. Gray, C. Landsea, G. Craig, J. Evans, Y. Kurihara, and C. Guard, 1994: Global climate change and tropical cyclones. *Bull. Amer. Meteor. Soc.*, 75, 2147–2157.

Liu, K.-b., and M. L. Fearn, 1993: Lake-sediment record of late Holocene hurricane activities from coastal Alabama. *Geology*, 21, 793–796.

Livezey, R. E., 1990: Variability of skill of long-range forecasts and implications for their use and value. *Bull. Amer. Meteor. Soc.*, 71, 300–309.

Loomis, E., 1876: *American Journal of Science*, 3rd ser., 15, 12–67.

Ludlum, D. M., 1963: *Early American Hurricanes, 1492–1870*, Amer. Meteor. Soc.

Lydolph, P. E., 1985: *The Climate of the Earth*, Rowman & Allanheld.

Maddox, R. A., 1980: Mesoscale convective complexes. *Bull. Amer. Meteor. Soc.*, 61, 1374–1387.

Mardia, K. V., J. T. Kent, and J. M. Bibby, 1979: *Multivariate Analysis*. Harcourt-Brace.

Marks, D. G., 1992: The beta and advection model for hurricane track forecasting. NOAA Tech. Memo. NWS NMC–70.

Marth, D., and M. J. Marth, 1998: *Florida Almanac: 1998–1999*, B. McGovern, Ed., Pelican Publishing Company.

Martinson, E. O., and M. A. Hamdan 1972: Maximum likelihood and some other asymptotically efficient estimators of correlation in two way contingency tables. *J. Statistical Computation and Simulation*, 1, 45–54.

Marzban, C., 1998: Bayesian probability and scalar performance measures in Gaussian models. *J. Appl. Meteor.*, 37, 72–82.

Mass, C. F., and D. A. Portman, 1989: Major volcanic eruptions and climate: A critical evaluation. *J. Climate*, 2, 566–593.

Maunder, W. J., 1986: *The Uncertainty Business*, Methuen and Co., Ltd.

McBride, J. L., 1995: Tropical cyclone formation. *Global Perspectives on Tropical Cyclones*, WMO Tech. Doc. No. 693.

McKenna, C. J., 1986: *The Economics of Uncertainty*. Wheatsheaf Books, Ltd., Sussex.

Mehta, V. M., 1998: Variability of the tropical ocean surface temperatures at decadal-multidecadal time scales, Part I: The Atlantic Ocean. *J. Climate*, 11, 2351–2375.

Meisner, B. N., 1995: An overview of NHC prediction models. NOAA Tech. Attach. SR/SSD 95–36.

Meldrum, C., 1872: On a periodicity in the frequency of cyclones in the Indian Ocean south of the equator. *Nature*, 6, 357–358.

Merrill, R. T., 1984: A comparison of large and small tropical cyclones. *Mon. Wea. Rev.*, 112, 1408–1418.

——, 1993: Tropical cyclone structure. *Global Guide to Tropical Cyclone Forecasting*, WMO, Geneva, 2.1–2.53.

Meyers, S., J. J. O'Brien, and E. Thelin, 1999: Reconstruction of monthly SST in the tropical Pacific Ocean during 1868–1993 using adaptive climate basis functions. *Mon. Wea. Rev.*, in press.

Michaels, J., 1973: Changes of the maximum winds in Atlantic tropical cyclones as deduced from central pressure changes. NOAA Tech. Memo., ERL WMPO-6.

Michaelsen, J., 1987: Cross-validation in statistical climate forecast models. *J. Climate Appl. Meteor.*, 26, 1589–1600.

Millás, J. C., 1968: *Hurricanes of the Caribbean and Adjacent Regions, 1492–1800.* Edwards-Brothers, Inc.

Miller, B. I., 1958: On the maximum intensity of hurricanes. *J. Meteor.*, 15, 184–195.

Miller, K. A., and M. W. Downton, 1993: The freeze risk to Florida citrus. Part I: Investment decisions. *J. Climate*, 6, 354–363.

Minar, N., R. Burkhart, C. Langton, and M. Askerazi, 1996: The Swarm simulation system: A toolkit for building multi-agent simulations. Santa Fe Institute Report, [Available from http://www.santafe.edu/projects/swarm/].

Mitchell, C. L., 1924: West Indian hurricanes and other tropical cyclones of the North Atlantic Ocean. *Mon. Wea. Rev.* Supplement No. 24, 1924, U.S. Weather Bureau.

Molinari, J., S. Skubis, and D. Vollaro, 1995: External influences on hurricane intensity, Part III: Potential vorticity evolution. *J. Atmos. Sci.*, 52, 3593–3606.

Montgomery, M. F., and B. F. Farrell, 1993: Tropical cyclone formation. *J. Atmos. Sci.*, 50, 285–310.

Moore, P. L., and W. R. Davis, 1951: A preseason hurricane of subtropical origin. *Mon. Wea. Rev.*, 79, 189–195.

Myers, K. D., 1995: *False Security: Greed and Deception in America's Multibillion-Dollar Insurance Industry*. Prometheus Books.

Namias, J., 1955: Secular fluctuations in vulnerability to tropical cyclones in and off New England. *Mon. Wea. Rev.*, 83, 155–162.

Namias, J., and G.E. Dunn, 1955: The weather and circulation of August 1955. *Mon. Wea. Rev.*, 83, 163–170.

Naranjo-Diaz, L. R., and A. Centella, 1998: Recent trends in the climate of Cuba. *Weather*, 53, 78–85.

Naughton, P. W., 1982: The Jamaican hurricane season—Changing the rhyme. *Caribbean J. Sci.*, 18, 107–111.

Neumann, C. J., 1972: An alternate to the HURRAN tropical cyclone forecast system. NOAA Tech. Memo. NWS SR–62.

——, 1975: A statistical study of tropical cyclone positioning errors with economic applications. NOAA Tech. Memo. NWS SR-82.

——, 1987: The National Hurricane Center risk analysis program (HURISK). NOAA Tech. Memo. NWS NHC-38.

Neumann, C. J., and C. J. McAdie, 1991: A revised National Hurricane Center NHC83 model (NHC90). NOAA Tech. Memo. NWS NHC–44.

——, 1997: The Atlantic tropical cyclone file: A critical need for a revision. Preprints, *22nd Conf. on Hurricanes and Tropical Meteorology*, Ft. Collins, CO, Amer. Meteor. Soc., 401–402.

Neumann, C. J., B. R. Jarvinen, C. J. McAdie, and J. D. Elms, 1993: *Tropical Cyclones of the North Atlantic Ocean, 1871–1992*. National Oceanic and Atmospheric Administration.

Nicholls, N., 1979: A possible method for predicting seasonal tropical cyclone activity in the Australian region. *Mon. Wea. Rev.*, 107, 1221–1224.

——, 1984: The Southern Oscillation, sea-surface temperature, and interannual fluctuations in Australian tropical cyclone activity. *J. Climatol.*, 4, 661–670.

——, 1992: Recent performance of a method for forecasting Australian seasonal tropical cyclone activity, *Australian Meteorological Magazine*, 40, 105–110.

Noonan, B., 1993: Catastrophes: The New Math. *Best's Review*, 83, 41–44.

Novlan, D. J., and W. M. Gray, 1974: Hurricane-spawned tornadoes. *Mon. Wea. Rev.*, 102, 476–488.

Nutter, F. W., 1994: The role of government in the United States in addressing natural catastrophes and environmental exposures. *The Geneva Papers on Risk and Insurance*, 19, 244–256.

O'Brien, D. P., and R. G. Currie, 1993: Observations of the 18.6 year cycle of air pressure and a theoretical model to explain certain aspects of this signal. *Climate Dynamics*, 8, 287–298.

Palmén, E., 1949: Origin and structure of high level cyclones south of the maximum westerlies. *J. Meteor.*, 6, 22–31.

Peixoto, J. P., and A. H. Oort, 1992: *Physics of Climate*. American Institute of Physics.

Penland, C., M. Ghil, and K. M. Weickmann, 1991: Adaptive filtering and maximum entropy spectra with application to changes in atmospheric angular momentum. *J. Geophys. Res.*, 96, 22659–22671.

Penland, C., and L. Matrosova, 1998: Prediction of tropical Atlantic sea surface temperatures using linear inverse modeling. *J. Climate*, 11, 483–496.

Petak, W. J., and A. A. Atkisson, 1982: *Natural Hazard Risk Assessment and Public Policy: Anticipating the Unexpected.* Springer-Verlag.

Pfeffer, R. L., and M. Challa, 1992: The role of environmental asymmetries in Atlantic hurricane formation. *J. Atmos. Sci.*, 49, 1051–1059.

Philander, S. G. H., 1983: El Niño-Southern Oscillation phenomena. *Nature*, 302, 295–301.

Pielke, R. A., Jr., 1995: "Hurricane Andrew: Mesoscale weather and societal responses," Environmental and Societal Impacts Group, National Center for Atmospheric Research, June 1, 1995.

——, 1996: "Societal Vulnerability to Tropical Cyclones," A discussion paper presented at the Climate Variability and Tropical Cyclone Prediction Meeting, 12–14 June 1996, Bermuda Biological Station for Research, Inc.

——, 1997: Reframing the U.S. hurricane problem. *Society and Natural Resources*, 10, 485–499.

Pielke, R. A., Jr., and R. A. Pielke, Sr., 1997a: Vulnerability to hurricanes along the U.S. Atlantic and Gulf coasts: Considerations of the use of long-term forecasts. *Hurricanes: Climate and Socioeconomic Impacts*, H. F. Diaz and R. S. Pulwarty, Eds., Springer-Verlag, 147–184.

——, 1997b: *Hurricanes: Their Nature and Impacts on Society*, John Wiley & Sons.

Pielke, R. A., Jr., J. Kimple, and the Sixth Prospectus Development Team, 1997: Societal aspects of weather: Report of the Sixth Prospectus Development Team of the U.S. Weather Research Program to NOAA and NSF. *Bull. Amer. Meteor. Soc.*, 78, 867–876.

Pielke, R. A., Jr., and C. W. Landsea, 1998: Normalized hurricane damages in the United States: 1925–95. *Wea. Forecasting*, 13, 621–631.

Poey, M. A., 1873: Sur les rapports entre les taches solaires et les ourages des Antilles de l'Atlantique-nord et de l'Ocean Indien sud. *Comptes Rendus*, 77, 1223–1226.

Porkess, R., 1991: *The Harper Collins Dictionary of Statistics*, HarperPerennial.

Powell, M. D., and S. H. Houston, 1996: Hurricane Andrew's landfall in south Florida. Part II: Surface wind fields and potential real-time applications. *Wea. Forecasting*, 11, 329–349.

Press, W. H., B. P. Flannery, S. A. Teukolsky, and W. T. Vetterling, 1989: *Numerical Recipes, The Art of Scientific Computing.* Cambridge University Press.

Prigogine, I., 1997: *The End of Certainty: Time, Chaos, and the New Laws of Nature.* The Free Press.

Pulwarty, R. S., and W. E. Riebsame, 1997: The political ecology of vulnerability to hurricane-related hazards. *Hurricanes: Climate and Socioeconomic Impacts*, H. F.

Diaz and R. S. Pulwarty, Eds., Springer-Verlag, 185–214.

Rampino, M. R., and S. Self, 1992: Volcanic winter and accelerated glaciation following the Toba super-eruption. *Nature*, 359, 50–52.

Raper, S. C. B., 1993: Observational data on the relationship between climate change and the frequency and magnitude of severe tropical storms. *Climate and Sea Level Change: Observations, Projections, and Implications*, R. A. Warrick, E. M. Barrow, and T. M. L. Wigley, Eds., Cambridge University Press, 192–212.

Rappaport, E. N., and J. J. Fernández-Partagás, 1995: The deadliest Atlantic tropical cyclones, 1492–1994. NOAA Tech. Memo. NWS NHC-47.

——, 1997: History of the deadliest Atlantic tropical cyclones since the discovery of the New World. *Hurricanes: Climate and Socioeconomic Impacts*, H. F. Diaz and R. S. Pulwarty, Eds., Springer-Verlag, 93–108.

Rasmusson, E. M., X. Wang, and C. F. Ropelewski, 1990: The biennial component of ENSO variability, *J. of Marine Systems*, 1, 71–96.

Ray, C. L., 1935: Relation of tropical cyclone frequency to summer pressures and ocean surface water temperatures. *Mon. Wea. Rev.*, 63, 10–12.

Reading, A. J., 1989: Caribbean tropical storm activity over the past four centuries. *Int. J. Climatol.*, 10, 365–376.

Riehl, H., 1954: *Tropical Meteorology*. McGraw-Hill Book Co.

——, 1956: Sea surface temperature anomalies and hurricanes. *Bull. Amer. Meteor. Soc.*, 37, 413–417.

Riehl, H., and R. J. Shafer, 1944: The recurvature of tropical storms. *J. Meteor.*, 1, 42–54.

Riehl, H., and N. M. Burgner, 1950: Further studies of the movement and formation of hurricanes and their forecasting. *Bull. Amer. Meteor. Soc.*, 31, 244–253.

Roebber, P. J., and L. F. Bosart, 1996: The complex relationship between forecast skill and forecast value: a real-world analysis. *Wea. Forecasting*, 11, 544–559.

Roth, R. J. Sr., 1997: Insurable risks, regulation, and the changing insurance environment. *Hurricanes: Climate and Socioeconomic Impacts*, H. F. Diaz and R. S. Pulwarty, Eds., Springer-Verlag, 261–272.

Roth, D. M., 1998: A historical study of tropical storms and hurricanes that have affected southwest Louisiana and southwest Texas. Tech. Attachment, U.S. Dept. of Commerce, NOAA, SR/SSD 98-16.

Saunders, M. A., and A. R. Harris, 1997: Statistical evidence links exceptional 1995 Atlantic season to record sea warming. *Geophys. Res. Lett.*, 24, 1255–1258.

Schneider, S. H., 1989: *Global Warming: Are We Entering The Greenhouse Century?* Sierra Club Books.

Schoner, R. W., 1968: Climatological regime of rainfall associated with hurricanes after landfall. Tech. Memo., U.S. Dept. of Commerce/Environmental Science Services Administration WBTM-ER-29.

Schwartz, G., 1978: Estimating the dimension of a model. *The Annals of Statistics*, 6, 461–464.

Shapiro, L. J., 1982: Hurricane climate fluctuations. Part II: Relation to large-scale circulation. *Mon. Wea. Rev.*, 110, 1014–1023.

——, 1987: Month-to-month variability of the Atlantic tropical circulation and its relationship to tropical storm formation. *Mon. Wea. Rev.*, 115, 2598–2614.

——, 1989: The relationship of the quasi-biennial oscillation to Atlantic tropical storm activity. *Mon. Wea. Rev.*, 117, 1545–1552.

Sheets, R. C., 1990: The National Hurricane Center: Past, present and future. *Wea. Forecasting*, 5, 185–232.

Simpson, R. H., 1952: Evoluation of the kona storm, a subtropical cyclone. *J. Meteor.*, 9, 24–35.

——, 1971: The decision process in hurricane forecasting. NOAA Tech. Memo. NWS SR-53.

——, 1998: Stepping stones in the evolution of a national hurricane policy. *Wea. Forecasting*, 13, 617–620.

Simpson, R. H., and M. B. Lawrence, 1971: Atlantic hurricane frequencies along the U.S. coastline. NOAA Tech. Memo. NWS TM SR-58.

Simpson, R. H., and H. Riehl, 1980: *The Hurricane and Its Impact.* 1st ed. Louisiana State University Press.

Simpson, R. H., N. L. Frank, D. Shideler, and H. M. Johnson, 1968: Atlantic tropical disturbances, 1967. *Mon. Wea. Rev.*, 96, 251–259.

——, 1969: Atlantic tropical disturbances of 1968. *Mon. Wea. Rev.*, 97, 240–255.

Smith, E. A., 1975: The McIDAS system. *IEEE Transactions: Geosciences*, GE-13, 123–128.

Solow, A. R., 1988: A bayesian approach to statistical inference about climate change. *J. Climate*, 1, 512–521.

Solow, A. R., and N. Nicholls, 1990: The relationship between the Southern Oscillation and tropical cyclone frequency in the Australian region. *J. Climate*, 3, 1097–1101.

Spiegler, D. B., 1971: The unnamed Atlantic tropical storms of 1970. *Mon. Wea. Rev.*, 99, 966–976.

Stewart, T. R., 1997: Forecast value: descriptive decision studies. *Economic Value of Weather and Climate Forecasts*, R. W. Katz and A. H. Murphy, Eds., Cambridge University Press, 222.

Street-Perrott, F. A., and R. A. Perrott, 1990: Abrupt climate fluctuations in the tropics: The influence of the Atlantic Ocean circulation. *Nature*, 343, 607–611.

Sugg, A. L., 1967: Economic aspects of hurricanes. *Mon. Wea. Rev.*, 95, 143–146.

Sun, D.-Z., and K. E. Trenberth, 1998: Coordinated heat removal from the equatorial Pacific during the 1986-87 El Niño. *Geophys. Res. Lett.*, 25, 2659–2662.

Sutton, R. T., and M. R. Allen, 1997: Decadal predictability of North Atlantic sea surface temperature and climate. *Nature*, 388, 563–567.

Tannehill, I. R., 1950: *Hurricanes: Their Nature and History–Particularly Those of the West Indies and the Southern Coasts of the United States.* Princeton University Press.

Theiler, J., S. Eubank, A. Longtin, B. Galdrikian, and J. D. Farmer, 1992: Testing for nonlinearity in time series: The method of surrogate data. *Physica D*, 58, 77–94.

Trenberth, K. E., 1997: Short-term climate variations: Recent accomplishments and issues for future progress. *Bull. Amer. Meteor. Soc.*, 78, 1081–1096.

Trenberth, K. E., and T. J. Hoar, 1996: The 1990-1995 El Niño-Southern Oscillation event: Longest on record. *Geophys. Res. Lett.*, 23, 57–60.

Tsonis, A. A., P. J. Roebber, and J. B. Elsner, 1998: A characteristic time scale in the global temperature record. *Geophys. Res. Lett.*, 25, 2821–2823.

Tucker, T., 1996: *Beware the Hurricane: The story of the cyclonic tropical storms that have struck Bermuda 1609–1995*. 4th ed. The Island Press, Ltd.

Tuleya, R. M., M. Bender, Y. Kurihara, and S. Lord, 1995: The GFDL multiply-nested moveable mesh hurricane system. NWS Tech. Proceed. Bull. No. 424. NWS Office of Meteorology.

U.S. Department of Commerce, 1990: *50 Years of Population Change along the Nation's Coasts: 1960–2010*. NOAA, National Ocean Service.

Vallis, G. K., 1986: El Niño: A chaotic dynamical system? *Science*, 232, 243–245.

Vautard, R., and M. Ghil, 1989: Singular spectrum analysis in nonlinear dynamics with applications to paleoclimatic time series. *Physica D*, 35, 395–424.

Walsh, J. E., and W. L. Chapman, 1990: Arctic contribution to upper-ocean variability in the North Atlantic. *J. Climate*, 3, 1462–1473.

Walsh, J. E., W. L. Chapman, and T.L. Shy, 1996: Recent decrease of sea-level pressure in the central Arctic. *J. Climate*, 9, 480–486.

Walsh, J. E., and R. Kleeman, 1997: Predicting decadal variations in Atlantic tropical cyclone numbers and Australian rainfall. *Geophys. Res. Lett.*, 24, 3249–3252.

Weaver, A., S. Aura, and P. Myers, 1994: Interdecadal variability in an idealized model of the North Atlantic. *J. Geophys. Res.*, 99, 12423–12441.

Wendland, W. M., 1977: Tropical storm frequencies related to sea surface temperatures. *J. Appl. Meteor.*, 16, 477–481.

Wilks, D. S., 1995: *Statistical Methods in the Atmospheric Sciences: An Introduction*. Academic Press.

——, 1997: Forecast value: prescriptive decision studies. *Economic Value of Weather and Climate Forecasts*, R. W. Katz and A. H. Murphy, Eds., Cambridge University Press.

Willett, H. C., 1955: A Study of the Tropical Hurricane Along the Atlantic and Gulf Coasts of the United States. Tech. Rep. to Inter-Regional Insurance Conference, New York.

Williams, J. M., and I. W. Duedall, 1997: *Florida Hurricanes and Tropical Storms, Revised Edition*. University Press of Florida.

Wilson, E. O., 1998: *Consilience: The Unity of Knowledge*. Knopf.

Wu, G., and N.-C. Lau, 1992: A GCM simulation of the relationship between tropical-storm formation and ENSO. *Mon. Wea. Rev.*, 120, 958–977.

Xu, Z., and C. J. Neumann, 1984: Frequency and motion of western North Pacific tropical cyclones. NOAA Tech. Memo. NWS NHC 23.

Zebrowski, E., Jr., 1997: *Perils of a Restless Planet*. Cambridge.

Zehr, R. M., 1992: Tropical cyclogenesis in the western North Pacific. NOAA Tech. Rep. NESDIS 61.

Index